Animal Models for Neurodegenerative Disease

RSC Drug Discovery Series

Editor-in-Chief
Professor David Thurston, *London School of Pharmacy, UK*

Series Editors:
Dr David Fox, *Pfizer Global Research and Development, Sandwich, UK*
Professor Salvatore Guccione, *University of Catania, Italy*
Professor Ana Martinez, *Instituto de Quimica Medica-CSIC, Spain*
Dr David Rotella, *Northeastern University, USA*

Advisor to the Board:
Professor Robin Ganellin, *University College London, UK*

How to obtain future titles on publication:
A standing order plan is available for this series. A standing order will bring delivery of each new volume immediately on publication.

For further information please contact:
Book Sales Department, Royal Society of Chemistry, Thomas Graham House, Science Park, Milton Road, Cambridge, CB4 0WF, UK
Telephone: +44 (0)1223 420066, Fax: +44 (0)1223 420247, Email: books@rsc.org
Visit our website at http://www.rsc.org/Shop/Books/

Animal Models for Neurodegenerative Disease

Edited by

Jesús Avila, Jose J. Lucas and Félix Hernández
Centro de Biología Molecular Severo Ochoa (CSIC-UAM), Madrid, Spain

RSCPublishing

RSC Drug Discovery Series No. 6

ISBN: 978-1-84973-184-3
ISSN: 2041-3203

A catalogue record for this book is available from the British Library

© Royal Society of Chemistry 2011

Published by The Royal Society of Chemistry,
Thomas Graham House, Science Park, Milton Road,
Cambridge CB4 0WF, UK

Registered Charity Number 207890

For further information see our web site at www.rsc.org

Preface

J. AVILA

Centro de Biología Molecular 'Severo Ochoa' (CSIC-UAM), Madrid, Spain

The aim of this book is to review how different animal models can mimic different neurodegenerative disorders and how these models can be used to test putative compounds that could modify these disorders.

Most prevalent neurological disorders like Alzheimer's disease, Parkinson's disease and stroke are uncommon in people under the age of 40 as aging is one of main risks for them. This is not the case for diseases like Huntington's disease, amyotrophic lateral sclerosis and prion diseases, which can appear at earlier ages. Animal models for all these disorders are presented in this book, together with a description of a technique to express exogenous genes in an animal model upon viral infection.

Alzheimer's disease (AD) was first described about a hundred years ago. In that first report Alois Alzheimer indicated the presence, in the brain of one patient, of two aberrant structures—senile plaques and neurofibrillary tangles.[1] We now know that the main component of senile plaques is beta amyloid peptide, a fragment of the amyloid precursor protein (APP)[2,3] resulting from cleavage of APP by two proteases (secretases), *i.e.* β and γ secretases.[4] Neurofibrillary tangles are aggregated filamentous structures, paired helical filaments,[5] which are polymers of the cytoskeletal protein tau[6,7] in phosphorylated form.[8]

In familial Alzheimer's disease, the cause for the onset of the disease is the presence of mutations in APP that facilitate the formation of beta amyloid peptide, or the presence of mutations in the protein presenilin-1 (or presenilin-2), the catalytic component of γ secretase complex, that result in an increase in γ secretase activity and the appearance of beta amyloid peptide[9] which could

RSC Drug Discovery Series No. 6
Animal Models for Neurodegenerative Disease
Edited by Jesús Avila, Jose J. Lucas and Félix Hernández
© Royal Society of Chemistry 2011
Published by the Royal Society of Chemistry, www.rsc.org

assemble into toxic peptide aggregates. In this book, there is a chapter[10] dealing with the generation of different animal models for Alzheimer's disease and another chapter focusing on those models involving amyloid/presenilin-1 pathology.[11] A recent report has described how other players could regulate beta amyloid aggregation. An example is the protein Reelin. Thus, the book contains a dedicated chapter about the role of Reelin in some aspects of AD pathology.[12]

The phosphorylation and aggregation of the main component, tau protein, of neurofibrillary tangles (the other aberrant structure present in AD) are described in two further chapters. In the first[13] animal models mimicking tau pathology in AD and in other disorders (tauopathies) are described; in the second,[14] the consequences of tau phosphorylation by one of the main tau kinases, GSK-3β (also known as tau kinase I) are discussed.

Another prevalent neurodegenerative disorder is Parkinson's disease (PD). The hallmark of this disorder is the loss of dopamine neurons in the substantia nigra and the appearance of cytoplasmic inclusions termed Lewy bodies of which the main component is the protein alpha synuclein.[15] There is a familial Parkinson disease caused by the mutations of different genes, one of them being that expressing α-synuclein. Other genes involved in familial PD are lrrk2, parkin, pink1 and dij1.[16] Animal models expressing these mutated genes are described in one of the chapters on PD.[16] In two additional chapters, other animal models and studies on neuroprotection in Parkinson's disease are indicated.[17,18]

Stroke is a leading cause of death and disability. Interruption of the blood supply to the entire brain or a region (ischemia) causes neuronal death and cessation of brain function. One chapter is dedicated to the analysis of animal models of stroke that have been developed to characterize the pathology of ischemia.[19]

Amyotrophic lateral sclerosis (ALS) is a degenerative disorder involving the death of motor neurons. Aberrant aggregates of proteins like those of SOD-1 or TDP-43 have been indicated in this disease.[21] There are also familial cases in this disease in which mutations in genes like SOD-1 result in the onset of the disease. Animal models expressing some of these mutated genes are discussed in another chapter.[20]

Huntington's disease is a neurodegenerative disorder also characterized by motor dysfunction.[21,22] The neuropathology of the disease involves striatal atrophy and the presence of aggregates of the protein huntingtin. Animal models for the disease are described.[23]

PrP related diseases are a group of neurodegenerative disorders known as transmissible spongiform encephalopathies or prion diseases.[24] Neuropathology of the disease includes neuronal loss and, in many cases, the presence of aggregates of prion protein deposits.[25] Different animal models for the disease are described.[26]

The last chapter describes the generation of virus-induced animal models.[27]

Finally, I will like to thank to the co-editors of this volume for their help and to Ms Nuria de la Torre for her excellent work in facilitating the edition of the book.

References

1. A. Alzheimer, *Psych. Genchtl. Med.*, 1907, **64**, 146.
2. C. L. Masters, G. Simms, N. A. Weinman, G. Multhaup, B. L. McDonald and K. Beyreuther, *Proc. Natl. Acad. Sci. U. S. A.*, 1985, **82**, 4245–4249.
3. G. G. Glenner and C. W. Wong, *Biochem. Biophys. Res. Commun.*, 1984, **122**, 1131–1135.
4. D. L. Price and S. S. Sisodia, *Annu. Rev. Neurosci.*, 1998, **21**, 479–505.
5. M. Kidd, *Nature*, 1963, **197**, 192–193.
6. I. Grundke-Iqbal, K. Iqbal, M. Quinlan, Y. C. Tung, M. S. Zaidi and H. M. Wisniewski, *J. Biol. Chem.*, 1986, **261**, 6084–6089.
7. E. Montejo de Garcini, L. Serrano and J. Avila, *Biochem. Biophys. Res. Commun.*, 1986, **141**, 790–796.
8. I. Grundke-Iqbal, K. Iqbal, Y. C. Tung, M. Quinlan, H. M. Wisniewski and L. I. Binder, *Proc. Natl. Acad. Sci. U. S. A.*, 1986, **83**, 4913–4917.
9. J. Hardy and D. J. Selkoe, *Science*, 2002, **297**, 353–356.
10. M. A. Chabrier, K. M. Neely, N. A. Castello and F. M. LaFerla, in: *Animal Models for Neurodegenerative Diseases*, J. Avila, J. J. Lucas and F. Hernandez (ed.), Royal Society of Chemistry, Cambridge, 2011, ch.1.
11. A. Takashima, in *Animal Models for Neurodegenerative Diseases*, ed. J. Avila, J. J. Lucas and F. Hernandez, Royal Society of Chemistry, Cambridge, 2011, ch.2.
12. E. Soriano, D. Rossi and L. Pujadas, in *Animal Models for Neurodegenerative Diseases*, ed. J. Avila, J. J. Lucas and F. Hernandez, Royal Society of Chemistry, Cambridge, 2011, ch.3.
13. J. Götz, L. M. Ittner, N. N. Götz, H. Lam and H. Nicholas, in *Animal models for Neurodegenerative Diseases*, ed. J. Avila, J. J. Lucas and F. Hernandez, Royal Society of Chemistry, Cambridge, 2011, ch.5.
14. F. Hernandez, in *Animal Models for Neurodegenerative Diseases*, ed. J. Avila, J. J. Lucas and F. Hernandez, Royal Society of Chemistry, Cambridge, 2011, ch.4.
15. M. G. Spillantini, R. A. Crowther, R. Jakes, M. Hasegawa and M. Goedert, *Proc. Natl. Acad. Sci. U. S. A.*, 1998, **95**, 6469–6473.
16. L. Stefanis and H. Rideout, in *Animal Models for Neurodegenerative Diseases*, ed. J. Avila, J. J. Lucas and F. Hernandez, Royal Society of Chemistry, Cambridge, 2011, ch.6.
17. J. G. Castaño T. Iglesias and J. G. de Yebenes, in *Animal Models for Neurodegenerative Diseases*, ed. J. Avila, J. J. Lucas and F. Hernandez, Royal Society of Chemistry, Cambridge, 2011, ch.7.
18. A. Pascual, J. Villadiego, M. Hidalgo-Figueroa, S. Mendez-Ferrer, R. Gomez-Diaz, J. J. Toledo-Aral and J. Lopez-Barneo, in *Animal Models for Neurodegenerative Diseases*, ed. J. Avila, J. J. Lucas and F. Hernandez, Royal Society of Chemistry, Cambridge, 2011, ch.8.
19. D. C. Henshall and R. P. Simon, in *Animal Models for Neurodegenerative Diseases*, ed. J. Avila, J. J. Lucas and F. Hernandez, Royal Society of Chemistry, Cambridge, 2011, ch.12.

20. R. Fujii and T. Takumi, in *Animal Models for Neurodegenerative Diseases*, ed. J. Avila, J. J. Lucas and F. Hernandez, Royal Society of Chemistry, Cambridge, 2011, ch.9.
21. G. A. Graveland, R. S. Williams and M. DiFiglia, *Science*, 1985, **227**, 770–773.
22. M. DiFiglia, E. Sapp, K. O. Chase, S. W. Davies, G. P. Bates, J. P. Vonsattel and N. Aronin, *Science*, 1997, **277**, 1990–1993.
23. Z. Ortega and J. J. Lucas, in *Animal Models for Neurodegenerative Diseases*, ed. J. Avila, J. J. Lucas and F. Hernandez, Royal Society of Chemistry, Cambridge, 2011, ch.10.
24. S. B. Prusiner, *N. Engl. J. Med.*, 2001, **344**, 1516–1526.
25. A. Aguzzi and A. M. Calella, *Physiol. Rev.*, 2009, **89**, 1105–1152.
26. M. Gasset and A. Aguzzi, in *Animal Models for Neurodegenerative Diseases*, ed. J. Avila, J. J. Lucas and F. Hernandez, Royal Society of Chemistry, Cambridge, 2011, ch.11.
27. K. Iqbal, X. Wang, J. Blanchard and I. Grundke-Iqbal, in *Animal Models for Neurodegenerative Diseases*, ed. J. Avila, J. J. Lucas and F. Hernandez, Royal Society of Chemistry, Cambridge, 2011, ch.13.

Contents

RSC Drug Discovery Series No. 6
Animal Models for Neurodegenerative Disease
Edited by Jesús Avila, Jose J. Lucas and Félix Hernández
© Royal Society of Chemistry 2011
Published by the Royal Society of Chemistry, www.rsc.org

Chapter 8   Neuroprotection in Parkinson's Disease   162
Alberto Pascual, Javier Villadiego, María Hidalgo-Figueroa,
Simón Méndez-Ferrer, Raquel Gómez-Díaz, Juan José
Toledo-Aral and José Lopez-Barneo

Chapter 9   Animal Models for ALS   177
Ritsuko Fujii and Toru Takumi

CHAPTER 1

The Contribution of Transgenic Models to the Understanding of Alzheimer's Disease Progression and Therapeutic Development

MEREDITH A. CHABRIER, KARA M. NEELY, NICHOLAS A. CASTELLO AND FRANK M. LAFERLA[*]

Department of Neurobiology and Behavior, Institute of Memory Impairments and Neurological Disorders, University of California Irvine, Irvine, CA, USA

1.1 Introduction

Alzheimer disease (AD), the most prevalent neurodegenerative disorder, is characterized by progressive memory loss and cognitive decline. Currently, there are over 35 million people throughout the world who suffer from AD and the prevalence is expected to increase by more than 50% by the year 2030.[1]

Over 100 rare mutations have been described in three AD-related genes that cause an autosomal dominant form of the disease, familial AD (fAD), which comprises less than 5% of all cases.[2,3] The vast majority of AD cases are sporadic (sAD) and the causes underlying these cases remain unknown. Neuropathologically, AD is characterized by the accumulation of amyloid-beta (Aβ) plaques and neurofibrillary tangles, in addition to widespread synaptic loss, inflammation and oxidative damage, and neuronal death. Curiously, the

RSC Drug Discovery Series No. 6
Animal Models for Neurodegenerative Disease
Edited by Jesús Avila, Jose J. Lucas and Félix Hernández
© Royal Society of Chemistry 2011
Published by the Royal Society of Chemistry, www.rsc.org

neuropathological and clinical phenotype is indistinguishable in both types of the disease, with the major difference being the age of onset, which occurs at less than 65 in fAD.[4]

As the cause of sAD is unknown, transgenic models have been based on the fAD component, utilizing fAD-associated genetic mutations to mimic specific elements of AD pathology, with the rationale that the events downstream of the initial trigger are quite similar. These genetic models have been invaluable in determining the molecular mechanisms of disease progression and for testing potential therapeutics. Although no single mouse model recapitulates all of the aspects of the disease spectrum, each model allows for in-depth analyses of one or two components of the disease, a feat not readily possible or ethical with human patients or samples.

1.2   Aβ and Tau Biology

Aβ is a central feature of AD pathology and is formed by the sequential cleavage of the amyloid precursor protein (APP) by β-secretase (BACE1) and γ-secretase, of which the catalytic component is either presenilin 1 or 2 (PS1 or PS2) (Figure 1.1A). The mutations found in fAD patients occur in APP, PS1 or PS2, and alter Aβ by either increasing its production, altering the ratio of Aβ42/Aβ40 or increasing its propensity to aggregate. Aβ42 has a higher propensity to aggregate and is considered the more pathological form.[5] Although insoluble Aβ plaques are the most prominent pathological feature of AD, human evidence shows that the plaque burden does not correlate with cognitive decline[6–10] and soluble forms of Aβ oligomers are now regarded as the primary pathological culprit,[11,12] raising the possibility that plaques might actually be protective by sequestering the toxic oligomeric Aβ species. Importantly, this finding is mirrored in studies with transgenic mice, which provides us with an important tool for investigating the pathological effects of different Aβ assemblies.[13,14]

Aβ oligomers are classified as either prefibrillar, which range in size from a small dimer to 75 kDa, or fibrillar which can assemble to a size greater than 500 kDa.[15,16] The leading hypothesis for AD pathogenesis is the Aβ cascade hypothesis, which states that Aβ is the primary trigger that initiates a cascade of disease effects such as the formation of neurofibrillary tangles and chronic inflammation, eventually culminating in neurodegeneration and dementia.[17,18]

Neurofibrillary tangles (NFTs) represent the second hallmark feature of AD pathology and consist of the microtubule binding protein tau. Tau is normally responsible for maintaining the structural integrity of cells, but in tauopathy disorders like AD, tau becomes aberrantly phosphorylated. Tau hyperphosphorylation renders it unable to bind to microtubules, causing their destablization and aggregation, eventually forming NFTs (Figure 1.1B). Transgenic models overexpressing mutant tau exhibit tangle formation, cognitive decline and neuronal loss.[19,20]

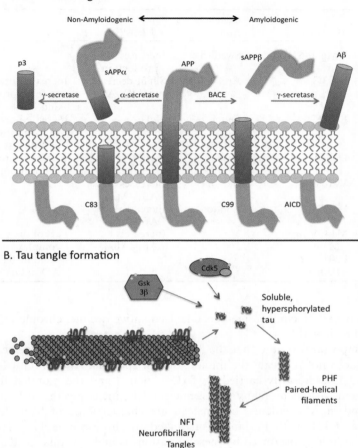

Figure 1.1 (**A**) APP processing. (**B**) Tau tangle formation.

1.3 Modelling Plaques and Tangles

Many transgenic models have successfully recapitulated amyloid pathology, generally by expressing mutated forms of APP and/or PS1. But models that develop both amyloid and tau pathology are rare, yet truly required for evaluation of the therapeutic efficacy of an anti-AD intervention. In 2003, our lab generated a triple-transgenic model of AD (3xTg-AD) that carries APPswe, tauP301L and PS1M146V mutant transgenes, which together promote the development of both Aβ and tau pathology.[21] This model has allowed investigation of these two major pathological hallmarks within the same mouse, and most importantly, provided insight into the interaction between Aβ and tau (discussed below). The 3xTg-AD model also recapitulates other important

Table 1.1 Mutations commonly used in AD transgenic models.

Protein	Mutation	Common Name	Effect on Aβ Levels	Notes
APP	KM670/671NL	Swedish	Increases Aβ production	
APP	E693G	Arctic	Slight decrease in $A\beta_{42}/A\beta_{40}$	Increased propensity to form protofibrils and fibrils
APP	E693K	Dutch	None	Increased propensity to form protofibrils and fibrils
APP	D694N	Iowa	None	Increased propensity to form fibrils
APP	V716V	Indiana	Increases $A\beta_{42}/A\beta_{40}$	
APP	V717I	London	Increases $A\beta_{42}/A\beta_{40}$	
PS1	M146V or M146L		Increases $A\beta_{42}/A\beta_{40}$	Part of the γ-secretase complex
Tau	P301L or P301S			Found in FTD, not AD patients

features of AD, including age-dependent cognitive decline, chronic inflammation and neuronal loss.[22–24]

It is important to note that the 3xTg-AD model, like all transgenic models of AD, does not replicate the initial cause of sporadic AD. Mutant forms of APP and PS1 are only carried by fAD patients, and the mutated forms of tau protein carried by many transgenic models, including the 3xTg-AD, are not found in AD patients, but instead are characteristic of frontotemporal dementia (FTD) with Parkinsonism linked to chromosome 17.[25] However, the tau pathology produced by these mutations is highly similar to that found in AD and has allowed investigation of the interactions between Aβ and tau.[24,26–31]

1.4 Mechanisms of Disease Progression

1.4.1 Aβ and Tau

Essential to our understanding of Aβ and tau interactions, studies with the 3xTg-AD model show that the accumulation of Aβ begins first and coincides with cognitive deficits, while tau pathology develops several months later.[21] Furthermore, characterization of the parental 3xTg-AD showed a significant increase in hyperphosphorylated tau compared with a derivative line without the PS1 mutation, suggesting that higher Aβ42 levels in the 3xTg-AD directly facilitates tau pathology.[32] Blocking Aβ accumulation by immunizing against Aβ also decreases tau pathology.[31,32] The mechanism for Aβ dependent tau pathology involves alterations in CHIP, a tau ubiquitin ligase[33,34] as increased Aβ causes decreased levels of CHIP concurrent with increased tau. Restoring CHIP levels ameliorated the Aβ-induced tau pathology.[32] This supports the amyloid cascade hypothesis and demonstrates that reducing Aβ levels can reverse tau pathology; it also implicates the ubiquitin proteasome system in disease progression.

Proteasome impairment has been implicated as an important mechanism for protein aggregation in AD. There is *in vitro* and *in vivo* evidence that Aβ blocks proteasome activity.[35–37] Given that tau is largely degraded by the proteasome, Aβ-mediated proteasome impairment leads to tau accumulation.[31,35] Using the 3xTg-AD model, our lab has shown that tau clearance follows Aβ clearance after Aβ immunotherapy, and that tau clearance is dependent on proteasome activity. Tau clearance is inhibited when the proteasome activity is blocked.[31] Furthermore we showed that Aβ oligomers, but not monomers, specifically target the proteasome, thereby facilitating tau accumulation.[35] In additional support of this mechanism of disease progression, Uch-L1, an enzyme responsible for ubiquitin recycling, is downregulated in the AD brain.[38] Increasing expression of Uch-L1 rescues synaptic and cognitive deficits[39,40] in the APP/PS1 model. This provides significant evidence that drugs targeting the ubiquitin–proteasome system may be a successful approach to AD therapy.

1.4.2 Inflammation

The immune system is chronically active in AD, leading to glial activation and the induction of pro-inflammatory cytokines near Aβ deposits.[41] Transgenic models have provided evidence that there are specific interactions between the inflammatory pathway, Aβ and tau that consequently lead to neuronal loss. Studies in the 3xTg-AD model have found that TNF-α and MCP-I expression are highly upregulated by three months of age, which corresponds with the time course of microglial activation.[42] The same study showed that neurons are a significant source of TNF-α, and chronic TNF-α exposure leads to neuronal cell death in transgenic but not non-transgenic mice.[43] Another study in APP transgenic models demonstrated that specific targeting of the CD40 ligand by genetic and immunological approaches lead to attenuation of Aβ pathology and a decrease in amyloidogenic processing.[44] While CD40 may be a future target for AD therapy, the effects on tau pathology are unknown. Together, these studies suggest that neuronal loss is caused by the upregulation of inflammatory proteins due to AD related pathology and targeting the immune system can attenuate disease progression.

Another pathway by which Aβ facilitates the development of tau pathology is through the activation of the CNS inflammatory system.[22,45] Induction of CNS inflammation in the 3xTg-AD model causes tau hyperphosphorylation, while not affecting APP levels.[22] The augmented tau pathology is due to increased formation of p25, which heightens the activity of Cdk5, a known kinase for tau phosphorylation (Figure 1.1B). Additionally, in the P301S tauopathy mouse model which accumulates tau pathology alone, microglial activation precedes tangle formation and neuronal loss.[45] Administration of an immunosuppressant, FK-506, attenuates tau pathology and increases the life-span of this model.[45] Further studies are needed to elucidate the precise mechanism of steps from microglia activation to tau pathology, as these could be possible therapeutic targets.

1.4.3 Risk Factors

As 95% of AD cases are of unknown origin, it is highly relevant to determine risk factors that may trigger or amplify disease pathology. There are many known risk factors of AD; here we discuss stress, diet, exercise, environmental factors and low cerebral blood oxygen levels.

Chronic stress exacerbates disease progression by increasing Aβ production[46,47] and tau accumulation.[47,48] Administration of dexamethasone, a potent glucocorticoid, to 3xTg-AD mice selectively increases APP and BACE (beta-site APP cleaving enzyme), causing Aβ levels to increase, leading to an increase in tau. To further show that the tau increases were due to increased Aβ levels, a control mouse with only the tauP301L and PS1M146V transgenes, without APP, was created. Dexamethasone treatment in tauP301L/PS1M146V mice did not cause an increase in tau pathology, suggesting that the increased tau accumulation is due to downstream effect of Aβ.[47] Further evidence from the 3xTg-AD model shows that corticosterone levels correlate with cognitive performance on stressful tasks such as Morris water maze and inhibitory avoidance.[49] These studies indicate that both therapeutically targeting stress-related hormones and/or lifestyle changes to reduce stress may slow AD disease progression.

Exercise and diet are known to improve cognitive function, possibly by increasing neurogenesis and synaptic plasticity.[50–53] Epidemiological studies show that physical activity inversely correlates with risk for dementia,[54] and specifically that AD patients were less active during their lifetimes than non-demented subjects.[55] To investigate the specific mechanisms that link exercise and cognition in AD, evidence from several different mouse models correlate voluntary exercise with enhanced learning and a decrease in amyloid burden.[56,57] Studies in the Tg2576 model have shown that these changes occur by altering APP processing in as little as one month of exercise.[56] Dietary supplementation with a variety of antioxidants, fatty acids or teas has also been shown to modulate Aβ[58,59] and improve learning and memory[53] in transgenic models. For example, three-month-old 3xTg-AD mice treated with DHA, an omega-3 polyunsaturated fatty acid, over a period of 12 months had significantly reduced intraneuronal accumulation of Aβ and tau.[58] However, it is important to note that this study shows that DHA is preventative for disease progression, not therapeutic. DHA given to mild to moderate AD patients in clinical trials did not significantly improve cognition.[60,61] Changes in diet should therefore be considered as a preventative measure to decrease risk for AD and not as way to ameliorate pathology.

Interestingly, hypoperfusion events are also linked to the onset and progression of AD pathology. Epidemiological data suggest that patients are 2–5 times more likely to develop AD following a stroke or other ischemic episode.[62] A number of studies examining hypoperfusion insults in AD models have shown robust increases in AD pathology, in part mediated *via* upregulation of BACE.[63] Surprisingly, even non-ischemic mild hypoperfusion insults produce significant increases in insoluble Aβ42 up to three weeks after the insult, as well as significantly altering tau phosphorylation in the 3xTg-AD model.[64] These data indicate a strong role for hypoperfusion on the development of AD pathology.

1.5 Therapeutic Development in Transgenic Models

Animal models have been extremely important for the discovery and development of novel AD therapeutics, particularly where studies in humans would be unethical, cost-prohibitive or too time-consuming. Here we highlight several promising treatment strategies that have been intensely investigated in transgenic mice. One such strategy is immunotherapy, which uses antibodies to target AD pathology and prevent or reverse disease progression. A second strategy uses histone deacetylase (HDAC) inhibitors to enhance synaptic plasticity in order to compensate for disease-associated cognitive deficits. A third strategy uses acetylcholinesterase inhibitors to compensate for aberrant acetylcholine signalling, which is an early and prominent feature of AD.[65,66]

1.5.1 Aβ Immunotherapy

Immunotherapy against Aβ, either by active or passive immunization, has been well established as an effective method to prevent Aβ aggregation. Early studies actively immunized PD-APP mice against Aβ and found a significant reduction in Aβ aggregation in both young, pre-pathological mice, as well as aged mice with established pathology.[67–69] Further studies in PD-APP mice found that passive delivery of Aβ antibodies to the periphery is also effective for reducing soluble and insoluble Aβ, and in Tg2576 mice led to a reversal in cognitive deficits.[70,71]

Studies from our lab provided insight into the effect of immunotherapy on both Aβ and tau pathology in the 3xTg-AD model. Administration of Aβ antibodies into the hippocampus lead to reductions in soluble and insoluble Aβ, as well as phosphorylated tau.[31] This finding supports the amyloid cascade hypothesis and suggests that targeting Aβ alone may be sufficient to affect multiple aspects of AD pathology. Interestingly, hyperphosphorylated tau was unaffected by immunotherapy, indicating that pathology may become more resistant to antibody-mediated clearance as the disease progresses. In a subsequent study, we demonstrated that clearance of soluble Aβ alone was insufficient to improve cognition in 3xTg-AD mice.[28] Concomitant reduction in soluble Aβ and tau was necessary to rescue memory impairments, which highlights the utility of a mouse that models both Aβ and tau pathology.

These findings in transgenic mice have resulted in numerous clinical trials of Aβ immunotherapy. One of the earlier and more well-known trials was of a synthetic Aβ42 peptide called AN1792, developed by Elan and Wyeth. Although the trial was stopped early due to complications in a small percentage of study participants, a follow-up study found that some participants with high antibody titres had a significantly reduced rate of cognitive decline.[72] Furthermore, neuropathological examination of many participants indicated a reduction of Aβ and tau load relative to controls.[73–75] Today, there are numerous clinical trials in progress for both active and passive immunotherapy against Aβ.

1.5.2   HDAC Inhibitors

Histone modification through histone acetyl transferases (HATs) and histone deacetylases (HDACs) plays an important role in gene transcription.[76] HDAC inhibitors, which increase acetylation of histones, have a beneficial effect on memory and provide neuroprotection in many models of neurodegenerative disease such as Huntington's disease and amyotrophic lateral sclerosis (ALS).[77–83] Given the ability of HDAC inhibitors to ameliorate the phenotype of synaptic dysfunction in other neurodegenerative disease models, their potential for AD treatment has recently been tested in AD models.

One of the first successful studies to explore the potential of HDAC inhibitors as an AD treatment was done in the 3xTg-AD model using nicotinamide, an inhibitor of the Class III, NAD+-dependent HDAC sirtuin family.[84] Nicotinamide treatment improved memory in hippocampal and amygdala-dependent tasks in pre-pathological animals. Although the treatment did not affect amyloid pathology, it decreased levels of pathological, phosphorylated tau. Reduction of pathological tau with nicotinamide caused an increase in microtubule-stability-associated proteins, which may allow neurons to maintain their structure and function. However, it is important to note that this effect of nicotinamide did not carry over to older animals with well-established pathology, demonstrating that nicotinamide should be used as a therapy for early stages of the disease.

The benefit of HDAC inhibitors for AD has been replicated in other models since the 3xTg-AD study. Treatment with sodium 4-phenylbutyrate, a class I HDAC inhibitor, in Tg2576 mice caused behavioural improvement (also by altering tau phosphorylation) and increased markers of synaptic plasticity.[85] Treatment with valproic acid, another HDAC Class I inhibitor, decreased Aβ production and improved memory deficits in APP23 and APP23/PS45 mice.[86] Other inhibitors of Class I HDACs were shown to improve contextual memory deficits in APPswe/PS1dE9 transgenic mice early in their pathological progression.[87] Taken together, these studies emphasize the promise of HDAC inhibitors as an AD treatment to improve cognition by stabilizing neuronal structure.

1.5.3 M1 Agonists

Acetylcholine is a critical neurotransmitter for systems that support learning and memory, and is one of the primary neurotransmitter systems affected by AD. Acetylcholinesterase inhibitors are the most widely used drugs to treat AD, though the currently available inhibitors confer only modest, temporary improvement for patients. Selective agonists to acetylcholine receptors present an alternative strategy to increase cholinergic signalling and provide important disease-modifying effects.

The M1 subtype of muscarinic acetylcholine receptors (mAchRs) are most abundant in the areas of the brain affected by AD pathology,[88,89] and evidence from animals and humans points to M1 agonists as promising drug candidates for AD treatment,[90–97] Most notably, studies in the 3xTg-AD mouse model show that M1 agonists not only have potential to increase cholinergic activity but can also modify disease pathology.[98] In this study, the selective M1 agonist, AF267B, was given to six-month-old animals for 10 weeks. Treated animals showed significant cognitive improvements, as well as decreased soluble and insoluble Aβ levels, with a concurrent decrease in tau pathology. Evidence from this and other studies highlights the potential of M1 agonists as an AD treatment that can not only improve cognition by increasing cholinergic signalling (like currently available treatments), but can also reduce both hallmark pathologies of AD.

1.6 Concluding Remarks

Here we discussed the importance of transgenic models in deciphering the multiple aspects of AD progression. Though the current animal models do not reproduce all the features of AD, they still provide valuable information about disease-related pathways and possible intervention points. Promising new therapeutics have been developed and continue to be refined in transgenic models and, in some cases, have already begun the transition into human clinical trials. Future efforts in transgenic modelling should focus on moving toward a more comprehensive model, which may improve the translation of therapeutics developed in animals. As the science of animal modelling continues to mature, better models will be created which will only accelerate the pace at which therapeutic discovery and development occurs.

References

1. L. E. Hebert, P. A. Scherr, J. L. Bienias, D. A. Bennett and D. A. Evans, *Arch. Neurol.*, 2003, **60**, 1119–1122.
2. M. Cruts and C. van Broeckhoven, *Ann. Med.*, 1998, **30**, 560–565.
3. R. E. Tanzi and L. Bertram, *Neuron*, 2001, **32**, 181–184.
4. D. J. Selkoe, *J. Clin. Invest.*, 2002, **110**, 1375–1381.
5. S. G. Younkin, *Ann. Neurol.*, 1995, **37**, 287–288.
6. R. D. Terry, E. Masliah, D. P. Salmon, N. Butters, R. DeTeresa, R. Hill, L. A. Hansen and R. Katzman, *Ann. Neurol.*, 1991, **30**, 572–580.

7. P. Delaère, C. Duyckaerts, Y. He, F. Piette and J. J. Hauw, *Acta Neuropathol.*, 1991, **81**, 328–335.

8. P. V. Arriagada, J. H. Growdon, E. T. Hedley-Whyte and B. T. Hyman, *Neurology*, 1992, **42**, 631–639.

9. D. W. Dickson, H. A. Crystal, L. A. Mattiace, D. M. Masur, A. D. Blau, P. Davies, S. H. Yen and M. K. Aronson, *Neurobiol. Aging*, **13**, 179–189.

10. L. Berg, D. W. McKeel, Jr., J. P. Miller, J. Baty and J. C. Morris, *Arch. Neurol.*, 1993, **50**, 349–358.

11. D. M. Walsh, B. P. Tseng, R. E. Rydel, M. B. Podlisny and D. J. Selkoe, *Biochemistry*, 2000, **39**, 10831–10839.

12. G. M. Shankar, S. Li, T. H. Mehta, A. Garcia-Munoz, N. E. Shepardson, I. Smith, F. M. Brett, M. A. Farrell, M. J. Rowan, C. A. Lemere, C. M. Regan, D. M. Walsh, B. L. Sabatini and D. J. Selkoe, *Nature Med.*, 2008, **14**, 837–842.

13. M. A. Westerman, D. Cooper-Blacketer, A. Mariash, L. Kotilinek, T. Kawarabayashi, L. H. Younkin, G. A. Carlson, S. G. Younkin and K. H. Ashe, *J. Neurosci.*, 2002, **22**, 1858–1867.

14. D. M. Walsh, I. Klyubin, J. V. Fadeeva, W. K. Cullen, R. Anwyl, M. S. Wolfe, M. J. Rowan and D. J. Selkoe, *Nature*, 2002, **416**, 535–539.

15. C. G. Glabe, *J. Biol. Chem.*, 2008, **283**, 29639–29643.

16. D. J. Selkoe, *Behav. Brain Res.*, 2008, **192**, 106–113.

17. J. Hardy and D. J. Selkoe, *Science (New York, N.Y.)*, 2002, **297**, 353–356.

18. J. Hardy and D. Allsop, *Trends Pharmacol. Sci.*, 1991, **12**, 383–388.

19. M. Ramsden, L. Kotilinek, C. Forster, J. Paulson, E. Mcgowan, K. Santacruz, A. Guimaraes, M. Yue, J. Lewis, G. Carlson, M. Hutton and K. H. Ashe, *J. Neurosci.*, 2005, **25**, 10637–10647.

20. K. Santacruz, J. Lewis, T. Spires, J. Paulson, L. Kotilinek, M. Ingelsson, A. Guimaraes, M. DeTure, M. Ramsden, E. McGowan, C. Forster, M. Yue, J. Orne, C. Janus, A. Mariash, M. Kuskowski, B. Hyman, M. Hutton and K. H. Ashe, *Science (New York, N.Y.)*, 2005, **309**, 476–481.

21. S. Oddo, A. Caccamo, J. D. Shepherd, M. P. Murphy, T. E. Golde, R. Kayed, R. Metherate, M. P. Mattson, Y. Akbari and F. M. LaFerla, *Neuron*, 2003, **39**, 409–421.

22. M. Kitazawa, S. Oddo, T. R. Yamasaki, K. N. Green and F. M. LaFerla, *J. Neurosci.*, 2005, **25**, 8843–8853.

23. M. Fuhrmann, T. Bittner, C. K. E. Jung, S. Burgold, R. M. Page, G. Mitteregger, C. Haass, F. M. LaFerla, H. Kretzschmar and J. Herms, *Nature Neurosci.*, 2010, **13**, 411–413.

24. L. M. Billings, S. Oddo, K. N. Green, J. L. McGaugh and F. M. LaFerla, *Neuron*, 2005, **45**, 675–688.

25. M. Hutton, C. L. Lendon, P. Rizzu, M. Baker, S. Froelich, H. Houlden, S. Pickering-Brown, S. Chakraverty, A. Isaacs, A. Grover, J. Hackett, J. Adamson, S. Lincoln, D. Dickson, P. Davies, R. C. Petersen, M. Stevens, E. de Graaff, E. Wauters, J. van Baren, M. Hillebrand, M. Joosse, J. M. Kwon, P. Nowotny, L. K. Che, J. Norton, J. C. Morris, L. A. Reed,

J. Trojanowski, H. Basun, L. Lannfelt, M. Neystat, S. Fahn, F. Dark, T. Tannenberg, P. R. Dodd, N. Hayward, J. B. Kwok, P. R. Schofield, A. Andreadis, J. Snowden, D. Craufurd, D. Neary, F. Owen, B. A. Oostra, J. Hardy, A. Goate, J. van Swieten, D. Mann, T. Lynch and P. Heutink, *Nature*, 1998, **393**, 702–705.

26. S. Oddo, A. Caccamo, L. Tran, M. P. Lambert, C. G. Glabe, W. L. Klein and F. M. Laferla, *J. Biol. Chem.*, 2006, **281**, 1599–1604.

27. L. M. Billings, K. N. Green, J. L. McGaugh and F. M. LaFerla, *J. Neurosci.*, 2007, **27**, 751–761.

28. S. Oddo, V. Vasilevko, A. Caccamo, M. Kitazawa, D. H. Cribbs and F. M. LaFerla, *J. Biol. Chem.*, 2006, **281**, 39413–39423.

29. S. Oddo, A. Caccamo, M. Kitazawa, B. P. Tseng and F. M. LaFerla, *Neurobiol. Aging*, 2003, **24**, 1063–1070.

30. S. Oddo, A. Caccamo, D. Cheng, B. Jouleh, R. Torp and F. M. LaFerla, *J. Neurochem.*, 2007, **102**, 1053–1063.

31. S. Oddo, L. Billings, J. P. Kesslak, D. H. Cribbs and F. M. LaFerla, *Neuron*, 2004, **43**, 321–332.

32. S. Oddo, A. Caccamo, B. Tseng, D. Cheng, V. Vasilevko, D. H. Cribbs and F. M. LaFerla, *J. Neurosci.*, 2008, **28**, 12163–12175.

33. C. A. Dickey, M. Yue, W. L. Lin, D. W. Dickson, J. H. Dunmore, W. C. Lee, C. Zehr, G. West, S. Cao, A. M. Clark, G. A. Caldwell, K. A. Caldwell, C. Eckman, C. Patterson, M. Hutton and L. Petrucelli, *J. Neurosci.*, 2006, **26**, 6985–6996.

34. C. A. Dickey, C. Patterson, D. Dickson and L. Petrucelli, *Trends Mol. Med.*, 2007, **13**, 32–38.

35. B. P. Tseng, K. N. Green, J. L. Chan, M. Blurton-Jones and F. M. LaFerla, *Neurobiol. Aging*, 2008, **29**, 1607–1618.

36. C. G. Almeida, R. H. Takahashi and G. K. Gouras, *J. Neurosci.*, 2006, **26**, 4277–4288.

37. S. Oh, H. S. Hong, E. Hwang, H. J. Sim, W. Lee, S. J. Shin and I. Mook-Jung, *Mech. Aging Dev.*, 2005, **126**, 1292–1299.

38. F. M. S. de Vrij, D. F. Fischer, F. W. van Leeuwen and E. M. Hol, *Prog. Neurobiol.*, 2004, **74**, 249–270.

39. D. L. Smith, J. Pozueta, B. Gong, O. Arancio and M. Shelanski, *Proc. Natl. Acad. Sci. U. S. A.*, 2009, **106**, 16877–16882.

40. B. Gong, Z. Cao, P. Zheng, O. V. Vitolo, S. Liu, A. Staniszewski, D. Moolman, H. Zhang, M. Shelanski and O. Arancio, *Cell*, 2006, **126**, 775–788.

41. E. G. McGeer and P. L. McGeer, *Exp. Gerontol.*, 1998, **33**, 371–378.

42. M. C. Janelsins, M. A. Mastrangelo, S. Oddo, F. M. LaFerla, H. J. Federoff and W. J. Bowers, *J. Neuroinflammation*, 2005, **2**, 23.

43. M. C. Janelsins, M. A. Mastrangelo, K. M. Park, K. L. Sudol, W. C. Narrow, S. Oddo, F. M. LaFerla, L. M. Callahan, H. J. Federoff and W. J. Bowers, *Am. J. Pathol.*, 2008, **173**, 1768–1782.

44. J. Tan, T. Town, F. Crawford, T. Mori, A. DelleDonne, R. Crescentini, D. Obregon, R. A. Flavell and M. J. Mullan, *Nature Neurosci.*, 2002, **5**, 1288–1293.

45. Y. Yoshiyama, M. Higuchi, B. Zhang, S.-M. Huang, N. Iwata, T. C. Saido, J. Maeda, T. Suhara, J. Q. Trojanowski and V. M.-Y. Lee, *Neuron*, 2007, **53**, 337–351.
46. G. Budas, C. M. Coughlan, J. R. Seckl and K. C. Breen, *Neurosci. Lett.*, 1999, **276**, 61–64.
47. K. N. Green, L. M. Billings, B. Roozendaal, J. L. McGaugh and F. M. LaFerla, *J. Neurosci.*, 2006, **26**, 9047–9056.
48. E. M. Elliott, M. P. Mattson, P. Vanderklish, G. Lynch, I. Chang and R. M. Sapolsky, *J. Neurochem.*, 1993, **61**, 57–67.
49. L. K. Clinton, L. M. Billings, K. N. Green, A. Caccamo, J. Ngo, S. Oddo, J. L. McGaugh and F. M. LaFerla, *Neurobiol. Dis.*, 2007, **28**, 76–82.
50. S. S. Vaynman, Z. Ying, D. Yin and F. Gomez-Pinilla, *Brain Res.*, 2006, **1070**, 124–130.
51. R. B. Wichi, K. De Angelis, L. Jones and M. C. Irigoyen, *Clinics (São Paulo)*, 2009, **64**, 253–258.
52. N. C. Berchtold, N. Castello and C. W. Cotman, *Neuroscience*, 2010, **167**, 588–597.
53. J. A. Joseph, N. A. Denisova, G. Arendash, M. Gordon, D. Diamond, B. Shukitt-Hale and D. Morgan, *Nutr. Neurosci.*, 2003, **6**, 153–162.
54. E. B. Larson, L. Wang, J. D. Bowen, W. C. McCormick, L. Teri, P. Crane and W. Kukull, *Ann. Intern. Med.*, 2006, **144**, 73–81.
55. R. P. Friedland, T. Fritsch, K. A. Smyth, E. Koss, A. J. Lerner, C. H. Chen, G. J. Petot and S. M. Debanne, *Proc. Natl. Acad. Sci. U. S. A.*, 2001, **98**, 3440–3445.
56. P. A. Adlard, V. M. Perreau, V. Pop and C. W. Cotman, *J. Neurosci.*, 2005, **25**, 4217–4221.
57. S. Mirochnic, S. Wolf, M. Staufenbiel and G. Kempermann, *Hippocampus*, 2009, **19**, 1008–1018.
58. K. N. Green, H. Martinez-Coria, H. Khashwji, E. B. Hall, K. A. Yurko-Mauro, L. Ellis and F. M. LaFerla, *J. Neurosci.*, 2007, **27**, 4385–4395.
59. K. Rezai-Zadeh, D. Shytle, N. Sun, T. Mori, H. Hou, D. Jeanniton, J. Ehrhart, K. Townsend, J. Zeng, D. Morgan, J. Hardy, T. Town and J. Tan, *J. Neurosci.*, 2005, **25**, 8807–8814.
60. S. Kotani, E. Sakaguchi, S. Warashina, N. Matsukawa, Y. Ishikura, Y. Kiso, M. Sakakibara, T. Yoshimoto, J. Guo and T. Yamashima, *Neurosci. Res.*, 2006, **56**, 159–164.
61. Y. Freund-Levi, H. Basun, T. Cederholm, G. Faxen-Irving, A. Garlind, M. Grut, I. Vedin, J. Palmblad, L. O. Wahlund and M. Eriksdotter-Jonhagen, *Int. J. Geriatr. Psychiatry*, 2008, **23**, 161–169.
62. J. T. Moroney, E. Bagiella, D. W. Desmond, M. C. Paik, Y. Stern and T. K. Tatemichi, *Stroke*, 1996, **27**, 1283–1289.
63. G. Tesco, Y. H. Koh, E. L. Kang, A. N. Cameron, S. Das, M. Sena-Esteves, M. Hiltunen, S.-H. Yang, Z. Zhong, Y. Shen, J. W. Simpkins and R. E. Tanzi, *Neuron*, 2007, **54**, 721–737.
64. M. A. Koike, K. N. Green, M. Blurton-Jones and F. M. Laferla, *Am. J. Pathol.*, 2010.

65. P. Davies and A. J. Maloney, *Lancet*, 1976, **2**, 1403.
66. J. L. Muir, *Pharmacol. Biochem. Behav.*, 1997, **56**, 687–696.
67. D. Schenk, R. Barbour, W. Dunn, G. Gordon, H. Grajeda, T. Guido, K. Hu, J. Huang, K. Johnson-Wood, K. Khan, D. Kholodenko, M. Lee, Z. Liao, I. Lieberburg, R. Motter, L. Mutter, F. Soriano, G. Shopp, N. Vasquez, C. Vandevert, S. Walker, M. Wogulis, T. Yednock, D. Games and P. Seubert, *Nature*, 1999, **400**, 173–177.
68. C. A. Lemere, R. Maron, E. T. Spooner, T. J. Grenfell, C. Mori, R. Desai, W. W. Hancock, H. L. Weiner and D. J. Selkoe, *Ann. N. Y. Acad. Sci.*, 2000, **920**, 328–331.
69. H. L. Weiner, C. A. Lemere, R. Maron, E. T. Spooner, T. J. Grenfell, C. Mori, S. Issazadeh, W. W. Hancock and D. J. Selkoe, *Ann. Neurol.*, 2000, **48**, 567–579.
70. F. Bard, C. Cannon, R. Barbour, R. L. Burke, D. Games, H. Grajeda, T. Guido, K. Hu, J. Huang, K. Johnson-Wood, K. Khan, D. Kholodenko, M. Lee, I. Lieberburg, R. Motter, M. Nguyen, F. Soriano, N. Vasquez, K. Weiss, B. Welch, P. Seubert, D. Schenk and T. Yednock, *Nature Med.*, 2000, **6**, 916–919.
71. L. A. Kotilinek, B. Bacskai, M. Westerman, T. Kawarabayashi, L. Younkin, B. T. Hyman, S. Younkin and K. H. Ashe, *J. Neurosci.*, 2002, **22**, 6331–6335.
72. B. Vellas, R. Black, L. J. Thal, N. C. Fox, M. Daniels, G. McLennan, C. Tompkins, C. Leibman, M. Pomfret and M. Grundman, *Curr. Alzheimer Res.*, 2009, **6**, 144–151.
73. E. Masliah, L. Hansen, A. Adame, L. Crews, F. Bard, C. Lee, P. Seubert, D. Games, L. Kirby and D. Schenk, *Neurology*, 2005, **64**, 129–131.
74. J. A. Nicoll, E. Barton, D. Boche, J. W. Neal, I. Ferrer, P. Thompson, C. Vlachouli, D. Wilkinson, A. Bayer, D. Games, P. Seubert, D. Schenk and C. Holmes, *J. Neuropathol. Exp. Neurol.*, 2006, **65**, 1040–1048.
75. J. P. Holmes, L. C. Benavides, J. D. Gates, M. G. Carmichael, M. T. Hueman, E. A. Mittendorf, J. L. Murray, A. Amin, D. Craig, E. von Hofe, S. Ponniah and G. E. Peoples, *J. Clin. Oncol.*, 2008, **26**, 3426–3433.
76. T. Kouzarides, *Cell*, 2007, **128**, 693–705.
77. R. M. Barrett and M. A. Wood, *Learn. Mem.*, 2008, **15**, 460–467.
78. A. McCampbell, A. A. Taye, L. Whitty, E. Penney, J. S. Steffan and K. H. Fischbeck, *Proc. Natl. Acad. Sci. U. S. A.*, 2001, **98**, 15179–15184.
79. J. S. Steffan, L. Bodai, J. Pallos, M. Poelman, A. McCampbell, B. L. Apostol, A. Kazantsev, E. Schmidt, Y. Z. Zhu, M. Greenwald, R. Kurokawa, D. E. Housman, G. R. Jackson, J. L. Marsh and L. M. Thompson, *Nature*, 2001, **413**, 739–743.
80. R. J. Ferrante, J. K. Kubilus, J. Lee, H. Ryu, A. Beesen, B. Zucker, K. Smith, N. W. Kowall, R. R. Ratan, R. Luthi-Carter and S. M. Hersch, *J. Neurosci.*, 2003, **23**, 9418–9427.
81. G. Gardian, L. Yang, C. Cleren, N. Y. Calingasan, P. Klivenyi and M. F. Beal, *Neuromolecular Med.*, 2004, **5**, 235–242.

82. S. Camelo, A. H. Iglesias, D. Hwang, B. Due, H. Ryu, K. Smith, S. G. Gray, J. Imitola, G. Duran, B. Assaf, B. Langley, S. J. Khoury, G. Stephanopoulos, U. De Girolami, R. R. Ratan, R. J. Ferrante and F. Dangond, *J. Neuroimmunol.*, 2005, **164**, 10–21.
83. H. Ryu, K. Smith, S. I. Camelo, I. Carreras, J. Lee, A. H. Iglesias, F. Dangond, K. A. Cormier, M. E. Cudkowicz, R. H. Brown and R. J. Ferrante, *J. Neurochem.*, 2005, **93**, 1087–1098.
84. K. N. Green, J. S. Steffan, H. Martinez-Coria, X. Sun, S. S. Schreiber, L. M. Thompson and F. M. LaFerla, *J. Neurosci.*, 2008, **28**, 11500–11510.
85. A. Ricobaraza, M. Cuadrado-Tejedor, A. Pérez-Mediavilla, D. Frechilla, J. Del Río and A. García-Osta, *Neuropsychopharmacology*, 2009, **34**, 1721–1732.
86. H. Qing, G. He, P. T. T. Ly, C. J. Fox, M. Staufenbiel, F. Cai, Z. Zhang, S. Wei, X. Sun, C.-H. Chen, W. Zhou, K. Wang and W. Song, *J. Exp. Med.*, 2008, **205**, 2781–2789.
87. M. Kilgore, C. A. Miller, D. M. Fass, K. M. Hennig, S. J. Haggarty, J. D. Sweatt and G. Rumbaugh, *Neuropsychopharmacology*, 2010, **35**, 870–880.
88. A. I. Levey, C. A. Kitt, W. F. Simonds, D. L. Price and M. R. Brann, *J. Neurosci.*, 1991, **11**, 3218–3226.
89. J. Wei, E. A. Walton, A. Milici and J. J. Buccafusco, *J. Neurochem.*, 1994, **63**, 815–821.
90. A. Fisher, Z. Pittel, R. Haring, N. Bar-Ner, M. Kliger-Spatz, N. Natan, I. Egozi, H. Sonego, I. Marcovitch and R. Brandeis, *J. Mol. Neurosci.*, 2003, **20**, 349–356.
91. C. Hock, A. Maddalena, I. Heuser, D. Naber, W. Oertel, H. von der Kammer, M. Wienrich, A. Raschig, M. Deng, J. H. Growdon and R. M. Nitsch, *Ann. N. Y. Acad. Sci.*, 2000, **920**, 285–291.
92. C. Hock, A. Maddalena, A. Raschig, F. Müller-Spahn, G. Eschweiler, K. Hager, I. Heuser, H. Hampel, T. Müller-Thomsen, W. Oertel, M. Wienrich, A. Signorell, C. Gonzalez-Agosti and R. M. Nitsch, *Amyloid*, 2003, **10**, 1–6.
93. H. Seo, A. W. Ferree and O. Isacson, *Eur. J. Neurosci.*, 2002, **15**, 498–506.
94. R. M. Nitsch, M. Deng, M. Tennis, D. Schoenfeld and J. H. Growdon, *Ann. Neurol.*, 2000, **48**, 913–918.
95. D. Müller, H. Wiegmann, U. Langer, S. Moltzen-Lenz and R. M. Nitsch, *J. Neural Transm.*, 1998, **105**, 1029–1043.
96. N. C. Bodick, W. W. Offen, A. I. Levey, N. R. Cutler, S. G. Gauthier, A. Satlin, H. E. Shannon, G. D. Tollefson, K. Rasmussen, F. P. Bymaster, D. J. Hurley, W. Z. Potter and S. M. Paul, *Arch. Neurol.*, 1997, **54**, 465–473.
97. A. Caccamo, A. Fisher and F. M. LaFerla, *Curr. Alzheimer Res.*, 2009, **6**, 112–117.
98. A. Caccamo, S. Oddo, L. M. Billings, K. N. Green, H. Martinez-Coria, A. Fisher and F. M. LaFerla, *Neuron*, 2006, **49**, 671–682.

CHAPTER 2

Animal Models of Amyloid/PS-1 Pathology

AKIHIKO TAKASHIMA

Laboratory for Alzheimer's Disease, Brain Science Institute, RIKEN, Saitama, Japan

2.1 β-Amyloid Generation

Extracellular amyloid β (Aβ) peptide is the major component of senile plaques, which have long been known to be a specific pathological marker of Alzheimer's disease (AD). This form of Aβ comprises about 40 amino acids and is derived from two successive cleavages of amyloid precursor protein (APP), which is a type 1 membrane protein. β-secretase, which has been cloned as a β-site APP-cleaving enzyme (BACE),[1–3] catalyses the initial cleavage. This cleavage of APP produces an APP C-terminus fragment comprising 99 amino acid residues (termed C99). C99 is further cleaved at the membrane-spanning region by γ-secretase—a complex of presenilin-1 (PS-1), Pen2, nicastrin and APH1—to generate Aβ.[4–10] Aβ is then transported to the extracellular space, where increasing concentrations initiate self-aggregation and the formation of β-pleated sheet structures.

Two classes of aggregated Aβ exist in the brain: Aβ oligomer and fibrillar Aβ. Aβ oligomer contributes to synaptotoxicity,[11–15] whereas fibrillar Aβ forms the classical 'senile' plaques, first observed by Alois Alzheimer. APP mutations have been found in cases of familial AD (FAD)[16–18] (Figure 2.1). Most of these mutations have been examined for their ability to generate Aβ by over-expressing mutant APP in cellular models, which either display increased levels

RSC Drug Discovery Series No. 6
Animal Models for Neurodegenerative Disease
Edited by Jesús Avila, Jose J. Lucas and Félix Hernández
© Royal Society of Chemistry 2011
Published by the Royal Society of Chemistry, www.rsc.org

of Aβ[19–21] or elevated ratios of Aβ42/Aβ40.[22,23] In the Arctic mutation,[24] modification occurs in the Aβ sequence; thus, it does not affect Aβ generation and the Aβ42/Aβ40 ratio. However, Aβ produced by this mutation has an increased probability of forming protofibrils,[25–27] which are more toxic than fibrillar Aβ. APP mutation studies support the Aβ hypothesis, which posits that increased Aβ levels, Aβ42/Aβ40 ratio, and toxic Aβ formation and aggregation underlie the development of AD.[28–30]

2.2   APP Transgenic Mouse Models

Motivated by the Aβ hypothesis and through the use of various APP mutations, promoters and vectors, many have attempted to generate realistic animal models of AD that exhibit AD pathologies such as Aβ deposition, neurofibrillary tangles (NFTs), neuronal and synaptic loss, and progressive memory impairment. Genes encoding three isoforms of human APP (hAPP) (695, 751 and 770) have been used as transgenes. The amino acid sequence of Aβ derived from mouse APP differs from that of hAPP; these differences affect the ability of Aβ to self-aggregate.[31,32] Thus, hAPP overexpression, hAPP gene knock-in and expression of yeast artificial chromosome (YAC) containing hAPP have been employed to model Aβ generation and deposition.[33]

Expression of hAPP751 containing the Kunitz serine protease inhibitor domain, driven by neuron specific enolase (NSE),[34] increases Aβ generation but results in little Aβ deposition. Expression of the amyloidogenic region of hAPP (104 amino acids of the *C*-terminal region), driven by a minigene of neurofilament light chain, causes a modest accumulation of Aβ in cytoplasm and neuropil.[35,36] This mouse model exhibited learning deficits and cell loss in hippocampal area CA1 concomitant with impairment in the maintenance of long-term potentiation (LTP).[36] However, because mice expressing wild-type APP or APP *C*-terminus show less extracellular Aβ deposition in the brain than mice expressing the FAD mutant form of APP, the latter mice are used as an AD mouse model. This is because extracellular Aβ deposition is similar to that observed in AD patients.

2.2.1   PDAPP Transgenic Mouse Model

This transgenic mouse was the first recognized AD model with AD-type pathological accumulation of Aβ peptide in the brain. This line was generated using a platelet-derived growth factor β-promoter (PDGF),[37] which drives a human APP minigene encoding alternative spliced hAPP. This sequence contains the FAD mutation V→F at APP position 717 (PDAPP).[38]

The PDAPP transgenic mouse expresses 4–6 fold higher levels of total APP mRNA compared with non-transgenic mice or humans, but expresses reduced levels of endogenous mouse APP mRNA.[39] In addition, these mice produce increased amounts of Aβ42, which is the predominant form of Aβ in brain plaques of AD cases observed at neuropathological evaluation. Age- and brain

VKMDAEFRHDSGYEVHHQKLVFFAEDVGSNKGAIIGLMVGGVVIATVIVITLVMLK

		APP T714I (Austrian)
APP670N/671L	A21G (Flemish)	APP T714A (Iranian)
(Swedish)	E22Q (Dutch)	APP V715M (French)
	E22K (Italian)	APP V715A (German)
H6R (English)	E22G (Arctic)	APP I716V (Florida)
D7N (Tottori)	E23N (Iowa)	APP I716T (Florida)
		APP V717I (London)
		APP V717L (London)
		APP V717F (Indiana)
		APP V717G (Indiana)

Figure 2.1 APP mutations. Shading indicates Aβ sequences.

region-dependent development of typical amyloid plaques, dystrophic neurites, loss of presynaptic terminals, astrocytosis and microgliosis occur in PDAPP mice. Extracellular amyloid fibrils (9–11 nm in diameter) are abundant and are strikingly similar to those observed in AD. Aβ deposition starts at about eight months of age in the cingulate cortex, and then progressively appears in the dentate gyrus, CA1 and entorhinal cortex by 12 months. By 22 months of age, the deposition of Aβ is very dense. However, a limitation of the PDAPP mouse model is that it does not exhibit neuronal loss[40] or NFTs,[41] which are other pathological markers of AD.

PDAPP transgenic mice learn the Morris water maze more slowly than non-transgenic mice, regardless of age. This behavioural task is often used to screen for hippocampal dysfunction and requires intact memory function. Learning capacity in the Morris water maze declines in an age-dependent manner in PDAPP transgenic mice. Aβ burden and learning capacity are significantly negatively correlated in PDAPP mice 16–22 months old. Other behavioural tests, such as object recognition memory, are normal.[42]

It was first reported that immunization with synthetic Aβ42 in PDAPP mice reduces Aβ plaque formation and related pathologies.[43] However, the quantitative relationship between Aβ burden and learning capacity remains uncertain in Aβ-immunized PDAPP mice.

2.2.2 Tg2576 Transgenic Mouse Model

Tg2576 mice[44] overexpress hAPP695 with the Swedish mutation (K670N/M671L) under the control of a hamster prion promoter. It was the first APP mouse line to show memory impairment in the Morris water maze task. Diffuse brain plaques and focal Aβ deposits are detectable at about 9–11 months. Tg2576 mice show neutrophil abnormalities, but they do not exhibit neuronal loss.[45]

In biochemical analyses, SDS-insoluble Aβ40 and Aβ42 are detectable at 6–7 months and increase exponentially in concentration with advancing age.

Unmodified full-length Aβ accumulates first, with N-terminal-truncated Aβ or modified Aβ developing only in old age.[46] However, far less Aβ accumulates in the brains of Tg2576 mice compared to the brains of AD patients. Coincident with the marked deposition of Aβ in the brain, there is a decrease in CSF Aβ and a substantial, highly significant decrease in plasma Aβ.[47] This reduction of CSF Aβ is exactly what is observed in AD patients, which is interpreted as indicating Aβ deposition has begun in the brain.

Tg2576 mice younger than six months do not show spatial reference learning and memory deficits in the Morris water maze. Mice that are six months or older have spatial reference learning and memory deficits, and become progressively impaired with age. Performance correlates well with the amount of SDS-insoluble brain Aβ This suggests that small Aβ aggregates induce neuronal dysfunction rather than Aβ deposition.[48] Because the amount of Aβ deposition does not correlate with the extent of neuronal loss and related brain dysfunction in AD patients, the suggestion that small Aβ aggregates induce neuronal dysfunction may explain how Aβ is involved in neuronal dysfunction in AD according to the Aβ hypothesis.[28-30]

To understand the mechanism of the SDS-insoluble, Aβ aggregate-induced memory deficit, Chapman and colleagues[49] investigated plasticity of synaptic transmission in CA1 and dentate gyrus (DG). Induction of LTP is significantly impaired in both of these areas of the hippocampus of old-aged Tg2576 mice, while no impairment is observed at younger ages (2–8 months). Impairment of LTP induction is well correlated with impaired performance of spatial working memory.

The temporal progression of morphological and functional deficits in Tg2576 mice was investigated by another group.[50] The earliest appearance of deficits occurred at about four months. A decrease in hippocampal spine density, LTP and hippocampal-dependent contextual fear memory was accompanied by an increase in the Aβ42/Aβ40 ratio. Astrocyte reactivity and microgliosis occurred at a later stage, which was also associated with Aβ deposition. Thus, before Aβ deposition is apparent, hippocampal-dependent memory is observed, presumably due to reduced synaptic plasticity but not to neuron loss.

It is unclear which Aβ aggregate underlies the age-dependent memory deficit in Tg2576 mice. Two approaches have been used to address this issue. Lesne and colleagues[51] investigated the temporal change of soluble Aβ oligomers and memory formation in Tg2576 mice and found that 56 kDa Aβ dodecamers (Aβ*56) were significantly inversely correlated with memory formation. Aβ monomers, trimers and hexamers were not related to memory formation. Also, injection of Aβ*56 into wild-type mice impaired memory formation.[51]

The other approach used electrophysiology, measuring hippocampal synaptic transmission. SDS-stable soluble Aβ oligomers impaired LTP induction[52-54] and enhanced long-term depression (LTD).[54,55] However, it remains unclear whether Aβ*56 in particular or some other Aβ oligomer is related to the impairment of LTP and enhancement of LTD.

Tg2576 mice develop many neurological features similar to that of AD, including amyloid plaques, dystrophic neuritis and inflammatory changes.[44,47,56] However, Tg2576 mice lack NFTs, significant neuronal loss and gross brain atrophy. These mice may be a good model for studying the preclinical stages of AD, when dementia is undetectable and neuronal loss is not apparent.

Immunization of Tg2576 mice with Aβ was investigated using various antibodies. Intraperitoneal injection of BAM-10, which recognizes the *N*-terminus of Aβ, fully reverses memory loss in Tg2576 mice, although it does not reduce Aβ deposition.[57] This result suggests that BAM-10 acts by neutralizing Aβ assemblies in the brain that impair cognitive function. Immunization with NAB61, which preferentially recognizes a conformational epitope present in dimeric, small oligomeric and higher-order Aβ species, resulted in significant improvement in spatial learning and memory. This was accompanied by a reduction of Aβ deposition, suggesting that oligomeric Aβ contributes to the memory loss and Aβ deposition.[58] Both antibodies commonly result in a reversal of memory impairment in Tg2576 mice, accompanied by a variable reduction of Aβ deposition.

Immunization of AD patients with Aβ does not halt the progression of dementia, although Aβ deposition is reduced. Although the Tg2576 mouse is considered a 'gold standard' of AD mouse models, it has limitations in capturing all the features of AD.

2.2.3 APP23 Transgenic Mouse Model

The APP23 transgenic mouse model was developed by Novartis Pharma, Inc. The APP23 line expresses hAPP751 with the Swedish mutation (K670N/M671L) under control of the mouse Thy-1.2 promoter.[59] It expresses sevenfold higher levels of hAPP compared with those of endogenous mouse APP. APP23 mice exhibit rare Congo-red positive Aβ plaques in the brain at six months, and Aβ plaques increase in size and number in neocortex and hippocampus with advancing age.

Two other hAPP transgenic mouse lines, APP17 and APP22, have been generated by the same group at Novartis.[59] APP17 mice express hAPP with the Swedish mutation, and APP22 mice express hAPP with the Swedish and London (V717I) mutations, both under control of the human Thy1 promoter. These two mouse lines express twofold more hAPP compared with that of endogenous mouse APP. APP17 mice do not show Aβ deposition in the brain until 24 months of age. APP22 mice mainly have diffuse plaques and a few Congo-red positive plaques in hippocampus and neocortex at 18 months. Compared with APP17 mice, APP22 mice express an increased level of hAPP, which is an important factor influencing Aβ deposition. Moreover, APP22 mice express the V717I London mutation, a more important factor affecting Aβ deposition. This mutation increases the Aβ42/40 ratio without changing the total Aβ level. APP17 mice exhibit brain Aβ deposition, whereas APP22 mice mainly exhibit diffuse plaques.

Detailed analysis of APP23 mice reveals that deposition of hAPP occurs preferentially in brain arterioles and capillaries. Within individual vessels, there is wide variability in the morphology of Aβ accumulation, ranging from a thin ring of amyloid in the vessel wall seen in cross-section to large plaque-like extrusions into the neutrophil. This cerebral amyloid angiopathy (CAA) is commonly observed with aging in many species, and is associated with local neuron loss, synaptic abnormalities, microglial activation and micro-haemorrhage.[60–62] The CAA in APP23 mice is strikingly similar to that observed in human aging and AD.

The Thy1 promoter causes expression of mutant hAPP and generates Aβ in neurons. Neuron-derived Aβ is deposited extracellularly and then accumulates in the vessel wall, leading to disruption of vessel walls. This, in turn leads to parenchymal haemorrhage,[61] which is observed as CAA-associated haemor-rhagic stroke in AD.

APP23 mice exhibit both behavioural impairments in the Morris water maze and in passive avoidance tasks at young ages,[63] both of which precede Aβ deposition.[64] Synaptic loss correlates with cognitive decline in AD patients.[65] However in APP23 mice, the number of synaptophysin-positive presynaptic terminals is maintained in the face of robust parenchymal Aβ deposition. Thus, Aβ deposition alone may not be sufficient to account for age-dependent behavioural changes in APP23.

Functional MRI studies suggest that the age-dependent dysfunction in APP23 mice may be attributed in part to compromised cerebrovascular reactivity.[66] Indeed, age-dependent cerebrovascular abnormalities and blood flow disturbances are found in APP23 mice.[67,68] Passive immunization of APP transgenic mice with anti-Aβ antibodies reduces Aβ deposition in the brain and improves cognition. On the down side, this kind of immunotherapy may also be accompanied by a significant increase in the frequency of intracerebral hae-morrhages.[69] Interestingly, approved drugs for symptomatic treatment of dementia in patients (*e.g.* galantamine, rivastigmine and donepezil) reduce cognitive impairment in APP23 mice,[70] suggesting that the APP23 mouse model may be a valuable tool for evaluating therapeutic drugs aimed at treating AD.

2.2.4 APP Dutch Transgenic Mouse Model

Cerebrovascular Aβ deposition is a common pathological feature of AD and related disorders. In particular, the Dutch E693Q and Iowa D694N mutations in the APP gene cause familial cerebrovascular amyloidosis with abundant diffuse Aβ plaques.[71–73] Cognitive deterioration due to white matter lesions is also present.

In APP Dutch transgenic mice, hAPP with the E693Q Dutch mutation is expressed by neuron-specific Thy1 promoter. Neuronal overexpression of human E693Q APP in APP Dutch mice causes extensive CAA, smooth muscle cell degeneration, haemorrhages and neuroinflammation, but results in no parenchymal amyloid deposition.[74–76] CAA, smooth muscle cell degeneration, haemorrhages, neuroinflammation and parenchymal amyloid deposition are present in humans with AD. CAA mainly consists of Aβ40. Crossing APP Dutch mice with mice overexpressing human PS-1 bearing the G384A mutation produces mice with an increased Aβ42/Aβ40 ratio. This double-transgenic mouse develops parenchymal amyloid deposits in the neocortex and hippocampus, although vascular amyloid is also present.[74] These observations show that Dutch mutant Aβ preferentially accumulates in blood vessels and that an increased Aβ42/Aβ4 ratio results in parenchymal Aβ deposition.

Crossing the APP Dutch mouse with the ApoE knockout mouse[77] produces mice with reduced fibrillar cerebral microvascular Aβ deposition and neuroinflammation, suggesting that ApoE may be involved in vascular Aβ deposition.

Behavioural tests with the APP Dutch mouse have been used to assess cognition. These mice are impaired in learning and memory of the Barnes' circle maze task, and performance correlates with early-onset accumulation of subicular microvascular amyloid and neuroinflammation.[78] Treatment of APP Dutch mice with the anti-inflammatory drug, minocycline, reduces neuroinflammation and results in improvement of behavioural performance without changing microvascular Aβ deposition.[79]

2.2.5 CRND8 Transgenic Mouse Model

The CRND8 transgenic mouse overexpresses an APP gene containing the Swedish (K670N and M671L) and the Indiana (V717F) FAD mutations. These mice exhibit an age-related increase in Aβ production and early onset of plaque deposition in neocortex and hippocampus.[80] Levels of hAPP expression are fivefold greater than endogenous APP levels. Up to eight weeks of age, CRND8 mice have elevated levels of Aβ40 and Aβ42, but no plaque deposition. From approximately nine weeks, plaque deposition begins and progresses such that, by 16 weeks, all CRND8 mice in a population show multiple brain plaque deposits.[80] This nature of Aβ deposition occurs quite early in CRND8 mice compared with the other hAPP mouse lines.

The high level of Aβ42 generation in CRND8 mice is associated with early impairment of learning in the Morris water maze, which is present by three months of age. In this mouse line, the behavioural deficit seems to coincide with the onset of Aβ deposition,[81] and the learning impairment in young mice is reversed by immunization against Aβ42.[82]

2.2.6 J20 Transgenic Mouse Model

In the J20 transgenic mouse, the PDGF β-chain promoter was used for direct neuronal expression of alternatively spliced minigenes encoding hAPP695, hAPP751 and hAPP770. The expressed APP gene carries the Swedish and Indiana mutations, as with the CRND8 mouse. Aβ brain plaques are present from 2–4 months of age in J20 transgenic mice, and high levels of Aβ42 result in age-dependent formation of amyloid plaques.[83] This pattern, however, is absent in expression-matched wild-type hAPP mice.[83] This suggests that high levels of Aβ42 are essential for plaque formation. However, both mouse lines have reduced synaptophysin immunoreactivity, which correlates with Aβ brain concentration but not with APP expression or Aβ plaque load. This suggests that an increase in Aβ concentration is a probable cause of the synaptic deficit.

In correspondence with the synaptic deficit, J20 mice are impaired in learning and memory of the Morris water maze.[84] When J20 mice are crossed with tau knockout heterozygote or homozygote mice, this memory deficit is absent. This crossbreeding does not affect Aβ deposition and neuritic dystrophy, indicating that tau plays an important role in the Aβ-induced synaptic deficit and consequent memory impairment. J20 mice do not show tau hyperphosphorylation.[84]

In cortical and hippocampal regions, J20 mice have spontaneous non-convulsive seizure activity that is associated with GABA-immunopositive fibre sprouting, enhanced synaptic inhibition and synaptic plasticity deficits in the dentate gyrus.[85,86] Because Aβ is a positive regulator of synaptic transmission,[85] hAPP mice may have an aberrant increase in neural excitability and simultaneous compensatory inhibition in the hippocampus through enhanced GABA signalling. This could lead to reduced LTP and memory impairment. GABA-A receptor inhibitor treatment restores LTP induction and restores memory performance without affecting Aβ deposition.[87]

2.2.7 APPSwDI/NOS2$^{-/-}$ Transgenic Mouse Model

It has been known for some time that human macrophages *in vitro* produce significantly less nitric oxide (NO) when stimulated under the same conditions that produce high NO levels in mouse macrophages.[88,89] Also, more recently, it has been shown that microglia isolated from mice expressing the human NOS2 gene have significantly lower NO production than wild-type mouse microglia under the same conditions.[90] Numerous studies have also shown that induced nitric oxide synthase (iNOS) activity can be neuroprotective.[91–93] This observation drives the hypothesis that hAPP transgenic mice may not show NFTs and neuronal loss because of iNOS activity.

To test this idea, Wilcock and colleagues[94] generated double transgenic mice by crossing APPSwDI (Swedish K760N/M671L, Dutch E693Q and Iowa D694N) transgenic mice with NOS2$^{-/-}$ mice. Although APPSwDI mice lack NFT-like pathology, APPSwDI/NOS2$^{-/-}$ double-transgenic mice display impaired spatial memory, neuronal loss and NFT-like pathology. Tg2576 APP transgenic mice crossed with NOS2 knockout (NOS2$^{-/-}$) mice produce mice that develop NFT-like pathology from endogenous mouse tau and show evidence of neurodegeneration.[95] These observations suggest that NOS2 may be a key factor in producing Aβ-induced NFT-like pathology from endogenous tau in hAPP mice.

Aβ-immunized APPSwDI/NOS2$^{-/-}$ mice,[96] which have predominantly vascular amyloid pathology, show a 30% decrease in brain Aβ and a 35–45% reduction in hyperphosphorylated tau. Neuron loss and cognitive deficits are partially reduced. In Aβ-immunized APPSw/NOS2$^{-/-}$ mice, brain Aβ is reduced by 65–85% and hyperphosphorylated tau by 50–60%. Furthermore, neuron loss is not present and memory deficits are fully reversed. Microhaemorrhage was observed in all Aβ-immunized APPSw/NOS2$^{-/-}$ mice, supporting the development of amyloid-lowering therapies for treatment of AD.

2.2.8 Adeno-associated Virus (AAV)-mediated Gene Transfer of BRI-Aβ

Transgenic models of AD utilize overexpression of mutant human APP to increase Aβ production and recapitulate AD cognitive deficits and pathologies. However, overexpression of APP results not only in increased production of both Aβ40 and Aβ42, but also in elevated levels of other APP fragments such as sAPP, *C*-terminal fragments of APP and AICD, which can have neuroprotective effects,[97] be neurotoxic,[98] or have signalling functions,[99] and which can influence learning and memory.[100–103]

Lawlor and colleagues[104] developed a novel expression system that selectively expresses Aβ40 or Aβ42 in the secretory pathway using AAV vectors encoding BRI-Aβ cDNAs, which are fusions between human Aβ peptides and the BRI protein (which is involved in amyloid deposition in British and Danish familial dementia).[105–107] These constructs achieve a high level of hippocampal expression and secretion of the specific-encoded Aβ peptide. BRI-Aβ42 expression and combination of BRI-Aβ40 and BRI-Aβ42 (BRI-Aβ40+Aβ42) expression result in elevated levels of detergent-insoluble Aβ. No significant increase in detergent-insoluble Aβ is seen in rat expressing BRI-Aβ40. Only BRI-Aβ40 + Aβ42 shows Aβ depositions, but they are not congophilic aggregations.[104]

2.2.9 Humanized Aβ Transgenic Mouse Models

2.2.9.1 Human APP Gene Knock-in Transgenic Mouse Models

Transgenic hAPP mouse lines overexpress mutant hAPP in the brain, which shows Aβ deposition, while wild-type hAPP mice exhibit much less Aβ deposition. In sporadic AD, Down's syndrome, and even in cases of normal human aging, extracellular Aβ deposits occur in the absence of the APP mutation. Therefore, the context of hAPP overexpression in rodents may be an artificial, unrealistic experimental approach to understanding the biological mechanisms behind AD brain pathologies. A difference in Aβ sequence between rodents and humans may make it difficult to form spontaneous Aβ deposition in rodents,[31,32] although aged monkey,[108] bear,[109] dogs[110] and cats[111] show spontaneous Aβ deposition.

Reaume and colleagues[112] generated a mouse line expressing humanized Aβ with the Swedish mutation, instead of mouse Aβ, by means of a gene targeting method. In these mice, human Aβ production is significantly increased to levels nine-fold greater than those in normal human brain. However, non-amyloidogenic processing is depressed, because of the APP-containing Swedish mutation. Even though humanized Aβ is generated by the Swedish mutation,

Aβ deposition requires an increased level of Aβ42, which can be accomplished by crossing these mice with mutant PS-1 mice.[113]

2.2.9.2   YAC hAPP Transgenic Mouse Model

The entire ~400 kbp human APP gene harbouring the FAD mutations is introduced into the germline of transgenic mice using a 650 kbp yeast artificial chromosome (YAC). This allows the proper spatial and temporal expression of mutant APP with appropriate splice donor and acceptor sites needed to generate the entire spectrum of alternatively spliced APP transcripts and encoded proteins.[114,115] YAC mutant hAPP mice (R1.40 mouse) show Aβ deposition at 26 months in hemizygous mice and 15 months in homozygous mice.[116]

Aged R1.40 transgenic and control mice were tested for spatial learning and memory in the Morris water maze and for working memory in the Y maze. Results from the Morris water maze demonstrated intact learning in the both control and R1.40 mice, but impairment in long-term retention only in the transgenic mice. This long-term memory deficit is not associated with Aβ deposition. However, age-related working memory impairment in the Y maze correlates with the presence of Aβ deposits.[117]

2.2.10   Amyloid Degradation

The two major endopeptidases involved in Aβ degradation are zinc metalloendopeptidases, also referred to as insulin degrading enzyme (IDE), and neprilysin (NEP). The overexpression of IDE (two-fold over endogenous concentrations) in mice overexpressing human APP harbouring both APP717 and APP670/671 missense mutations results in a 50% decrease in both soluble and insoluble Aβ.[118] In addition, in APP/IDE double-transgenic mice, the Aβ plaque burden is diminished by 50%. Heterozygous IDE–/+ mice exhibit Aβ levels that are intermediate between those of wild-type control mice and IDE–/– mice. Because naturally occurring oligomers (dimers and trimers) of secreted Aβ in culture medium are resistant to IDE, while Aβ monomers in the same sample are avidly degraded by the enzyme,[119,120] IDE cannot rescue Aβ-induced LTP deficits.[53] Studies examining viral expression of NEP in primary neurons,[121] studies of knockout mice[122] and studies of chronic transgenic overexpression show identical results to those described above for IDE. Although overexpression of NEP in J20 transgenic mice reduces Aβ deposition, the memory deficit persists because NEP cannot remove Ab*58.[123]

2.3   Other Transgenic Mouse Models

2.3.1   Physiological and Pathological Roles of PS-1

Through genetic linkage analysis of FAD patients, missense mutations of the PS-1 gene on chromosome 14 have been discovered. Several point mutations of the PS-2 gene on chromosome 1 have also been found in FAD patients.[124] Coexpression of various mutants of PS-1 and APP C100 cDNAs in COS7 cells increases the Aβ42/Aβ40 ratio.[125] PS-1 gene-deficient fibroblasts exhibit increased amounts of β-secretase-cleaved APP-C terminal fragments, and they do not generate Aβ,[126] suggesting that PS-1 is involved in γ-secretase activity.

Biochemical studies reveal that presenilins are part of a larger γ-secretase complex, associated with nicastrin, Pen2 and APH1. The γ-secretase complex cleaves several type I membrane proteins, including APP at the intramembrane region.[127] In the γ-secretase complex, presenilins are stabilized as *N*- and *C*-terminus cleaved forms, suggesting that cleaved PS-1 serves as an active form in the γ-secretase complex. Some mutants of PS-1 reduce the generation of cleaved *N*- and *C*-terminus PS-1,[128] suggesting that mutant PS-1 may reduce γ-secretase activity. Indeed, PS-1 mutant cDNA expression affects Aβ40 more than Aβ42. Consequently, mutant PS-1 increases the ratio of Aβ42/Aβ40.[129,130]

PS-1 also reportedly has non-γ-secretase functions. PS-1 is associated with GSK-3β and its substrate β-catenin and tau.[131,132] Mutant PS-1 activates GSK-3β and enhances phosphorylation of β-catenin and tau. Phosphorylated β-catenin is degraded rapidly by proteasomes, which may explain why PS-1 knockout mice have an increased probability of developing skin cancer.[133] β-catenin is involved in cell proliferation.

Loss of PS-1 function, either by genetic knockout or by introducing mutations causing FAD, results in Ca^{2+} overloading in the endoplasmic reticulum.[134–137] Inositol 1,4,5-triphosphate (InsP3)-mediated intracellular Ca^{2+} release[135,138] and capacitative calcium entry (CCE)[134] disturb PS-1 mutations, which may lead to impairments of synaptic plasticity. Indeed, mice lacking presenilin in the forebrain exhibit age-related impairments in LTP and in hippocampal-dependent memory. These mice eventually show synaptic and neuronal degeneration.[139]

2.3.2 PS-1 Transgenic Mouse Model

Mutant PS-1 increases the ratio of Aβ42/Aβ40, although the total amount of Aβ generated is reduced. Several lines of mice derived from crossing hAPP transgenic mice with PS-1 mutants have been produced. Duff and colleagues[140] generated the PSAPP mouse by crossing hAPP transgenic mice (Tg2576) with wild-type and mutant PS-1 transgenic mice. Analysis showed that overexpression of mutant PS-1 selectively increases brain Aβ42(43), and mutant PS-1 accelerates Aβ deposition at six months of age. Wild-type mice do not show this pattern. Tg2576 mice show Aβ deposition at nine months.

Both Tg2576 mice and double transgenic mice show reduced spontaneous alternation in the Y maze before substantial Aβ deposition is apparent, suggesting that the PS-1 mutation accelerates the behavioural deficit and Aβ deposition by increasing the ratio of Aβ42/Aβ40.[141] The acceleration of Aβ deposition and behavioural deficit in double-transgenic mice may model early onset AD.

The PS-1 mutant knock-in (PS1 KI) mouse also shows an increased ratio of Aβ42/Aβ40, but no deposition of mouse Aβ. This line is generated by introducing the FAD mutation of presenilin into the mouse PS gene. Histological and quantitative RT-PCR analyses of PS1 KI mice showed normal gross hippocampal morphology and unaltered expression of three genes involved in inflammatory responses. These mice show hippocampal-dependent spatial memory impairments caused by the mutation (PS1 M146V) and have age-related deterioration of memory.[142]

Another mutant (I213T) PS1 KI mouse exhibits NFT-like tau pathology in the hippocampus[143] in old age, which suggests that the PS-1 mutation may be involved not only in Aβ deposition but also in NFT formation.

Mice with PS-1 mutations (M146L and G183V) have Pick's bodies,[144,145] a neuropathological marker of frontotemporal dementia (FTD). They do not show amyloid plaques. The literature suggests that PS-1 mutations result in an overlapping continuum of the clinical and neuropathological features of AD and FTD. With PS-1 mutations, AD or FTD development may depend on the degree of functional loss of the PS-1 gene and the resultant tau pathophysiology.

M233T/L235P knock-in mutations in PS-1 mice overexpress a mutated hAPP. These mice are termed APP(SL)/PS1 KI. They show a loss of parvalbumin-immunoreactive neurons (40–50%) in the CA1-2 region and a loss of calretinin-immunoreactive neurons (37–52%) in the dentate gyrus and hilus, suggesting that this mutation results in hippocampal interneuron loss, a feature also seen in AD brain.[146]

The PS-1 gene knockout is lethal due to skeletal and central nervous system (CNS) deficits. However, through loxP/Cre-recombinase-mediated deletion, it is possible to delete the PS-1 gene from postnatal neurons. This conditional PS-1 knockout mouse is viable and can be crossed with the mutant hAPP (V717I) mouse. The absence of PS-1 in neurons effectively prevents amyloid pathology, even in mice that are 18 months old. This also rescues hippocampus LTP impairment in progeny mice expressing only hAPP (V717I). However, the absence of PS-1 failed to rescue the object recognition impairment of parent hAPP mice.[147] This suggests that therapies based on γ-secretase inhibitors will be limited for treatment of AD.

2.3.3   3xTg Mouse Model

LaFerla's group[148] at the University of California, Irvine, has generated triple transgenic (3xTg) mice expressing hAPP (K670N/M671L), PS-1 (M146V) and Tau (P301L) (see Chapter 1). In these mice, Aβ deposits appear first in neocortex and later in hippocampus with aging. Tau pathology is first apparent in the hippocampus and then later in neocortex. Despite equivalent overexpression of the human beta-APP and human tau transgenes, Aβ deposition develops prior to the tangle pathology. Aβ immunization reduces not only extracellular Aβ plaques but also early tau pathology.[149] This observation strongly supports the Aβ cascade hypothesis.

In behavioural tests of two-month-old mice, these mice are cognitively unimpaired but brain pathology is absent. The earliest cognitive impairment manifests at four months as a deficit in long-term retention of a task. This deficit correlates with the accumulation of intraneuronal Aβ in the hippocampus and amygdala. Plaque and tangle pathology is absent at this age. Aβ immunization reduces intraneuronal Aβ accumulation and rescues the early behavioural impairment on a hippocampal-dependent task,[150] suggesting that intraneuronal Aβ accumulation is the cause of the early-appearing cognitive deficits in this mouse.

Some potential therapeutic drugs of AD have been tested using 3xTg mice. Chronic nicotine administration causes upregulation of nicotinic receptors, which correlates with a marked increase in the aggregation and phosphorylation state of tau, without changing Aβ.[151] The selective M1 muscarinic agonist AF267B rescues the cognitive deficits in a spatial task and inhibits Aβ deposition and tau pathology, suggesting that M1 agonists may be potent candidate drugs for treating AD.[152] Lithium, a GSK-3 inhibitor, reduces tau phosphorylation but not Aβ accumulation.[153] It also does not affect the working memory deficit.[154] However, this result is controversial when the GSK-3 inhibitor, NP12, is used in the APP/tau Tg mice.[155]

Memantine, a *N*-methyl-D-aspartate (NMDA) receptor antagonist that is approved for the treatment of moderate to severe AD, restores cognition and significantly reduces the levels of insoluble Aβ and hyperphosphorylated tau in 3xTg mice.[156] Prophylactic treatment of young 3xTg mice with ibuprofen reduces intraneuronal oligomeric Aβ, reduces cognitive deficits, and results in no hyperphosphorylated tau immunoreactivity.[157] Ibuprofen treatment in mild to moderate AD patients does not improve cognition.[158]

2.4 hAPP Mouse Models and Treatment of AD

The effectiveness of Aβ immunization was first reported using the PDAPP mouse. Aβ immunization removed Aβ deposits from brain parenchyma.[43] Subsequent reports investigated the recovery of memory deficits in Aβ-immunized hAPP mice. A ∼ 50% reduction in dense-core Aβ plaques in the brain of CRND8 mice was sufficient to significantly improve memory performance in the Morris water maze. Active Aβ immunization in Tg2576/PS-1 mutant mice also resulted in restoration of memory performance in the radial arm water maze test.[159] Passive immunization in Tg2576 mice with BAM-10 antibody recovered memory performance and was associated with no apparent Aβ deposition.[57]

Based on the above hAPP mouse studies, clinical trials of Aβ immunization were started. The trial was stopped, however, because of oedema associated with inflammation. Postmortem evaluation of the brains of some patients was performed. Atrophy of the cerebral cortex and white matter was present and was associated with ventricular enlargement. Immunized patients showed features that are not normally seen in AD and which bear remarkable similarities to features of immunized aged PDAPP mice. Both have a low density of Aβ plaques in extensive areas of the cerebral cortex.[160]

Although immunization of AD patients with Aβ42 (AN1792) is associated with a long-term reduction in Aβ load and a variable degree of plaque removal compared with unimmunized control patients, plaque removal was not sufficient to halt progressive neurodegeneration in AD patients. This is in contrast to immunized hAPP mice, in which the memory impairment is ameliorated and Aβ load is reduced.[161]

From Aβ immunization studies, including hAPP mouse model and patient experiments, Aβ immunization commonly reduces Aβ deposition, but the effect of Aβ immunization on cognition differs between the two experiments. Several reasons may explain this difference.[162] Although NFTs are prominent features in AD patients, most hAPP mouse lines do not show NFTs and neuronal loss, except the 3xTg mouse and hAPP/NOS$^{-/-}$ mouse. Aβ immunization of both 3xTg mice and hAPP/NOS$^{-/-}$ mice reduced tau pathology and recovered impaired cognitive function, which may mean that APP overexpression-associated overproduction of Aβ mice is involved in tau pathology and impaired cognition in the hAPP mouse model.

Recently, the physiological role of Aβ was revealed.[86] Increased extracellular concentration of endogenous Aβ enhances the probability of transmitter release, and decreased extracellular concentration (through experimental addition of Aβ antibody) reduces probability of transmitter release. Both cases do not respond to electrical stimulation, suggesting that an appropriate concentration of Aβ is required to maintain normal synaptic activity. If this is the case, the total Aβ level in the hAPP mouse may return to normal endogenous Aβ levels during Aβ immunization. This, in turn, could lead to an alleviation of Aβ pathology and recovery of cognitive deficits in the immunized hAPP mouse. Because AD patients do not overproduce Aβ, Aβ immunization reduces endogenous brain Aβ levels, which may lead to deficits of synaptic activity similar to what occurs with Aβ overproduction in some transgenic mice. Thus, the failure of Aβ immunization in AD patients may be because Aβ concentrations were reduced too much.

In the search for effective therapeutics for the treatment or prevention of AD, we will do well to pay increased attention to the entire spectrum of differences between hAPP mouse models and AD patients.

References

1. S. Sinha, J. P. Anderson, R. Barbour, G. S. Basi, R. Caccavello, D. Davis, M. Doan, H. F. Dovey, N. Frigon, J. Hong, K. Jacobson-Croak, N. Jewett, P. Keim, J. Knops, I. Lieberburg, M. Power, H. Tan, G. Tatsuno, J. Tung, D. Schenk, P. Seubert, S. M. Suomensaari, S. Wang, D. Walker, J. Zhao, L. McConlogue and V. John, *Nature*, 1999, **402**(6761), 537–540.

2. R. Vassar, B. D. Bennett, S. Babu-Khan, S. Kahn, E. A. Mendiaz, P. Denis, D. B. Teplow, S. Ross, P. Amarante, R. Loeloff, Y. Luo, S. Fisher, J. Fuller, S. Edenson, J. Lile, M. A. Jarosinski, A. L. Biere,

E. Curran, T. Burgess, J. C. Louis, F. Collins, J. Treanor, G. Rogers and M. Citron, *Science*, 1999, **286**(5440), 735–741.

3. L. B. Yang, K. Lindholm, R. Yan, M. Citron, W. Xia, X. L. Yang, T. Beach, L. Sue, P. Wong, D. Price, R. Li and Y. Shen, *Nat. Med.*, 2003, **9**(1), 3–4.

4. A. Capell, H. Steiner, H. Romig, S. Keck, M. Baader, M. G. Grim, R. Baumeister and C. Haass, *Nat. Cell Biol.*, 2000, **2**, 205–211.

5. B. De Strooper, *Neuron*, 2003, **38**(1), 9–12.

6. B. De Strooper, W. Annaert, P. Cupers, P. Saftig, K. Craessaerts, J. S. Mumm, E. H. Schroeter, V. Schrijvers, M. S. Wolfe, W. J. Ray, A. Goate and R. Kopan, *Nature*, 1999, **398**(6727), 518–522.

7. W. P. Esler, W. T. Kimberly, B. L. Ostaszewski, T. S. Diehl, C. L. Moore, J. Y. Tsai, T. Rahmati, W. Xia, D. J. Selkoe and M. S. Wolfe, *Nat. Cell Biol.*, 2000, **2**, 428–434.

8. N. Takasugi, T. Tomita, I. Hayashi, M. Tsuruoka, M. Niimura, Y. Takahashi, G. Thinakaran and T. Iwatsubo, *Nature*, 2003, **422**(6930), 438–441.

9. M. S. Wolfe, W. Xia, B. L. Ostaszewski, T. S. Diehl, W. T. Kimberly and D. J. Selkoe, *Nature*, 1999, **398**, 513–517.

10. D. Edbauer, E. Winkler, J. T. Regula, B. Pesold, H. Steiner and C. Haass, *Nat. Cell Biol.*, 2003, **5**(5), 486–488.

11. A. Deshpande, H. Kawai, R. Metherate, C. G. Glabe and J. Busciglio, *J. Neurosci.*, 2009, **29**(13), 4004–4015.

12. P. N. Lacor, M. C. Buniel, L. Chang, S. J. Fernandez, Y. Gong, K. L. Viola, M. P. Lambert, P. T. Velasco, E. H. Bigio, C. E. Finch, G. A. Krafft and W. L. Klein, *J. Neurosci.*, 2004, **24**(45), 10191–10200.

13. P. N. Lacor, M. C. Buniel, P. W. Furlow, A. S. Clemente, P. T. Velasco, M. Wood, K. L. Viola and W. L. Klein, *J. Neurosci.*, 2007, **27**(4), 796–807.

14. V. Nimmrich, C. Grimm, A. Draguhn, S. Barghorn, A. Lehmann, H. Schoemaker, H. Hillen, G. Gross, U. Ebert and C. Bruehl, *J. Neurosci.*, 2008, **28**(4), 788–797.

15. P. J. Shughrue, P. J. Acton, R. S. Breese, W. Q. Zhao, E. Chen-Dodson, R. W. Hepler, A. L. Wolfe, M. Matthews, G. J. Heidecker, J. G. Joyce, S. A. Villarreal and G. G. Kinney, *Neurobiol. Aging*, **31**(2), 189–202.

16. A. Goate, M. C. Chartier-Harlin, M. Mullan, J. Brown, F. Crawford, L. Fidani, L. Giuffra, A. Haynes, N. Irving, L. James, R. mant, P. newton, K. Rooke, P. Roques, C. Talbot, M. Pericak-Vance, A. Roses, R. Williamson, M. N. Roser, M. Owen and J. Hardy, *Nature*, 1991, **349**(6311), 704–706.

17. L. Lannfelt, N. Bogdanovic, H. Appelgren, K. Axelman, L. Lilius, G. Hansson, D. Schenk, J. Hardy and B. Winblad, *Neurosci. Lett.*, 1994, **168**(1–2), 254–256.

18. E. Almqvist, S. Lake, K. Axelman, K. Johansson and B. Winblad, *J. Neural Transm. Park. Dis. Dement. Sect.*, 1993, **6**, 151–156.

19. K. M. Felsenstein, L. W. Hunihan and S. B. Roberts, *Nat. Genet.*, 1994, **6**(3), 251–255.

20. C. Haass, C. A. Lemere, A. Capell, M. Citron, P. Seubert, D. Schenk, L. Lannfelt and D. J. Selkoe, *Nat. Med.*, 1995, **1**(12), 1291–1296.

21. M. Citron, T. Oltersdorf, C. Haass, L. McConlogue, A. Y. Hung, P. Seubert, C. Vigo-Pelfrey, I. Lieberburg and D. J. Selkoe, *Nature*, 1992, **360**(6405), 672–674.

22. B. Zhao, S. S. Sisodia and J. W. Kusiak, *J. Neurosci. Res.*, 1995, **40**(2), 261–268.

23. A. Tamaoka, A. Odaka, Y. Ishibashi, M. Usami, N. Sahara, N. Suzuki, N. Nukina, H. Mizusawa, S. Shoji and I. Kanazawa, *J. Biol. Chem.*, 1994, **269**(52), 32721–32724.

24. C. Nilsberth, A. Westlind-Danielsson, C. B. Eckman, M. M. Condron, K. Axelman, C. Forsell, C. Stenh, J. Luthman, D. B. Teplow, S. G. Younkin, J. Naslund and L. Lannfelt, *Nat. Neurosci.*, 2001, **4**(9), 887–893.

25. C. Sahlin, A. Lord, K. Magnusson, H. Englund, C. G. Almeida, P. Greengard, F. Nyberg, G. K. Gouras, L. Lannfelt and L. N. Nilsson, *J. Neurochem.*, 2007, **101**(3), 854–862.

26. A. Lord, H. Kalimo, C. Eckman, X. Q. Zhang, L. Lannfelt and L. N. Nilsson, *Neurobiol. Aging*, 2006, **27**(1), 67–77.

27. H. A. Lashuel, D. M. Hartley, B. M. Petre, J. S. Wall, M. N. Simon, T. Walz and P. T. Lansbury Jr., *J. Mol. Biol.*, 2003, **332**(4), 795–808.

28. J. Hardy, *J. Neurochem.*, 2009, **110**(4), 1129–1134.

29. J. Hardy and D. J. Selkoe, *Science*, 2002, **297**(5580), 353–356.

30. J. A. Hardy and G. A. Higgins, *Science*, 1992, **256**(5054), 184–185.

31. S. T. Liu, G. Howlett and C. J. Barrow, *Biochemistry*, 1999, **38**(29), 9373–9378.

32. L. Otvos Jr., G. I. Szendrei, V. M. Lee and H. H. Mantsch, *Eur. J. Biochem.*, 1993, **211**(1-2), 249–257.

33. J. Gotz and L. M. Ittner, *Nat. Rev. Neurosci.*, 2008, **9**(7), 532–544.

34. D. Quon, Y. Wang, R. Catalano, J. M. Scardina, K. Murakami and B. Cordell, *Nature*, 1991, **352**(6332), 239–241.

35. A. Kammesheidt, F. M. Boyce, A. F. Spanoyannis, B. J. Cummings, M. Ortegon, C. Cotman, J. L. Vaught and R. L. Neve, *Proc. Natl. Acad. Sci. U. S. A.*, 1992, **89**(22), 10857–10861.

36. J. Nalbantoglu, G. Tirado-Santiago, A. Lahsaini, J. Poirier, O. Goncalves, G. Verge, F. Momoli, S. A. Welner, G. Massicotte, J. P. Julien and M. L. Shapiro, *Nature*, 1997, **387**(6632), 500–505.

37. D. Games, D. Adams, R. Alessandrini, R. Barbour, P. Berthelette, C. Blackwell, T. Carr, J. Clemens, T. Donaldson, F. Gillespie, T. Guido, S. Hagopian, K. Johnson, K. Khan, M. Lee, P. Leibowitz, I. Lieberburg, S. Little, E. Masliah, L. McConiogue, M. Montoya-Zavala, L. Mucke, L. Paganini, E. Penniman, M. Power, D. Schenk, P. Seubert, B. Snyder, F. Soriano, H. Tan, J. Vitalo, S. Wadsworth, B. Wolozin and J. Zhao, *Nature*, 1995, **373**(6514), 523–527.

38. J. Murrell, M. Farlow, B. Ghetti and M. D. Benson, *Science*, 1991, **254**(5028), 97–99.

39. E. M. Rockenstein, L. McConlogue, H. Tan, M. Power, E. Masliah and L. Mucke, *J. Biol. Chem.*, 1995, **270**(47), 28257–28267.
40. M. C. Irizarry, F. Soriano, M. McNamara, K. J. Page, D. Schenk, D. Games and B. T. Hyman, *J. Neurosci.*, 1997, **17**(18), 7053–7059.
41. E. Masliah, A. Sisk, M. Mallory and D. Games, *J. Neuropathol. Exp. Neurol.*, 2001, **60**(4), 357–368.
42. G. Chen, K. S. Chen, J. Knox, J. Inglis, A. Bernard, S. J. Martin, A. Justice, L. McConlogue, D. Games, S. B. Freedman and R. G. Morris, *Nature*, 2000, **408**(6815), 975–979.
43. D. Schenk, R. Barbour, W. Dunn, G. Gordon, H. Grajeda, T. Guido, K. Hu, J. Huang, K. Johnson-Wood, K. Khan, D. Kholodenko, M. Lee, Z. Liao, I. Lieberburg, R. Motter, L. Mutter, F. Soriano, G. Shopp, N. Vasquez, C. Vandevert, S. Walker, M. Wogulis, T. Yednock, D. Games and P. Seubert, *Nature*, 1999, **400**(6740), 173–177.
44. K. Hsiao, P. Chapman, S. Nilsen, C. Eckman, Y. Harigaya, S. Younkin, F. Yang and G. Cole, *Science*, 1996, **274**(5284), 99–102.
45. M. C. Irizarry, M. McNamara, K. Fedorchak, K. Hsiao and B. T. Hyman, *J. Neuropathol. Exp. Neurol.*, 1997, **56**(9), 965–973.
46. W. Kalback, M. D. Watson, T. A. Kokjohn, Y. M. Kuo, N. Weiss, D. C. Luehrs, J. Lopez, D. Brune, S. S. Sisodia, M. Staufenbiel, M. Emmerling and A. E. Roher, *Biochemistry*, 2002, **41**(3), 922–928.
47. T. Kawarabayashi, L. H. Younkin, T. C. Saido, M. Shoji, K. H. Ashe and S. G. Younkin, *J. Neurosci.*, 2001, **21**(2), 372–381.
48. M. A. Westerman, D. Cooper-Blacketer, A. Mariash, L. Kotilinek, T. Kawarabayashi, L. H. Younkin, G. A. Carlson, S. G. Younkin and K. H. Ashe, *J. Neurosci.*, 2002, **22**(5), 1858–1867.
49. P. F. Chapman, G. L. White, M. W. Jones, D. Cooper-Blacketer, V. J. Marshall, M. Irizarry, L. Younkin, M. A. Good, T. V. Bliss, B. T. Hyman, S. G. Younkin and K. K. Hsiao, *Nat. Neurosci.*, 1999, **2**(3), 271–276.
50. J. S. Jacobsen, C. C. Wu, J. M. Redwine, T. A. Comery, R. Arias, M. Bowlby, R. Martone, J. H. Morrison, M. N. Pangalos, P. H. Reinhart and F. E. Bloom, *Proc. Natl. Acad. Sci. U. S. A.*, 2006, **103**(13), 5161–5166.
51. S. Lesne, M. T. Koh, L. Kotilinek, R. Kayed, C. G. Glabe, A. Yang, M. Gallagher and K. H. Ashe, *Nature*, 2006, **440**(7082), 352–357.
52. J. P. Cleary, D. M. Walsh, J. J. Hofmeister, G. M. Shankar, M. A. Kuskowski, D. J. Selkoe and K. H. Ashe, *Nat. Neurosci.*, 2005, **8**(1), 79–84.
53. D. M. Walsh, I. Klyubin, J. V. Fadeeva, W. K. Cullen, R. Anwyl, M. S. Wolfe, M. J. Rowan and D. J. Selkoe, *Nature*, 2002, **416**(6880), 535–539.
54. G. M. Shankar, S. Li, T. H. Mehta, A. Garcia-Munoz, N. E. Shepardson, I. Smith, F. M. Brett, M. A. Farrell, M. J. Rowan, C. A. Lemere, C. M. Regan, D. M. Walsh, B. L. Sabatini and D. J. Selkoe, *Nat. Med.*, 2008, **14**(8), 837–842.

55. S. Li, S. Hong, N. E. Shepardson, D. M. Walsh, G. M. Shankar and D. Selkoe, *Neuron*, 2009, **62**(6), 788–801.

56. W. C. Benzing, J. R. Wujek, E. K. Ward, D. Shaffer, K. H. Ashe, S. G. Younkin and K. R. Brunden, *Neurobiol. Aging*, 1999, **20**(6), 581–589.

57. L. A. Kotilinek, B. Bacskai, M. Westerman, T. Kawarabayashi, L. Younkin, B. T. Hyman, S. Younkin and K. H. Ashe, *J. Neurosci.*, 2002, **22**(15), 6331–6335.

58. E. B. Lee, L. Z. Leng, B. Zhang, L. Kwong, J. Q. Trojanowski, T. Abel and V. M. Lee, *J. Biol. Chem.*, 2006, **281**(7), 4292–4299.

59. C. Sturchler-Pierrat, D. Abramowski, M. Duke, K. H. Wiederhold, C. Mistl, S. Rothacher, B. Ledermann, K. Burki, P. Frey, P. A. Paganetti, C. Waridel, M. E. Calhoun, M. Jucker, A. Probst, M. Staufenbiel and B. Sommer, *Proc. Natl. Acad. Sci. U. S. A.*, 1997, **94**(24), 13287–13292.

60. L. Bondolfi, M. Calhoun, F. Ermini, H. G. Kuhn, K. H. Wiederhold, L. Walker, M. Staufenbiel and M. Jucker, *J. Neurosci.*, 2002, **22**(2), 515–522.

61. D. T. Winkler, L. Bondolfi, M. C. Herzig, L. Jann, M. E. Calhoun, K. H. Wiederhold, M. Tolnay, M. Staufenbiel and M. Jucker, *J. Neurosci.*, 2001, **21**(5), 1619–1627.

62. M. E. Calhoun, P. Burgermeister, A. L. Phinney, M. Stalder, M. Tolnay, K. H. Wiederhold, D. Abramowski, C. Sturchler-Pierrat, B. Sommer, M. Staufenbiel and M. Jucker, *Proc. Natl. Acad. Sci. U. S. A.*, 1999, **96**(24), 14088–14093.

63. P. H. Kelly, L. Bondolfi, D. Hunziker, H. P. Schlecht, K. Carver, E. Maguire, D. Abramowski, K. H. Wiederhold, C. Sturchler-Pierrat, M. Jucker, R. Bergmann, M. Staufenbiel and B. Sommer, *Neurobiol. Aging*, 2003, **24**(2), 365–378.

64. D. Van Dam, R. D'Hooge, M. Staufenbiel, C. Van Ginneken, F. Van Meir and P. P. De Deyn, *Eur. J. Neurosci.*, 2003, **17**(2), 388–396.

65. R. D. Terry, E. Masliah, D. P. Salmon, N. Butters, R. DeTeresa, R. Hill, L. A. Hansen and R. Katzman, *Ann. Neurol.*, 1991, **30**(4), 572–580.

66. T. Mueggler, D. Baumann, M. Rausch, M. Staufenbiel and M. Rudin, *J. Neurosci.*, 2003, **23**(23), 8231–8236.

67. N. Beckmann, A. Schuler, T. Mueggler, E. P. Meyer, K. H. Wiederhold, M. Staufenbiel and T. Krucker, *J. Neurosci.*, 2003, **23**(24), 8453–8459.

68. D. R. Thal, E. Capetillo-Zarate, S. Larionov, M. Staufenbiel, S. Zurbruegg and N. Beckmann, *Neurobiol. Aging*, 2009, **30**(12), 1936–1948.

69. G. J. Burbach, A. Vlachos, E. Ghebremedhin, D. Del Turco, J. Coomaraswamy, M. Staufenbiel, M. Jucker and T. Deller, *Neurobiol. Aging*, 2007, **28**(2), 202–212.

70. D. Van Dam and P. P. De Deyn, *Eur. Neuropsychopharmacol.*, 2006, **16**(1), 59–69.

71. C. Van Broeckhoven, J. Haan, E. Bakker, J. A. Hardy, W. Van Hul, A. Wehnert, M. Vegter-Van der Vlis and R. A. Roos, *Science*, 1990, **248**(4959), 1120–1122.

72. E. Levy, M. D. Carman, I. J. Fernandez-Madrid, M. D. Power, I. Lieberburg, S. G. van Duinen, G. T. Bots, W. Luyendijk and B. Frangione, *Science*, 1990, **248**(4959), 1124–1126.
73. M. Bornebroek, J. Haan, M. A. van Buchem, J. B. Lanser, M. A. de Vries-vd Weerd, M. Zoeteweij and R. A. Roos, *Arch. Neurol.*, 1996, **53**(1), 43–48.
74. M. C. Herzig, D. T. Winkler, P. Burgermeister, M. Pfeifer, E. Kohler, S. D. Schmidt, S. Danner, D. Abramowski, C. Sturchler-Pierrat, K. Burki, S. G. van Duinen, M. L. Maat-Schieman, M. Staufenbiel, P. M. Mathews and M. Jucker, *Nat. Neurosci.*, 2004, **7**(9), 954–960.
75. J. Davis, F. Xu, R. Deane, G. Romanov, M. L. Previti, K. Zeigler, B. V. Zlokovic and W. E. Van Nostrand, *J. Biol. Chem.*, 2004, **279**(19), 20296–20306.
76. J. Miao, F. Xu, J. Davis, I. Otte-Holler, M. M. Verbeek and W. E. Van Nostrand, *Am. J. Pathol.*, 2005, **167**(2), 505–515.
77. J. Miao, M. P. Vitek, F. Xu, M. L. Previti, J. Davis and W. E. Van Nostrand, *J. Neurosci.*, 2005, **25**(27), 6271–6277.
78. F. Xu, A. M. Grande, J. K. Robinson, M. L. Previti, M. Vasek, J. Davis and W. E. Van Nostrand, *Neuroscience*, 2007, **146**(1), 98–107.
79. R. Fan, F. Xu, M. L. Previti, J. Davis, A. M. Grande, J. K. Robinson and W. E. Van Nostrand, *J. Neurosci.*, 2007, **27**(12), 3057–3063.
80. M. A. Chishti, D. S. Yang, C. Janus, A. L. Phinney, P. Horne, J. Pearson, R. Strome, N. Zuker, J. Loukides, J. French, S. Turner, G. Lozza, M. Grilli, S. Kunicki, C. Morissette, J. Paquette, F. Gervais, C. Bergeron, P. E. Fraser, G. A. Carlson, P. S. George-Hyslop and D. Westaway, *J. Biol. Chem.*, 2001, **276**(24), 21562–21570.
81. L. A. Hyde, T. M. Kazdoba, M. Grilli, G. Lozza, R. Brusa, Q. Zhang, G. T. Wong, M. F. McCool, L. Zhang, E. M. Parker and G. A. Higgins, *Behav. Brain Res.*, 2005, **160**(2), 344–355.
82. C. Janus, J. Pearson, J. McLaurin, P. M. Mathews, Y. Jiang, S. D. Schmidt, M. A. Chishti, P. Horne, D. Heslin, J. French, H. T. Mount, R. A. Nixon, M. Mercken, C. Bergeron, P. E. Fraser, P. St George-Hyslop and D. Westaway, *Nature*, 2000, **408**(6815), 979–982.
83. L. Mucke, E. Masliah, G. Q. Yu, M. Mallory, E. M. Rockenstein, G. Tatsuno, K. Hu, D. Kholodenko, K. Johnson-Wood and L. McConlogue, *J. Neurosci.*, 2000, **20**(11), 4050–4058.
84. E. D. Roberson, K. Scearce-Levie, J. J. Palop, F. Yan, I. H. Cheng, T. Wu, H. Gerstein, G. Q. Yu and L. Mucke, *Science*, 2007, **316**(5825), 750–754.
85. J. J. Palop, J. Chin, E. D. Roberson, J. Wang, M. T. Thwin, N. Bien-Ly, J. Yoo, K. O. Ho, G. Q. Yu, A. Kreitzer, S. Finkbeiner, J. L. Noebels and L. Mucke, *Neuron*, 2007, **55**(5), 697–711.
86. E. Abramov, I. Dolev, H. Fogel, G. D. Ciccotosto, E. Ruff and I. Slutsky, *Nat. Neurosci.*, 2009, **12**(12), 1567–1576.
87. Y. Yoshiike, T. Kimura, S. Yamashita, H. Furudate, T. Mizoroki, M. Murayama and A. Takashima, *PLoS One*, 2008, **3**(8), e3029.

88. C. Colton, S. Wilt, D. Gilbert, O. Chernyshev, J. Snell and M. Dubois-Dalcq, *Mol. Chem. Neuropathol.*, 1996, **28**(1–3), 15–20.
89. J. B. Weinberg, M. A. Misukonis, P. J. Shami, S. N. Mason, D. L. Sauls, W. A. Dittman, E. R. Wood, G. K. Smith, B. McDonald and K. E. Bachus, *Blood*, 1995, **86**(3), 1184–1195.
90. M. P. Vitek, C. Brown, Q. Xu, H. Dawson, N. Mitsuda and C. A. Colton, *Antioxid. Redox Signal*, 2006, **8**(5-6), 893–901.
91. K. D. Kroncke, C. V. Suschek and V. Kolb-Bachofen, *Antioxid. Redox Signal*, 2000, **2**(3), 585–605.
92. X. Q. Tang, H. M. Yu, J. L. Zhi, Y. Cui, E. H. Tang, J. Q. Feng and P. X. Chen, *Life Sci.*, 2006, **79**(9), 870–876.
93. E. Ciani, S. Guidi, R. Bartesaghi and A. Contestabile, *J. Neurochem.*, 2002, **82**(5), 1282–1289.
94. D. M. Wilcock, M. R. Lewis, W. E. Van Nostrand, J. Davis, M. L. Previti, N. Gharkholonarehe, M. P. Vitek and C. A. Colton, *J. Neurosci.*, 2008, **28**(7), 1537–1545.
95. C. A. Colton, M. P. Vitek, D. A. Wink, Q. Xu, V. Cantillana, M. L. Previti, W. E. Van Nostrand, J. B. Weinberg and H. Dawson, *Proc. Natl. Acad. Sci. U. S. A.*, 2006, **103**(34), 12867–12872.
96. D. M. Wilcock, N. Gharkholonarehe, W. E. Van Nostrand, J. Davis, M. P. Vitek and C. A. Colton, *J. Neurosci.*, 2009, **29**(25), 7957–7965.
97. K. Furukawa, B. L. Sopher, R. E. Rydel, J. G. Begley, D. G. Pham, G. M. Martin, M. Fox and M. P. Mattson, *J. Neurochem.*, 1996, **67**(5), 1882–1896.
98. D. C. Lu, S. Rabizadeh, S. Chandra, R. F. Shayya, L. M. Ellerby, X. Ye, G. S. Salvesen, E. H. Koo and D. E. Bredesen, *Nat. Med.*, 2000, **6**(4), 397–404.
99. F. M. LaFerla, *Nat. Rev. Neurosci.*, 2002, **3**(11), 862–872.
100. J. C. Dodart, C. Mathis and A. Ungerer, *Rev. Neurosci.*, 2000, **11**(2–3), 75–93.
101. P. R. Turner, K. O'Connor, W. P. Tate and W. C. Abraham, *Prog. Neurobiol.*, 2003, **70**(1), 1–32.
102. A. Bour, S. Little, J. C. Dodart, C. Kelche and C. Mathis, *Neurobiol. Learn. Mem.*, 2004, **81**(1), 27–38.
103. V. Galvan, O. F. Gorostiza, S. Banwait, M. Ataie, A. V. Logvinova, S. Sitaraman, E. Carlson, S. A. Sagi, N. Chevallier, K. Jin, D. A. Greenberg and D. E. Bredesen, *Proc. Natl. Acad. Sci. U. S. A.*, 2006, **103**(18), 7130–7135.
104. P. A. Lawlor, R. J. Bland, P. Das, R. W. Price, V. Holloway, L. Smithson, B. L. Dicker, M. J. During, D. Young and T. E. Golde, *Mol. Neurodegener.*, 2007, **2**, 11.
105. P. A. Lewis, S. Piper, M. Baker, L. Onstead, M. P. Murphy, J. Hardy, R. Wang, E. McGowan and T. E. Golde, *Biochim. Biophys. Acta*, 2001, **1537**(1), 58–62.
106. R. Vidal, B. Frangione, A. Rostagno, S. Mead, T. Revesz, G. Plant and J. Ghiso, *Nature*, 1999, **399**(6738), 776–781.

107. R. Vidal, T. Revesz, A. Rostagno, E. Kim, J. L. Holton, T. Bek, M. Bojsen-Moller, H. Braendgaard, G. Plant, J. Ghiso and B. Frangione, *Proc. Natl. Acad. Sci. U. S. A.*, 2000, **97**(9), 4920–4925.
108. D. L. Price, L. J. Martin, S. S. Sisodia, M. V. Wagster, E. H. Koo, L. C. Walker, V. E. Koliatsos and L. C. Cork, *Brain Pathol.*, 1991, **1**(4), 287–296.
109. K. Uchida, T. Yoshino, R. Yamaguchi, S. Tateyama, Y. Kimoto, H. Nakayama and N. Goto, *Vet. Pathol.*, 1995, **32**(4), 412–414.
110. B. J. Cummings, J. H. Su, C. W. Cotman, R. White and M. J. Russell, *Neurobiol. Aging*, 1993, **14**(6), 547–560.
111. B. J. Cummings, T. Satou, E. Head, N. W. Milgram, G. M. Cole, M. J. Savage, M. B. Podlisny, D. J. Selkoe, R. Siman, B. D. Greenberg and C. W. Cotman, *Neurobiol. Aging*, 1996, **17**(4), 653–659.
112. A. G. Reaume, D. S. Howland, S. P. Trusko, M. J. Savage, D. M. Lang, B. D. Greenberg, R. Siman and R. W. Scott, *J. Biol. Chem.*, 1996, **271**(38), 23380–23388.
113. R. Siman, A. G. Reaume, M. J. Savage, S. Trusko, Y. G. Lin, R. W. Scott and D. G. Flood, *J. Neurosci.*, 2000, **20**(23), 8717–8726.
114. B. T. Lamb, S. S. Sisodia, A. M. Lawler, H. H. Slunt, C. A. Kitt, W. G. Kearns, P. L. Pearson, D. L. Price and J. D. Gearhart, *Nat. Genet.*, 1993, **5**(1), 22–30.
115. B. T. Lamb, L. M. Call, H. H. Slunt, K. A. Bardel, A. M. Lawler, C. B. Eckman, S. G. Younkin, G. Holtz, S. L. Wagner, D. L. Price, S. S. Sisodia and J. D. Gearhart, *Hum. Mol. Genet.*, 1997, **6**(9), 1535–1541.
116. B. T. Lamb, K. A. Bardel, L. S. Kulnane, J. J. Anderson, G. Holtz, S. L. Wagner, S. S. Sisodia and E. J. Hoeger, *Nat. Neurosci.*, 1999, **2**(8), 695–697.
117. B. J. Hock, L. M. Lattal, L. S. Kulnane, T. Abel and B. T. Lamb, *Curr. Aging Sci.*, 2009, **2**(3), 205–213.
118. M. A. Leissring, W. Farris, A. Y. Chang, D. M. Walsh, X. Wu, X. Sun, M. P. Frosch and D. J. Selkoe, *Neuron*, 2003, **40**(6), 1087–1093.
119. W. Q. Qiu, D. M. Walsh, Z. Ye, K. Vekrellis, J. Zhang, M. B. Podlisny, M. R. Rosner, A. Safavi, L. B. Hersh and D. J. Selkoe, *J. Biol. Chem.*, 1998, **273**(49), 32730–32738.
120. K. Vekrellis, Z. Ye, W. Q. Qiu, D. Walsh, D. Hartley, V. Chesneau, M. R. Rosner and D. J. Selkoe, *J. Neurosci.*, 2000, **20**(5), 1657–1665.
121. E. Hama, K. Shirotani, H. Masumoto, Y. Sekine-Aizawa, H. Aizawa and T. C. Saido, *J. Biochem.*, 2001, **130**(6), 721–726.
122. N. Iwata, S. Tsubuki, Y. Takaki, K. Shirotani, B. Lu, N. P. Gerard, C. Gerard, E. Hama, H. J. Lee and T. C. Saido, *Science*, 2001, **292**(5521), 1550–1552.
123. W. J. Meilandt, M. Cisse, K. Ho, T. Wu, L. A. Esposito, K. Scearce-Levie, I. H. Cheng, G. Q. Yu and L. Mucke, *J. Neurosci.*, 2009, **29**(7), 1977–1986.
124. B. A. Bergmans and B. De Strooper, *Lancet Neurol.*, **9**(2), 215–226.

125. O. Murayama, T. Tomita, N. Nihonmatsu, M. Murayama, X. Sun, T. Honda, T. Iwatsubo and A. Takashima, *Neurosci. Lett.*, 1999, **265**(1), 61–63.

126. B. De Strooper, P. Saftig, K. Craessaerts, H. Vanderstichele, G. Guhde, W. Annaert, K. Von Figura and F. Van Leuven, *Nature*, 1998, **391**(6665), 387–390.

127. T. Wakabayashi and B. De Strooper, *Physiology (Bethesda)*, 2008, **23**, 194–204.

128. O. Murayama, M. Murayama, T. Honda, X. Sun, N. Nihonmatsu and A. Takashima, *Prog. Neuropsychopharmacol. Biol. Psychiatry*, 1999, **23**(5), 905–913.

129. M. Shimojo, N. Sahara, M. Murayama, H. Ichinose and A. Takashima, *Neurosci. Res.*, 2007, **57**(3), 446–453.

130. M. Shimojo, N. Sahara, T. Mizoroki, S. Funamoto, M. Morishima-Kawashima, T. Kudo, M. Takeda, Y. Ihara, H. Ichinose and A. Takashima, *J. Biol. Chem.*, 2008, **283**(24), 16488–16496.

131. A. Takashima, M. Murayama, O. Murayama, T. Kohno, T. Honda, K. Yasutake, N. Nihonmatsu, M. Mercken, H. Yamaguchi, S. Sugihara and B. Wolozin, *Proc. Natl. Acad. Sci. U. S. A.*, 1998, **95**(16), 9637–9641.

132. M. Murayama, S. Tanaka, J. Palacino, O. Murayama, T. Honda, X. Sun, K. Yasutake, N. Nihonmatsu, B. Wolozin and A. Takashima, *FEBS Lett.*, 1998, **433**(1–2), 73–77.

133. X. Xia, S. Qian, S. Soriano, Y. Wu, A. M. Fletcher, X. J. Wang, E. H. Koo, X. Wu and H. Zheng, *Proc. Natl. Acad. Sci. U. S. A.*, 2001, **98**(19), 10863–10868.

134. M. A. Leissring, Y. Akbari, C. M. Fanger, M. D. Cahalan, M. P. Mattson and F. M. LaFerla, *J. Cell Biol.*, 2000, **149**(4), 793–798.

135. M. A. Leissring, I. Parker and F. M. LaFerla, *J. Biol. Chem.*, 1999, **274**(46), 32535–32538.

136. O. Nelson, H. Tu, T. Lei, M. Bentahir, B. de Strooper and I. Bezprozvanny, *J. Clin. Invest.*, 2007, **117**(5), 1230–1239.

137. H. Tu, O. Nelson, A. Bezprozvanny, Z. Wang, S. F. Lee, Y. H. Hao, L. Serneels, B. De Strooper, G. Yu and I. Bezprozvanny, *Cell*, 2006, **126**(5), 981–993.

138. M. A. Leissring, B. A. Paul, I. Parker, C. W. Cotman and F. M. LaFerla, *J. Neurochem.*, 1999, **72**(3), 1061–1068.

139. C. A. Saura, S. Y. Choi, V. Beglopoulos, S. Malkani, D. Zhang, B. S. Shankaranarayana Rao, S. Chattarji, R. J. Kelleher 3rd, E. R. Kandel, K. Duff, A. Kirkwood and J. Shen, *Neuron*, 2004, **42**(1), 23–36.

140. K. Duff, C. Eckman, C. Zehr, X. Yu, C. M. Prada, J. Perez-tur, M. Hutton, L. Buee, Y. Harigaya, D. Yager, D. Morgan, M. N. Gordon, L. Holcomb, L. Refolo, B. Zenk, J. Hardy and S. Younkin, *Nature*, 1996, **383**(6602), 710–713.

141. L. Holcomb, M. N. Gordon, E. McGowan, X. Yu, S. Benkovic, P. Jantzen, K. Wright, I. Saad, R. Mueller, D. Morgan, S. Sanders,

C. Zehr, K. O'Campo, J. Hardy, C. M. Prada, C. Eckman, S. Younkin, K. Hsiao and K. Duff, *Nat. Med.*, 1998, **4**(1), 97–100.

142. X. Sun, V. Beglopoulos, M. P. Mattson and J. Shen, *Neurodegener. Dis.*, 2005, **2**(1), 6–15.

143. K. Tanemura, D. H. Chui, T. Fukuda, M. Murayama, J. M. Park, T. Akagi, Y. Tatebayashi, T. Miyasaka, T. Kimura, T. Hashikawa, Y. Nakano, T. Kudo, M. Takeda and A. Takashima, *J. Biol. Chem.*, 2006, **281**(8), 5037–5041.

144. G. M. Halliday, Y. J. Song, G. Lepar, W. S. Brooks, J. B. Kwok, C. Kersaitis, G. Gregory, C. E. Shepherd, F. Rahimi, P. R. Schofield and J. J. Kril, *Ann. Neurol.*, 2005, **57**(1), 139–143.

145. M. F. Mendez and A. McMurtray, *Am. J. Alzheimers Dis. Other Demen.*, 2006, **21**(4), 281–286.

146. H. Takahashi, I. Brasnjevic, B. P. Rutten, N. Van Der Kolk, D. P. Perl, C. Bouras, H. W. Steinbusch, C. Schmitz, P. R. Hof and D. L. Dickstein, *Brain Struct. Funct.*, 2010, **214**(2–3), 145–160.

147. I. Dewachter, D. Reverse, N. Caluwaerts, L. Ris, C. Kuiperi, C. Van den Haute, K. Spittaels, L. Umans, L. Serneels, E. Thiry, D. Moechars, M. Mercken, E. Godaux and F. Van Leuven, *J. Neurosci.*, 2002, **22**(9), 3445–3453.

148. S. Oddo, A. Caccamo, M. Kitazawa, B. P. Tseng and F. M. LaFerla, *Neurobiol. Aging*, 2003, **24**(8), 1063–1070.

149. S. Oddo, L. Billings, J. P. Kesslak, D. H. Cribbs and F. M. LaFerla, *Neuron*, 2004, **43**(3), 321–32.

150. L. M. Billings, S. Oddo, K. N. Green, J. L. McGaugh and F. M. LaFerla, *Neuron*, 2005, **45**(5), 675–688.

151. S. Oddo, A. Caccamo, K. N. Green, K. Liang, L. Tran, Y. Chen, F. M. Leslie and F. M. LaFerla, *Proc. Natl. Acad. Sci. U. S. A.*, 2005, **102**(8), 3046–3051.

152. A. Caccamo, S. Oddo, L. M. Billings, K. N. Green, H. Martinez-Coria, A. Fisher and F. M. LaFerla, *Neuron*, 2006, **49**(5), 671–682.

153. A. Caccamo, A. Fisher and F. M. LaFerla, *Curr. Alzheimer Res.*, 2009, **6**(2), 112–117.

154. A. Caccamo, S. Oddo, L. X. Tran and F. M. LaFerla, *Am. J. Pathol.*, 2007, **170**(5), 1669–1675.

155. L. Sereno, M. Coma, M. Rodriguez, P. Sanchez-Ferrer, M. B. Sanchez, I. Gich, J. M. Agullo, M. Perez, J. Avila, C. Guardia-Laguarta, J. Clarimon, A. Lleo and T. Gomez-Isla, *Neurobiol. Dis.*, 2009, **35**(3), 359–367.

156. H. Martinez-Coria, K. N. Green, L. M. Billings, M. Kitazawa, M. Albrecht, G. Rammes, C. G. Parsons, S. Gupta, P. Banerjee and F. M. LaFerla, *Am. J. Pathol.*, **176**(2), 870–880.

157. A. C. McKee, I. Carreras, L. Hossain, H. Ryu, W. L. Klein, S. Oddo, F. M. LaFerla, B. G. Jenkins, N. W. Kowall and A. Dedeoglu, *Brain Res.*, 2008, **1207**, 225–236.

158. P. Pasqualetti, C. Bonomini, G. Dal Forno, L. Paulon, E. Sinforiani, C. Marra, O. Zanetti and P. M. Rossini, *Aging Clin. Exp. Res.*, 2009, **21**(2), 102–110.

159. D. Morgan, D. M. Diamond, P. E. Gottschall, K. E. Ugen, C. Dickey, J. Hardy, K. Duff, P. Jantzen, G. DiCarlo, D. Wilcock, K. Connor, J. Hatcher, C. Hope, M. Gordon and G. W. Arendash, *Nature*, 2000, **408**(6815), 982–985.

160. J. A. Nicoll, D. Wilkinson, C. Holmes, P. Steart, H. Markham and R. O. Weller, *Nat. Med.*, 2003, **9**(4), 448–452.

161. C. Holmes, D. Boche, D. Wilkinson, G. Yadegarfar, V. Hopkins, A. Bayer, R. W. Jones, R. Bullock, S. Love, J. W. Neal, E. Zotova and J. A. Nicoll, *Lancet*, 2008, **372**(9634), 216–223.

162. P. H. St George-Hyslop and J. C. Morris, *Lancet*, 2008, **372**(9634), 180–182.

CHAPTER 3

The Ying and Yang of the Reelin Signalling Pathway in Alzheimer's Disease Pathology

EDUARDO SORIANO*, DANIELA ROSSI AND LLUÍS PUJADAS

Development of Neurobiology and Regeneration Laboratory, Institute for Research in Biomedicine, Barcelona; Department of Cell Biology, University of Barcelona, and Centro de Investigación en Red sobre Enfermedades Neurodegenerativas (CIBERNED-ISCIII), 08028 Barcelona, Spain

3.1 The Extracellular Matrix Protein Reelin in the Developing and Adult Brain

Reelin is a large extracellular protein crucial for neuronal migration and brain development. Furthermore, there is increasing evidence that Reelin controls synaptogenesis and synaptic maturation during development.[1–5] Reelin acts through the receptors apolipoprotein E receptor 2 (ApoER2) and very-low lipoprotein receptor (VLDLR), which trigger a complex signalling cascade involving members of the Src kinase family, the adaptor Dab1, the PI3K, Erk1/2 and GSK-3β kinases, and CrkL, among others.[6–13] It is important to remark that *reeler*, Dab1 and ApoER2/VLDLR $(-/-)$ mice are phenotypically almost identical, which highlights the essential role of Dab1 and Reelin receptors in the Reelin pathway.

In the adult brain, Reelin is expressed by subsets of neurons in the cerebral cortex and in other regions, particularly by GABAergic interneurons.[1]

RSC Drug Discovery Series No. 6
Animal Models for Neurodegenerative Disease
Edited by Jesús Avila, Jose J. Lucas and Félix Hernández
© Royal Society of Chemistry 2011
Published by the Royal Society of Chemistry, www.rsc.org

Although the role of Reelin in the adult brain is not well understood, researchers have shown that this protein is expressed in synaptic contacts and that neurons deficient in ApoER2 and VLDLR receptors have impaired long-term potentiation (LTP).[14] Moreover, Reelin was recently shown to participate in the composition, recruitment and trafficking of N-methyl-D-aspartate (NMDA) receptor subunits;[15–17] in the generation of dendrites, and in the formation of dendritic spines.[18,19] Furthermore, Reelin signalling antagonizes β-amyloid effects at hippocampal synapses.[20] These studies suggest that Reelin is involved in the correct formation of cortical synapses. This protein also acts as a detachment factor in the migration of subventricular zone (SVZ)-derived neurons in the rostral migratory stream to the olfactory bulb; however, changes in Reelin levels do not alter neurogenesis rates in the SVZ.[21,22]

3.2 Generation of Conditional Transgenic Mice Overexpressing Reelin in the Adult Forebrain: A Tool to Dissect Reelin Functions in the Adult Brain

Most studies on Reelin functions in the adult brain have employed analyses of the Reelin deficient *reeler* mice (or Dab1 or ApoER2/VLDLR (−/−) mice) and heterozygous *reeler* mice as a model of Reelin haploinsufficiency. Studies in *reeler* mice (or Dab1 or ApoER2/VLDLR (−/−) mice) are hampered by these animals' dramatic defects in brain lamination and organization (and behavioural deficits), which raises the question as to whether defects in *reeler* mouse function are secondary to neuronal mispositioning. To unravel the function of Reelin in the adult brain, we generated a gain-of-function transgenic mouse model that conditionally overexpresses Reelin, specifically in the postnatal and adult forebrain, under the control of the calcium/calmodulin-dependent kinase II promoter (pCaMKII-Reelin-OE; Tg1/Tg2)[22] (Figure 3.1).

Our data provided evidence that overexpression of Reelin increases adult hippocampal neurogenesis and impairs the migration and positioning of adult-generated hippocampal neurons. Furthermore, hippocampal overexpression of Reelin caused an increase in synaptic contacts and hypertrophy of dendritic spines (see Figure 3.2); in addition, induction of LTP in the CA3–CA1 synapse in alert-behaving mice showed that Reelin overexpression evokes a dramatic increase in LTP responses.[22] Thus, Reelin levels in the adult brain regulate neurogenesis and migration, as well as the structural and functional properties of synapses, which in turn implies that Reelin controls developmental processes that remain active in the adult brain.

Because Reelin is essential in brain development, we mainly addressed the impact of Reelin protein levels on developmental-like processes that remain active in the adult brain, namely adult neurogenesis, neural migration and synaptic plasticity. Our data show that Reelin levels do not produce changes in the adult neurogenesis in the SVZ. In contrast, the pattern of neuronal migration generated in the SVZ is altered in *Tg1/Tg2* mice. Recent data have shown that *in vitro* incubation with recombinant Reelin causes a detachment of

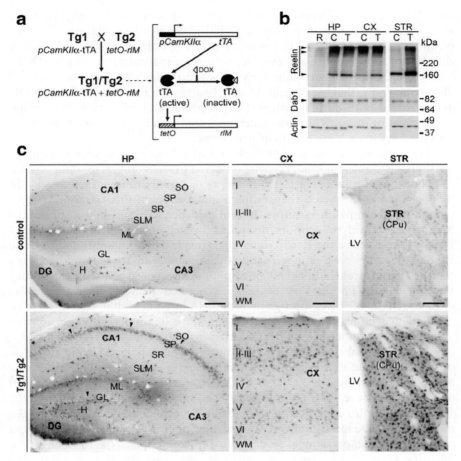

Figure 3.1 Generation and characterization of conditional *Tg1/Tg2* transgenic mice. (**a**) Transgenic mice that overexpress Reelin were based on the Tet-off regulated binary system: the *Tg1* transgene contains the *tTA* transactivator set under the control of *pCaMKIα*, while the *Tg2* transgene contains *rlM* controlled by the *tetO* promoter. Double transgenic mice (*Tg1/Tg2*) express Reelin in neurons expressing CaMKIIα; transgene expression can be switched off by doxycycline administration, which inactivates tTA transactivator. (**b**) Western blot experiments showing Reelin, Dab1 and actin protein levels in protein lysates obtained from control and *Tg1/Tg2* adult hippocampus (HP), cerebral cortex (CX) and striatum (STR). Reelin levels are increased in *Tg1/Tg2* (T) mice compared with control littermates (C) or *reeler* animals (R). Dab1 levels are slightly reduced in *Tg1/Tg2* mice. (**c**) Immunohistochemical detection of Reelin shows that expression of this protein in control adult mice is restricted to a subset of interneurons distributed throughout the cortex and hippocampal layers, while the striatum shows a diffuse staining (top). In *Tg1/Tg2* mice, overexpression of Reelin is observed in hippocampal pyramidal cells and in granule cells of the dentate gyrus (arrows in bottom left panel); Reelin is also expressed in neocortical pyramidal cells (bottom middle panel) and in striatal neurons (bottom right

neuroblast migrating chains from the RMS in a process that is interpreted as a switch from tangential migration to radial glia-controlled migration.[9,21] The phenotype observed in *Tg1/Tg2* mice is consistent with this vision and suggests that Reelin is required to achieve successful neuronal migration in the RMS.

The hippocampal neurogenic niche is active throughout the entire life.[23,24] Our data show that neurogenesis and neuronal incorporation in the DG are increased both during postnatal development and adult stages in Reelin-overexpressing mice. These data are consistent with studies in *reeler* mice, which have shown a lower rate of neurogenesis.[25]

Our results support an effect of Reelin on adult neurogenesis which cannot be attributable to the abnormal layering observed in *reeler* mice. Our data provide evidence that Reelin increases the numbers of proliferating progenitors in the hippocampal SGZ, thereby suggesting a direct effect of this extracellular protein on hippocampal progenitors. Recent evidence indicates that a substantial number of neural progenitors disappear by cell death during adult life.[26] Finally, and because the Reelin signalling cascade activates both the ERK and PI3K pathways,[9,13] we cannot exclude the possibility that Reelin may increase the survival of progenitors. This possibility is also consistent with the finding that differences in numbers of newly generated neurons are much more dramatic in old *Tg1/Tg2* transgenic mice than in young adult animals.[22]

Our data demonstrate that overexpression of Reelin results in a dispersion and abnormal positioning of adult-generated neurons within the GL. Given that Reelin deficiency also leads to abnormal neuronal migration and lamination,[27-29] our results suggest that Reelin levels control neuronal migration. However, it is also possible that the ectopic expression of Reelin in the granular layer (GL) of adult *Tg1/Tg2* mice contributes to the altered migration observed in the DG.

Previous studies have shown that Reelin-deficient *reeler* and heterozygous mice display a reduced number of dendritic spines. Moreover, incubation of cultured *reeler* hippocampal neurons with Reelin increases the density of dendritic spines.[19] We have shown that Reelin overexpression in adult mice does not alter the number of dendritic spines in the hippocampus but does affect their structural properties. This apparent discrepancy could be explained in several ways. First, both *reeler* and heterozygous mice have increased Dab1 levels, an adaptor required for most Reelin functions including dendritogenesis.[12,30-32] This is particularly evident in *reeler* mice, which express

---

panel). Abbreviations: I-VI, cortical layers; CA1–CA3, hippocampal regions; CPu, caudateputamen nucleous; CX, cortex; DG, dentate gyrus; DOX, doxycycline; GL, granular layer; H, hilus; HP, hippocampus; LV, lateral ventricle; ML, molecular layer; pCaMKIIα, calcium/calmodulin-dependent kinase IIα promoter; rlM, myc-tagged Reelin cDNA; SLM, stratum lacunosum moleculare; SO, stratum oriens; SP, stratum piramidale; SR, stratum radiatum; STR, striatum; tetO, tetracyclin operator; tTA, tetracycline-controlled transactivator; WM, white matter. Scale bars: c (left panels): 200 μm; c (medium and right panels): 100 μm. (Subjected to copyright from *The Journal of Neuroscience*).

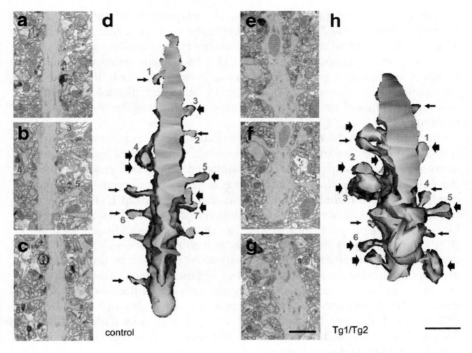

Figure 3.2 Three-dimensional serial electron microscopic reconstructions of dendrites in the SR of CA1 region in control and *Tg1/Tg2* mice, illustrating hypertrophy of dendritic spines in *Tg1/Tg2* mice. (**a–c, e–g**) Examples of serial electron micrographs in which the dendritic shaft and spines originating from it have been colored in green; numbers pointing to examples of reconstructed dendritic spines in d and h. (**d, h**) Three-dimensional reconstructions of the identified dendritic segments shown in (**a–c, e–g**); note the large hypertrophy of spines in (**h**), compared to controls (**d**). Dendritic spine heads with sizes below and above 0.4 μm in diameter are represented by small and large arrows, respectively. Synaptic contacts are drawn in red. Dendritic spines in electron micrographs are labeled by numbers. Scale bars: **a–c, e–g**: 1 μm; **d, h**: 1 μm. (Subjected to copyright from *The Journal of Neuroscience*).

a seven-fold increase in Dab1 protein,[30] and it is therefore likely that application of Reelin to *reeler* neurons leads to hyperactivation of the Reelin signalling pathway. Second, it is also possible that normal Reelin levels are required to achieve standard numbers of dendritic spines and that, above a certain threshold, Reelin is not sufficient to modify these numbers. Our results show that Reelin overexpression in a wild-type background *in vivo* does not alter spine density in adult hippocampal neurons. Finally, our finding that doxycycline treatment of *Tg1/Tg2* mice strongly reverses the synaptic phenotype in adult mice indicates that this phenotype depends on acute overexpression of Reelin.

Perhaps the most interesting synaptic phenotypic feature of Reelin-overexpressing *Tg1/Tg2* mice is the formation of very large, hypertrophied dendritic spines. Most of these spines display typical mushroom-type shapes and two or more synaptic active zones. This feature correlated with increased numbers of synapses in all hippocampal layers. Persistent dendritic spine enlargement is commonly associated with increased physiological efficacy and stable LTP, and the latter is thought to underlie long-lasting memory and learning.[33,34]

One of the questions raised by these structural features is how they are correlated with the functional electrophysiological properties of hippocampal neurons. To address this question, we monitored the physiological properties of the CA3–CA1 connection in adult mice *in vivo* under specific experimental conditions, which included intensity tests, electrical-evoked LTP and classical conditioning. Our electrophysiological data revealed that Reelin-overexpressing hippocampal neurons do not exhibit the shift from paired-pulse facilitation (PPF) to paired-pulse depression (PPD) at high stimulation intensities. When evoked in behaving animals, PPD is likely to be the result of the activation of disynaptic inhibitory circuits or to reflect synaptic fatigue caused by massive exocytosis of the releasable pool of synaptic vesicles. In Reelin-overexpressing mice, electrophysiological saturation was also reached although the shift from PPF to PPD did not typically occur even at high stimulation intensities. This finding suggests that transgenic neurons respond better to repetitive stimulation than wild-type neurons.

Interestingly, the present study in alert-behaving mice revealed that LTP responses are increased two-fold in the hippocampus of Reelin-overexpressing hippocampi. Moreover, because high-frequency stimulation (HFS) induction of LTP is likely to overstimulate neurons in a non-physiological way, we also monitored patterns of hippocampal activity during a classical eye blink conditioning using a hippocampal-dependent trace paradigm.[35] Again, this study showed that during classical conditioning, Reelin-overexpressing neurons are more prone to be activated and to elicit increased electrophysiological responses than those of controls. In both experimental situations (LTP and classical conditioning), persistent robust responses were recorded in *Tg1/Tg2* mice over several days, as compared to control litter mates. Thus, our results indicate that the modelling of Reelin expression enhances physiological plasticity in the CA3–CA1 synapse both during early and late LTP and across the conditioning sessions. This observation implies that, in both paradigms, activation of the Reelin system is sufficient to drive the structural and molecular changes responsible for short-term and long-term changes in synaptic plasticity.

The data are consistent with those reported in previous studies showing that ApoER2/VLDLR-deficient mice display reduced LTP in slices *in vitro*, which suggests that the effects described herein are mediated by these receptors.[14,36] Interestingly Reelin has been found to control NMDA and AMPA subunit receptor composition and trafficking during development,[15–17] and both types of ionotropic glutamate receptors are essential for LTP induction.[37,38]

Furthermore, GSK-3β, a downstream effector of the Reelin signalling cascade, has been found recently to be involved in LTP.[39] Taken together, our data suggest that extracellular Reelin acts through the ApoER2/VLDLR receptors to trigger a signalling cascade which is sufficient to induce marked LTP physiological alterations *in vivo*. Moreover, we have shown that Reelin over-expression also leads to increased LTP responses over several days, which suggests that this extracellular protein controls gene expression and protein synthesis required for late phases of LTP.[37,38]

Altogether, our results indicate that in the adult brain Reelin levels are crucial for the modulation of neurogenesis and migration, particularly in the hippocampus, as well as for the modulation of the structural and functional plastic properties of adult synapses, including induction and maintenance of LTP. Thus, Reelin, a protein with pivotal roles in normal development, also controls plasticity processes in the adult brain that are reminiscent of developmental processes.

3.3 Potential Role of Reelin in Alzheimer's Disease: A Unifying Hypothesis Linking Reelin to Amyloid Deposits, Tangles and Stress

Alzheimer's disease (AD) is a devastating disorder that affects a large segment of the senior population. Over the past two decades the proteins APP, ApoER2, ApoE, GSK-3 and tau have all been imputed in the pathogenesis of AD. For instance, ApoE4 haplotype is associated with increased AD risk[40] and mutations in APP, presenilin I and II and tau genes are responsible for roughly 6% of AD cases and tauopathy-type dementia.[41] Moreover, analyses from recent Genome-Wide Association Studies (GWAS) have suggested that specific haplotypes are linked to sporadic AD; these include the *clusterin* gene (ApoJ), a ligand of the ApoER2/VLDR receptor system which is related to APP/β-amyloid metabolism.[42] But although the spatiotemporal development and main pathological hallmarks of the disease are relatively well-known, the exact mechanisms that trigger sporadic AD have yet to be determined.

Recent work supports a link between Reelin and AD. First, Reelin and β-amyloid colocalize in plaques in AD patients and in animal models.[43,44] Secondly, AD patients have been reported to exhibit abnormal Reelin expression.[45,46] Thirdly, the Reelin signalling pathway interacts with APP processing: activation of this pathway triggers phosphorylation of Dab1, which in turn binds to the *C*-terminal domain of APP, promoting both αAPP cleavage (producing non-amyloidogenic peptides instead of the β-cleavage leading to the toxic β-42).[47] Fourthly, Reelin antagonizes the blocking of LTP by β-42 amyloid *in vitro*.[20] And lastly, the Reelin signalling pathway activates phosphorylation of GSK-3β-Ser, thereby resulting in decreased phosphorylation of GSK-3β targets, including tau protein.[8,13] Furthermore, a recent study has shown that overexpression of the receptor LDLR prevents the formation of amyloid in transgenic mice.[48]

The data above, together with the fact that Reelin has positive effects on synaptic remodelling and potentiation and in adult hippocampal neurogenesis (which is essential for learning and memory), suggest that the Reelin pathway could be considered protective in AD, and contrariwise, that deficiencies in this pathway could lead to an elevated risk of AD. Our preliminary data support these hypotheses. For instance, in biochemical experiments using purified Reelin and β-42 peptide, the two proteins co-aggregate and Reelin modifies β-amyloid formation (L. Pujadas *et al.* unpublished data). Furthermore, we recently found that overexpression of Reelin in the adult forebrain results in increased GSK-3β-phospho-Ser and decreased tau phosphorylation levels (D. Rossi *et al.* unpublished data). Moreover, two important hallmarks corroborating a protective role for Reelin against AD have recently been observed. First, during normal aging, extracellular Reelin deposits initially appear in the hippocampal region (including the entorhinal cortex) and then in neocortical areas,[44] paralleling the regional development of AD in humans. Secondly, sequential analysis of Reelin+ deposits suggest that these derive from Reelin+ beaded dendrites generated by the natural death of Reelin + neurons (Figure 3.3).

Another issue that must be tackled is to identify the factors (including environmental ones) that facilitate and trigger the development of sporadic AD. Non-genetic environmental factors are thought to be critically involved in AD. For example, the results from clinical studies and murine model experiments support the notion that chronic stress correlates to both the rate of AD incidence and to the presence of β-amyloid.[49] However, the physiological consequences of chronic stress in the context of AD are unknown. Indeed, we recently found that corticosterone-induced stress in mice leads to a massive appearance of extracellular Reelin deposits in the hippocampus, suggesting that stress might amplify β-amyloid deposition *via* Reelin deposit formation (unpublished data).

Considering the findings described above, we propose the following temporal sequence of cellular events implying that Reelin is functionally paramount in the pathogenesis of sporadic AD (Figure 3.4):

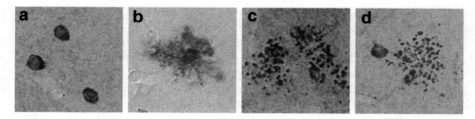

Figure 3.3   Reelin-immunostained hippocampal tissue sections from old wild-type mice suggest that Reelin+ extracellular deposits arise from the death of Reelin+ neurons. (**a**) Examples of Reelin+ neurons stained with G10 antibody. (**b**) Example of an individual Reelin+ neuron with degenerating appearance. (**c, d**) Accumulation of Reelin-containing deposits occur in regions where Reelin+ cells are present.

(a) Activation of the Reelin pathway in normal brain function leads to decreases in amyloid β-42 production, GSK-3β activity and tau phosphorylation, thereby protecting the individual from two pathological hallmarks of AD.

(b) During normal aging, the natural death of Reelin+ neurons results in decreased Reelin signalling, and consequently, increased amyloid β-42 production and tau phosphorylation; concomitantly, the death of Reelin+ neurons leads to accumulation of extracellular Reelin-deposits, which in turn could nucleate amyloid β-42 deposition.

(c) The processes described in (b) would be accelerated upon environmental insults (*e.g.* stress) and would act as a triggering or amplifying mechanism in the pathogenesis of sporadic AD.

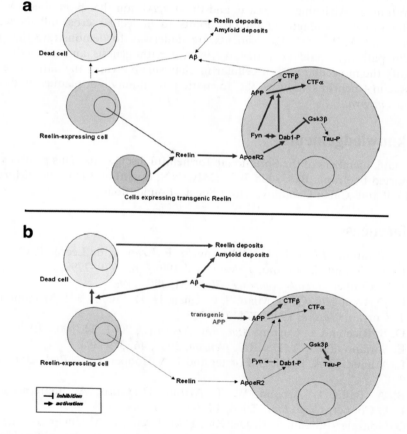

Figure 3.4 Schematic diagram showing AD-related pathways regulated by Reelin. (**a**) Reelin decreases amyloid production, GSK-3β activity and Tau phosphorylation. (**b**) Death of Reelin+ neurons results in increased amyloid production and Tau phosphorylation, and amyloid deposition.

Thus, given that Reelin levels control adult neurogenesis and synaptic plasticity (which are essential for learning and memory), we are also investigating how Reelin levels correlate to these plasticity processes and to AD progression.

3.4 Conclusions

The above data indicate that adult brain Reelin levels are crucial for the modulation of neurogenesis and migration, as well as for the modulation of the structural and functional properties of synapses, including induction and maintenance of LTP. Thus, Reelin, a protein with pivotal roles in normal development, also controls plasticity processes in the adult brain that are reminiscent of developmental processes. Moreover, the modulation of adult neurogenesis by Reelin as well as the recent findings implicating Reelin alterations in Alzheimer's disease, and the observation that β-amyloid species inhibit synaptic strength and LTP, points to molecular cross-talk between Reelin and APP.[20,43–45] Experiments are underway to demonstrate that the Reelin pathway could be protective in AD pathogenesis and, secondly, to identify the molecular and/or cellular mechanisms by which the natural death of Reelin + neurons induces the formation of β-amyloid plaques and tau tangles *in vivo*.

Acknowledgements

We thank members of the Soriano lab for helpful discussions. This project was supported by grants BFU2008-3980 (MICINN, Spain) and from the 'Marató de TV3' and 'Caixa Catalunya-Obra Social' Foundations.

References

1. S. Alcantara, M. Ruiz, G. D'Arcangelo, F. Ezan, L. de Lecea, T. Curran, C. Sotelo and E. Soriano, *J. Neurosci.*, 1998, **18**, 7779–7799.
2. J. A. Cooper, *Trends Neurosci.*, 2008, **31**, 113–119.
3. G. D'Arcangelo, G. G. Miao, S. C. Chen, H. D. Soares, J. I. Morgan and T. Curran, *Nature*, 1995, **374**, 719–723.
4. D. S. Rice and T. Curran, *Ann. Rev. Neurosci.*, 2001, **24**, 1005–1039.
5. E. Soriano and J. A. Del Rio, *Neuron*, 2005, **46**, 389–394.
6. L. Arnaud, B. A. Ballif, E. Forster and J. A. Cooper, *Curr. Biol.*, 2003, **13**, 9–17.
7. B. A. Ballif, L. Arnaud, W. T. Arthur, D. Guris, A. Imamoto and J. A. Cooper, *Curr. Biol.*, 2004, **14**, 606–610.
8. C. Gonzalez-Billault, J. A. Del Rio, J. M. Urena, E. M. Jimenez-Mateos, M. J. Barallobre, M. Pascual, L. Pujadas, S. Simo, A. L. Torre, R. Gavin, F. Wandosell, E. Soriano and J. Avila, *Cereb. Cortex*, 2005, **15**, 1134–1145.
9. S. Simo, L. Pujadas, M. F. Segura, A. La Torre, J. A. Del Rio, J. M. Urena, J. X. Comella and E. Soriano, *Cereb. Cortex*, 2007, **17**, 294–303.

10. B. W. Howell, R. Hawkes, P. Soriano and J. A. Cooper, *Nature*, 1997, **389**, 733–737.
11. T. Hiesberger, M. Trommsdorff, B. W. Howell, A. Goffinet, M. C. Mumby, J. A. Cooper and J. Herz, *Neuron*, 1999, **24**, 481–489.
12. B. W. Howell, T. M. Herrick and J. A. Cooper, *Genes Dev.*, 1999, **13**, 643–648.
13. U. Beffert, G. Morfini, H. H. Bock, H. Reyna, S. T. Brady and J. Herz, *J. Biol. Chem.*, 2002, **277**, 49958–49964.
14. U. Beffert, E. J. Weeber, A. Durudas, S. Qiu, I. Masiulis, J. D. Sweatt, W. P. Li, G. Adelmann, M. Frotscher, R. E. Hammer and J. Herz, *Neuron*, 2005, **47**, 567–579.
15. S. Qiu, L. F. Zhao, K. M. Korwek and E. J. Weeber, *J. Neurosci.*, 2006, **26**, 12943–12955.
16. Y. Chen, U. Beffert, M. Ertunc, T. S. Tang, E. T. Kavalali, I. Bezprozvanny and J. Herz, *J. Neurosci.*, 2005, **25**, 8209–8216.
17. L. Groc, D. Choquet, F. A. Stephenson, D. Verrier, O. J. Manzoni and P. Chavis, *J. Neurosci.*, 2007, **27**, 10165–10175.
18. T. Matsuki, A. Pramatarova and B. W. Howell, *J. Cell Sci.*, 2008, **121**, 1869–1875.
19. S. Niu, O. Yabut and G. D'Arcangelo, *J. Neurosci.*, 2008, **28**, 10339–10348.
20. M. S. Durakoglugil, Y. Chen, C. L. White, E. T. Kavalali and J. Herz, *Proc. Natl. Acad. Sci. U. S. A.*, 2009, **106**, 15938–15943.
21. I. Hack, M. Bancila, K. Loulier, P. Carroll and H. Cremer, *Nat. Neurosci.*, 2002, **5**, 939–945.
22. L. Pujadas, A. Gruart, C. Bosch, L. Delgado, C. M. Teixeira, D. Rossi, L. de Lecea, A. Martinez, J. M. Delgado-Garcia and E. Soriano, *J. Neurosci.*, 2010, **30**, 4636–4649.
23. N. Kee, C. M. Teixeira, A. H. Wang and P. W. Frankland, *Nat. Neurosci.*, 2007, **10**, 355–362.
24. C. Zhao, W. Deng and F. H. Gage, *Cell*, 2008, **132**, 645–660.
25. S. J. Won, S. H. Kim, L. Xie, Y. Wang, X. O. Mao, K. Jin and D. A. Greenberg, *Exp.Neurol.*, 2006, **198**, 250–259.
26. D. Dupret, A. Fabre, M. D. Dobrossy, A. Panatier, J. J. Rodriguez, S. Lamarque, V. Lemaire, S. H. Oliet, P. V. Piazza and D. N. Abrous, *PLoS Biol.*, 2007, **5**, e214.
27. A. Drakew, T. Deller, B. Heimrich, C. Gebhardt, D. Del Turco, A. Tielsch, E. Forster, J. Herz and M. Frotscher, *Exp. Neurol.*, 2002, **176**, 12–24.
28. S. Zhao, X. Chai, E. Forster and M. Frotscher, *Development*, 2004, **131**, 5117–5125.
29. C. Gong, T. W. Wang, H. S. Huang and J. M. Parent, *J. Neurosci.*, 2007, **27**, 1803–1811.
30. D. S. Rice, M. Sheldon, G. D'Arcangelo, K. Nakajima, D. Goldowitz and T. Curran, *Development*, 1998, **125**, 3719–3729.
31. S. Niu, A. Renfro, C. C. Quattrocchi, M. Sheldon and G. D'Arcangelo, *Neuron*, 2004, **41**, 71–84.
32. E. C. Olson, S. Kim and C. A. Walsh, *J. Neurosci.*, 2006, **26**, 1767–1775.

33. R. Yuste and T. Bonhoeffer, *Ann. Rev. Neurosci.*, 2001, **24**, 1071–1089.
34. Y. Yang, X. B. Wang, M. Frerking and Q. Zhou, *J. Neurosci.*, 2008, **28**, 5740–5751.
35. A. Gruart, M. D. Munoz and J. M. Delgado-Garcia, *J. Neurosci.*, 2006, **26**, 1077–1087.
36. E. J. Weeber, U. Beffert, C. Jones, J. M. Christian, E. Forster, J. D. Sweatt and J. Herz, *J. Biol. Chem.*, 2002, **277**, 39944–39952.
37. M. W. Jones, M. L. Errington, P. J. French, A. Fine, T. V. Bliss, S. Garel, P. Charnay, B. Bozon, S. Laroche and S. Davis, *Nat. Neurosci.*, 2001, **4**, 289–296.
38. W. Ju, W. Morishita, J. Tsui, G. Gaietta, T. J. Deerinck, S. R. Adams, C. C. Garner, R. Y. Tsien, M. H. Ellisman and R. C. Malenka, *Nat. Neurosci.*, 2004, **7**, 244–253.
39. S. Peineau, C. Taghibiglou, C. Bradley, T. P. Wong, L. Liu, J. Lu, E. Lo, D. Wu, E. Saule, T. Bouschet, P. Matthews, J. T. Isaac, Z. A. Bortolotto, Y. T. Wang and G. L. Collingridge, *Neuron*, 2007, **53**, 703–717.
40. J. Kim, J. M. Basak and D. M. Holtzman, *Neuron*, 2009, **63**, 287–303.
41. H. W. Querfurth and F. M. LaFerla, *N. Eng. J. Med.*, **362**, 329–344.
42. D. Harold, R. Abraham, P. Hollingworth, R. Sims, A. Gerrish, M. L. Hamshere, J. S. Pahwa, V. Moskvina, K. Dowzell, A. Williams, N. Jones, C. Thomas, A. Stretton, A. R. Morgan, S. Lovestone, J. Powell, P. Proitsi, M. K. Lupton, C. Brayne, D. C. Rubinsztein, M. Gill, B. Lawlor, A. Lynch, K. Morgan, K. S. Brown, P. A. Passmore, D. Craig, B. McGuinness, S. Todd, C. Holmes, D. Mann, A. D. Smith, S. Love, P. G. Kehoe, J. Hardy, S. Mead, N. Fox, M. Rossor, J. Collinge, W. Maier, F. Jessen, B. Schurmann, H. van den Bussche, I. Heuser, J. Kornhuber, J. Wiltfang, M. Dichgans, L. Frolich, H. Hampel, M. Hull, D. Rujescu, A. M. Goate, J. S. Kauwe, C. Cruchaga, P. Nowotny, J. C. Morris, K. Mayo, K. Sleegers, K. Bettens, S. Engelborghs, P. P. De Deyn, C. Van Broeckhoven, G. Livingston, N. J. Bass, H. Gurling, A. McQuillin, R. Gwilliam, P. Deloukas, A. Al-Chalabi, C. E. Shaw, M. Tsolaki, A. B. Singleton, R. Guerreiro, T. W. Muhleisen, M. M. Nothen, S. Moebus, K. H. Jockel, N. Klopp, H. E. Wichmann, M. M. Carrasquillo, V. S. Pankratz, S. G. Younkin, P. A. Holmans, M. O'Donovan, M. J. Owen and J. Williams, *Nat. Genet.*, 2009, **41**, 1088–1093.
43. H. S. Hoe, K. J. Lee, R. S. Carney, J. Lee, A. Markova, J. Y. Lee, B. W. Howell, B. T. Hyman, D. T. Pak, G. Bu and G. W. Rebeck, *J. Neurosci.*, 2009, **29**, 7459–7473.
44. I. Knuesel, M. Nyffeler, C. Mormede, M. Muhia, U. Meyer, S. Pietropaolo, B. K. Yee, C. R. Pryce, F. M. LaFerla, A. Marighetto and J. Feldon, *Neurobiol. Aging*, 2009, **30**, 697–716.
45. A. Botella-Lopez, F. Burgaya, R. Gavin, M. S. Garcia-Ayllon, E. Gomez-Tortosa, J. Pena-Casanova, J. M. Urena, J. A. Del Rio, R. Blesa, E. Soriano and J. Saez-Valero, *Proc. Natl. Acad. Sci. U. S. A.*, 2006, **103**, 5573–5578.

46. J. Chin, C. M. Massaro, J. J. Palop, M. T. Thwin, G. Q. Yu, N. Bien-Ly, A. Bender and L. Mucke, *J. Neurosci.*, 2007, **27**, 2727–2733.
47. H. S. Hoe, T. S. Tran, Y. Matsuoka, B. W. Howell and G. W. Rebeck, *J. Biol. Chem.*, 2006, **281**, 35176–35185.
48. H. M. Kim, T. Qu, V. Kriho, P. Lacor, N. Smalheiser, G. D. Pappas, A. Guidotti, E. Costa and K. Sugaya, *Proc. Natl. Acad. Sci. U. S. A.*, 2002, **99**, 4020–4025.
49. Y. H. Jeong, C. H. Park, J. Yoo, K. Y. Shin, S. M. Ahn, H. S. Kim, S. H. Lee, P. C. Emson and Y. H. Suh, *FASEB J.*, 2006, **20**, 729–731.

CHAPTER 4

Transgenic Mice Overexpressing GSK-3β as Animal Models for Alzheimer's Disease

FÉLIX HERNÁNDEZ

Centro de Biología Molecular 'Severo Ochoa' (CBMSO), Universidad Autónoma de Madrid, Madrid, Spain

4.1 Introduction

Alzheimer's disease (AD) is a dementia which gives rise to progressive cognitive impairments including memory failure. Two major neuropathological abnormalities are present in the brain of patients with AD; the extracellular senile plaques (SPs) and the intracellular neurofibrillary tangles (NFTs). SPs are made up of amyloid deposits composed of aggregates of β-amyloid peptide, a fragment of an integral membrane protein called amyloid precursor protein (APP). NFTs are aggregates of paired helical filaments (PHFs) whose main component is the hyperphosphorylated form of the microtubule associated protein called tau. It now seems clear that tau changes in AD are secondary to β-amyloid deposition[1,2] and that tau is essential for β-amyloid induced toxicity.[3] However, the intracellular connection between β-amyloid and tau changes in AD is still a matter for debate, and GSK-3β has emerged as one of the possible mediators (Figure 4.1).

In a pioneer study looking for tau kinases, Ishiguro et al.[4] found a key kinase, associated to microtubules, which they described as like tau kinase I. This kinase was later recognized as glycogen synthase kinase-3 (GSK-3),[5]

RSC Drug Discovery Series No. 6
Animal Models for Neurodegenerative Disease
Edited by Jesús Avila, Jose J. Lucas and Félix Hernández
© Royal Society of Chemistry 2011
Published by the Royal Society of Chemistry, www.rsc.org

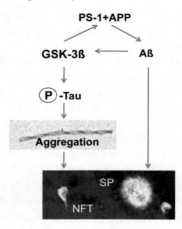

Figure 4.1   Interaction of GSK-3β with the cellular components related to the neuropathology of AD such as amyloid precursor protein, β-amyloid peptide,
presenilins and tau protein.

a kinase involved in the regulation of glucose metabolism.[6] In vertebrates, two
closely related isoforms (GSK-3α and GSK-3β) are present.[7]

GSK-3β has been proposed as the link between the two neuropathological
hallmarks of AD, the extracellular β-amyloid and NFTs. Deregulation of
GSK-3 activity in neurons has been postulated as a key feature in Alzheimer's
disease pathogenesis. Exposure of cortical and hippocampal primary neuronal
cultures to β-amyloid induces activation of GSK-3β,[8] tau hyperphosphorylation[9,10] and cell death.[9,11] Blockade of GSK-3β expression by antisense
oligonucleotides[11] or its activity by lithium addition[12] inhibits β-amyloid
induced neurodegeneration of cortical and hippocampal primary cultures.
Furthermore, interaction of GSK-3β isoform has been described with many of
the cellular components related to the neuropathology of AD such as:

- APP;
- the β-amyloid peptide;
- the metabolic pathway leading to acetylcholine synthesis;
- the presenilins, which are mutated in many cases of familial AD;
- tau protein.

GSK-3β has been shown to phosphorylate tau in most sites hyperphosphorylated in NFTs, both in transfected cells[13] and *in vivo*.[14,15] Increased levels
of active GSK-3β have been found in AD brains and GSK-3β accumulates in
the cytoplasm of pre-tangle neurons and can be found associated to NFTs.[16–18]

4.2   Glycogen Synthase Kinase-3

GSK-3 plays important roles in embryonic development, cell differentiation,
microtubule dynamics, cell cycle division, cell adhesion, glucose metabolism,

apoptosis and neurogenesis.[19–21] In addition, a growing amount of experimental data has demonstrated its involvement in some pathological processes such as Alzheimer's disease.

The GSK-3 protein (E.C. 2.7.1.[37]) is a proline-directed serine/threonine protein kinase that was originally identified and named for its ability to phosphorylate, among other kinases, the enzyme glycogen synthase. The central nervous system (CNS) is the tissue with the highest GSK-3 level.[7] In mammals, two GSK-3 isoenzymes (α and β) have been described.[7] They share 95% amino acid identity; the N-terminal and C-terminal ends are the regions with less homology. Both isoenzymes are products of two independent genes. *GSK-3α* gene encodes a 51 kDa protein, while *GSK-3β* gene encodes a 47 kDa protein (mapped to chromosome 19q12.3 and 3q13.3, respectively[22,23]). A new alternative splicing isoform of GSK-3β with an additional 13 amino acid insert (exon 8A) in the catalytic domain has been described.[24] *GSK-3β* gene has been analysed[25] and its promoter studied to identify variations that could be associated with an abnormal function.[26] However, these studies have concluded that aberrant tau phosphorylation is not due to mutations in the GSK-3β promoter.

GSK-3 is regulated at several levels. Activation of Tyr216 by phosphorylation in GSK-3β or Tyr276 in GSK-3α is necessary for its activity.[27] Mutation of that residue reduces the activity.[27] From its crystal structure, it has been proposed that unphosphorylated Tyr216/276 blocks the access of primed substrates. The structure of phosphorylated GSK-3β has been published[28] showing that phosphorylated Tyr216 suffers a conformational change that allows the primed substrate to bind the enzyme.

Which is the kinase that phosphorylated Tyr216? In *Dictyostelium,* the Zaphod kinase activates GSK-3 by Tyr-phosphorylation.[29] In mammals, experimental data suggest that Fyn tyrosine kinase[30] or some related tyrosine kinase are involved. However, it seems that phosphorylation in that residue is the result of an autophosphorylation event.[31] In postmortem samples from AD patients, active GSK-3β colocalizes with hyperhosphorylated tau.[32] To support these data, an accumulation of active GSK-3β has been observed in dystrophic neuritis close to amyloid deposits in transgenic mice overexpressing mutant APP.[33]

Inhibition of the enzyme is mediated by two different types of mechanisms.[19] The first mechanism is used by insulin and growth factors, and is mediated mainly by PKB which phosphorylates Ser9 in GSK-3β and Ser21 in GSK-3α. Structural studies have suggested a model for the inhibition.[34,35] Thus, the phosphorylated Ser9/21 binds as a competitive pseudosubstrate to the primed-binding site, inhibiting the binding of the protein to be phosphorylated. In this way, it has been suggested that β-amyloid could facilitate the phosphorylation of tau protein as β-amyloid could act as an antagonist of insulin receptor[36] facilitating the activation of GSK-3 through inhibition of Ser-9/21 phosphorylation. Interestingly, GSK-3β phosphorylated at Ser9, which inactivates GSK-3, is found in the majority of neurons with neurofibrillary tangles and dystrophic neurites of senile plaques in AD and other phospho-tau-containing neurons in other tauopathies (for a review see ref. 37).

Another important mechanism controlling GSK-3 function is through sequestration. Thus, subcellular localization of GSK-3 regulates substrate accessibility. GSK-3 is observed in cytosol, mitochondria and the nucleus, and also found enriched in specific neuronal compartments such as dendritic spines. Therefore, maintaining GSK-3 to specific compartments may affect responsiveness to different signals.[20]

On the other hand, GSK-3 can be regulated by binding to other proteins in a multiprotein complex. For example, GSK-3 can associate with p53[38] or 14-3-3.[39,40] Another protein able to modulate GSK-3 activity is DISC1 (Disrupted In Schizophrenia 1) implicated in schizophrenia susceptibility. DISC1 inhibits GSK-3β activity through direct physical interaction.[41]

However, the most studied pathways involving GSK-3 complex formation is regulated by the Wnt pathway. GSK-3 is present in a multiprotein complex formed by axin and APC where GSK-3 is able to phosphorylates β-catenin targeting it for proteasome degradation.[42] Wnt proteins bind to frizzled receptor activating the dishevelled protein. This inhibits GSK-3 activity by disrupting the multiprotein complex. As a consequence, β-catenin is accumulated and translocated into the nucleus where it activates the transcription through binding to transcription factors.

Many GSK-3 substrates require a priming phosphorylation by a priming kinase on a Ser or Thr residue four amino acids *C*-terminal to the phosphorylatable Ser or Thr residue. The crystal structure of human GSK-3β has provided a model for the binding of prephosphorylated substrates to the kinase.[34,35] Thus, primed Ser/Thr are recognized by a positively charged 'binding pocket' which facilitates the binding of primed substrates. Knowledge of the priming kinase is of obvious interest. Kinases like CDK5[43,44] PAR1[45] casein kinase I[46] or PKA[47] could act like priming kinases for a further GSK-3 phosphorylation.

Finally, an additional way for the regulation of GSK-3β phosphorylation has been recently proposed.[48] This regulation involves the removal by calpain of a fragment from the *N*-terminal region of GSK-3, including the regulatory serines 9/21. After removal of that fragment GSK-3 becomes activated.

4.3 Alzheimer's Disease and Glycogen Synthase Kinase-3

Increased levels of active GSK-3β have been found in AD brains and GSK-3β accumulates in the cytoplasm of pre-tangle neurons and can be found associated to NFTs.[16–18] AD has been classified in two types: AD of familiar origin (FAD, monogenic) or sporadic AD (polygenic).

Three genes have been found to be the cause of the onset of FAD. The first one that was described codified for APP, the other two for two proteins known as presenilin-1 (PS1) and presenilin-2 (PS2). GSK-3 is involved in APP processing. Thus, GSK-3β phosphorylates APP increasing its intracellular processing.[49,50] Furthermore, inhibition of GSK-3β reduces the production of β-amyloid.[51,52] One of the functions of PS1 and PS2, the other proteins linked to FAD, is to facilitate the cleavage of APP yielding β-amyloid.[53,54] Additionally, it was found that aggregates of β-amyloid induce tau phosphorylation[9] and that GSK-3β was essential for the β-amyloid induced neurotoxicity.[11]

These and other studies were the basis of the so-called 'amyloid hyppothesis'.[55,56] This hypothesis suggested that the origin of AD must be in the appearance of β-amyloid peptide. β-Amyloid production is facilitated by mutations associated to FAD in APP or PS (a gain of function). β-Amyloid will aggregate and those aggregates facilitate tau phosphorylation, and as consequence of that phosphorylation, tau protein polymerizes into PHFs and later on aggregates into NFTs. If that is the case, presenilin (PS) mutations yielding increasing amounts of β-amyloid peptide will induce a faster onset of FAD. However, this is not the case[57] as the amount of β-amyloid generated in cells transfected with different PS1 mutations did not correlate with the starting age of FAD caused by that mutation. To support this view, a PS1 mutation has been described in a patient with Pick-type tauopathy without extracellular β-amyloid deposits.[58] Thus, preselinin proteins could have another function that could be important for AD onset. One of these functions was described by Weihl *et al.*,[59] indicating that PS1 downregulates PKB. Afterwards, Baki *et al.*[60] described how PS1 activates PI3K, inhibiting GSK-3 activity and tau hyperphosphorylation. Therefore, it can be suggested that some PS1 mutations will result in a lack of PI3K activation (loss of function) which will result in tau phosphorylation. To summarize, it can be speculated that there exist two pathways from preselinin to induce AD. In one case, a gain of function will result in the appearance of β-amyloid and, afterwards, tau phosphorylation by GSK-3; in the other case there will be a loss of function but also in tau phosphorylation by GSK-3. Thus, a common feature for FAD from PS1 mutations will be the appearance of tau phosphorylated by GSK-3. Other functions in which presenilins are involved are related to autophagy.[61]

Phosphorylation of tau has been proposed as an important non-genetic factor involved in tauopathies such as AD. In addition to some residues present in the microtubule binding domain (MBD), tau molecules present in PHFs are hyperphosphorylated at sites flanking the MBD. More than 30 sites could be modified by phosphorylation in tau protein. These sites have been divided in non-proline kinase directed sites and proline kinase directed sites;[62] it is mainly the second class that are preferentially modified by GSK-3.

Phospho-tau isolated from the brain of AD patients contains more primed than unprimed GSK-3 modified sites.[63] Several of these sites have been characterized and antibodies that specifically recognize those phosphosites have been generated. The contribution of α- and β-isoforms requires further analysis and it is necessary to understand the contribution of both isoforms to tau phosphorylation. In fact, the β-isoform generated by splicing has less affinity for tau protein than the low molecular weight β-isoform.[24] The role of phosphorylation in the self-assembly of tau is a fundamental question in the study of AD and other tauopathies. It has been suggested that phosphorylation of some specific tau sites may be a prerequisite for its assembly.[64,65] GSK-3β is one of the best candidate enzymes for generating the hyperphosphorylated tau that is characteristic of PHFs.

4.4 Mouse Models Overexpressing Glycogen Synthase Kinase-3

One of the strategies for generating animal models of tauopathies has been to overexpress the kinases responsible for tau hyperphosphorylation. GSK-3 and CDK5 are two of the main kinases phosphorylating tau protein in AD. Several genetic approaches have been used to generate mouse strains with altered kinase activity.

CDK5 is able to phosphorylate tau protein and is associated with NFTs. CDK5 was originally identified as a tau kinase[66,67] and has been shown to be associated with the early stages of NFTs.[68,69] Triple transgenic mice overexpressing human p35, its regulatory subunit, Cdk5 and tau do not display an increase in tau phosphorylation despite significant increases in CDK5 activity in these mice.[70] These findings indicate that p35/Cdk5 does not efficiently phosphorylate tau. This notion is further corroborated by the report that recombinant p25 (the truncated form of p35)/CDK5 phosphorylates tau with higher efficiency than p35/CDK5 *in vitro*.[71] In addition, an study describing the expression of human mutant P301L tau with p25 in double transgenic mice shows increased tau phosphorylation and aggregation into filaments.[44] The insoluble tau was associated with active GSK-3, suggesting that CDK5 can be one of the main priming kinases that allows GSK-3 to phosphorylate tau protein.

Several genetic approaches have been used to generate mouse strains with altered GSK-3 activity. The first lines of GSK-3β transgenic animals described used either ubiquitous (murine sarcoma virus) or CNS-specific promoters (murine neurofilament light chain) to drive expression.[72] Although slight increases in tau phosphorylation at the AT8 epitope were detected, overexpression was not observed using transgenes encoding either wild-type GSK-3β or a mutant GSK-3β(S9A) that is not inhibited by phosphorylation. The authors speculated that the low levels of kinase expression obtained were probably because high levels of GSK-3 are lethal. Conversely, as it will be discussed below, GSK-3β knock-out mice die during embryonic life.[73]

With the lethality of GSK-3β as well as the known role of GSK-3β in development taken into account, a further GSK-3β transgenic mouse was generated using the conditional tetracycline-regulated system under the control of CaM kinase IIα-promoter to achieve substantial overexpression of the kinase.[74] Animal models of neurodegeneration using the tetracycline-regulated system have the advantage of being able to perform reversibility studies. In this transgenic mouse (Tet/GSK-3β), GSK-3β overexpression was restricted to certain cortical neurons and hippocampal neurons. *In vivo* overexpression of GSK-3β resulted in an increase in the phosphorylation of tau in Tet/GSK-3β animals, as detected with antibodies raised against the phosphorylated tau modified in AD. Hyperphosphorylated tau was found in the somatodendritic compartment. However, the change in the subcellular localization of tau observed was exclusively due to the hyperphosphorylation of tau by GSK-3β since there was no change in the overall levels of tau. Somatodendritic hyperphosphorylated tau in the conditional Tet/GSK-3β transgenic mice produced an increase in the levels of tau not bound to microtubules in this neuronal compartment.

These data suggest that the increased phosphorylation of tau results in a decrease in the affinity of tau for microtubules, reproducing two of the characteristics of AD tau: hyperphosphorylation and decreased interaction with microtubules. However, the aberrant tau aggregations found in AD were not observed in the Tet/GSK-3β transgenic mice. Tet/GSK-3β mice also show neuronal stress and neuronal death as revealed by the presence of reactive glia, terminal deoxynucleotidyl transferase dUTP nick end labelling (TUNEL) and cleaved caspase-3 staining in the hippocampal dentate gyrus. These data are in agreement with the known role of GSK-3β in the survival pathway, as well as supporting the neuroprotective effect of lithium, a relatively specific GSK-3β inhibitor.

All these *in vivo* genetic approaches support the involvement of GSK-3β in tau phosphorylation in the nervous system, suggesting that GSK-3β deregulation might be relevant to the pathogenesis of AD. Hyperphosphorylation of hippocampal tau in transgenic Tet/GSK-3 mice, despite the lack of filament formation, results in a behavioural impairment that can be measured in the Morris water maze test[75] and in the object recognition test.[76] Thus hyperphosphorylation of tau could be sufficient to provoke some of the pathological manifestations found in AD.

The implication of wild-type GSK-3β in tau pathology was further analysed by combining this transgenic model with transgenic mice expressing tau with a triple FTDP-17 mutation which develop prefibrillar tau-aggregates (VLW mice).[77] Tet/GSK-3β/VLW transgenic mice show a higher tau hyperphosphorylation in hippocampal neurons compared with Tet/GSK-3β and this is accompanied by thioflavin-*S* or Thiazin Red staining (Figure 4.2), and formation of filaments similar in width to those found in tauopathies. This model demonstrates that an increase in GSK-3β activity is a key factor able to induce tau aggregation (sarkosyl-insoluble material as well as tau filaments).

The main way to regulate GSK-3 is through phosphorylation of *N*-terminal regulatory serines. To study the effect of a permanent GSK-3 activation, homozygous GSK-3α/β(S21A/S21A/S9A/S9A) knock-in mice have been generated. This model has similar basal levels of mutant forms to wild-type mice. Interestingly, these mice reproduce and develop normally.[78] Analysing this model it was demonstrated that insulin regulates muscle glycogen synthase mainly through GSK-3β rather than GSK-3α.

However, the main implication of this study is in relation to the physiological role of *N*-terminal phosphorylation showing that, although it has some implications, they are not especially relevant. Some of these effects are related to neurogenesis in the CNS. This model shows altered neurogenesis as well as deficient production of the supporting growth factor, vascular endothelial

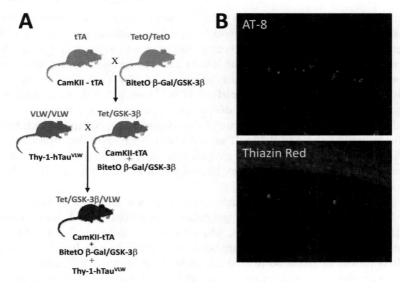

Figure 4.2 (A) Generation of Tet/GSK-3β/VLW triple transgenic mice. Tet/GSK-3β mice are generated by crossing mice expressing tTA under control of the CamKIIα promoter (tTA) with mice that have incorporated the BitetO construct in their genome (TetO). Triple transgenic mice are generated by crossing Tet/GSK-3β mice with homozygotes VLW mice carrying three FTDP-17 linked mutations (G272V, P301L and R406W). (B) Tau pathology in Tet/GSK-3β/VLW mice. Double-labelling of tau pathology in the hippocampus (CA1) from 15 months old Tet/GSK-3β/VLW mice with AT8 antibody (green) and Thiazin Red (aggregated protein, red).

growth factor (VEGF) in the hippocampus.[79] Given that administration of fluoxetine with lithium increases proliferation in the dentate gyrus of wild-type but not in GSK-3α/β(S21A/S21A/S9A/S9A) knock-in mice, it was suggested that impaired neurogenesis seems to contribute to psychiatric diseases and likely to Alzheimer's disease.[79] In this respect, GSK-3 has been described as a master regulator of neuronal progenitor homeostasis during embryonic development.[21] In the specific case of adult neurogenesis taking place in the subgranular zone of the dentate gyrus, a role for tau phosphorylation (by GSK-3) has been found in the migration and differentiation of neuronal precursors into mature neurons.[80,81]

Another approach has consisted of expressing mutant GSK-3β(S9A) under the control of Thy-1 promoter. These animals present a reduction in brain volume, mainly at the cerebral cortex, and an increase in tau phosphorylation but only in older transgenic mice (6–7 months).[82,83] These mutant GSK-3β(S9A) transgenic mice performed normally in the Morris water maze test, but interestingly, when cross-bred with transgenic mice that overexpress the longest human protein tau isoform, the number of axonal enlargements present and the motor impairment typical of these tau transgenic animals were reduced.[83,84] In these mice, an increase in the phosphorylation of human tau was observed, although neither an increase in insoluble tau aggregates nor the presence of paired helical filaments was observed. The most likely explanation for this is that an excess of tau bound to microtubules inhibits axonal transport. This is in agreement with the fact that the axonopathy in tau-overexpressing mice occurs preferentially in neurons with long axons such as brainstem and spinal cord motor neurons. Because phosphorylation of tau by GSK-3β decreases the affinity of tau for microtubules, Spittaels *et al.*[84] concluded that the most likely explanation for the improvement of this axonopathy and the motor problems found in htau40 mice was the rescue of the axonal transport that was inhibited by the excess of the MAP tau. The same group crossed this mutant GSK-3β(S9A) mice with a tau model overexpressing Tau(P301L),[85] observing an increase in forebrain tau pathology although the bigenic mice survived longer than the parental Tau(P301L).

From these mouse models it seems clear that not all neurons behave similarly. Thus, in Tet/GSK-3β/VLW, granular neurons from the dentate gyrus die and the dentate gyrus degenerate, while pyramidal neurons from CA1 accumulate phosphorylated tau protein and form NFT-like inclusions; in GSK-3β(S9A) model neurons with long axons such as spinal cord motor neurons GSK-3 can rescue tau pathology, although in all of them tau is phosphorylated in AD epitopes. In a similar approach, Li *et al.* also used the mutant GSK-3β (S9A) under the control of a neuron-specific human PDGFβ chain promoter.[86] This promoter drove GSK-3β expression mainly in the cortex and hippocampus of the transgenic mice, with the result being tau phosphorylation.

In addition to crossing these GSK-3β models with transgenic mouse overexpressing different tau isoforms, a transgenic model overexpressing wild-type GSK-3β was crossed with mice overexpressing APP with FAD mutations. Thus, mice expressing V717I and Swedish K679M/N671L mutations were crossed with transgenic mice overexpressing wild-type GSK-3β.[87] These mice showed an improvement in Morris water maze compared with single mutant APP mice, although they had a reduction in tau phosphorylation and a decrease in APP phosphorylation and Aβ levels. The authors found that, although GSK-3β was overexpressed, the increase resulted in an increase in the inhibitory domain of the kinase. This was the reason to name, in a confusing way, that animal model as a dominant negative GSK-3β model. Inactivation of GSK-3 by the use of inhibitors, like lithium, results in an increase in phosphorylation at serines 21 (GSK-3α) and 9 (GSK-3β).[88] This *N*-terminal phosphorylation has been explained by indirect inhibitory effect of lithium on a protein inhibitor (I2PP1) of protein phosphatase 1 (PP1), an enzyme that dephosphorylates the phospho-GSK-3 isoform, like phosphatase 2 (PP2).[89] Overall the paper demonstrates that GSK-3 is able to modulate the amyloid aspect of AD *in vivo*.

4.5   Animal Models with Reduced Glycogen Synthase Kinase-3 Activity

Several genetic approaches have been used to generate mice with a decrease in GSK-3 levels. The first approach was to generate GSK-3β knock-out mice.

These mice were not viable and died *in utero*, demonstrating first that GSK-3β is essential for development and second that GSK-3α does not compensate GSK-3β.[73] Thus, disruption of the *GSK-3β* gene that results in embryonic lethality is caused by liver degeneration during gestation, a phenotype consistent with excessive tumor necrosis factor (TNF) toxicity. This model could not offer information with respect to the CNS. However, heterozygote GSK-3β knockout (GSK-3β +/−) mice are viable and fertile. But although these heterozygous mice perform the Morris water maze without any problem, they show impaired memory reconsilidation.[90]

Mice without GSK-3α are viable and show an increase in liver glycogen.[91] With respect to the CNS ,these animals show an increase in cerebellar volume, although the rest is normal, suggesting a role for GSK-3α in this area. In addition, GSK-3α knock-out mice present alterations in a wide range of behaviour tests involving this isoenzyme in normal brain function.[92]

Deletion of GSK-3α/β was carried out in order to delete GSK-3 specifically from neurons. Thus, mice with a GSK-3α full background carrying the construction GSK-3β(loxP/loxP) were crossed with nestin-Cre line.[21] These animals showed bigger heads and died at P0. The main finding was a massive hyperproliferation of neural progenitors by expansion of the radial progenitor pool. These effects were linked with β-catenin dysregulation, Sonic Hedgehog, Notch and fibroblast growth factor signalling. This model mainly demonstrates the role of GSK-3β in brain development.[41,93,94] However, it has recently been demonstrated that GSK-3β is involved not only in embryonic development but also in adult neurogenesis. Thus, GSK-3 is able to phosphorylate tau protein in doublecortin positive cells in adult dentate gyrus.[80,81] In addition, the GSK-3α/β-(S21A/S21A/S9A/S9A) knock-in mice described previously[79] present a drastic impairment in adult neurogenesis *in vivo*. In good agreement, adult neurogenesis is also altered in Tet/GSK-3β mice.[95]

Another approach to reduce GSK-3 activity has been to generate real DN-GSK-3β mice, more precisely the K85R mutant form of GSK-3β to study permanent GSK-3 inhibition.[96] The neurological consequences were impaired motor coordination and an increase of neuronal apoptosis. These consequences were restored after shut-down expression (doxycycline administration).

4.6 GSK-3 Inhibition as a Therapy for Alzheimer's Disease

The previous data allow a discussion of the potential use of GSK-3 inhibition as a therapy for AD. This hypothesis has been tested in the conditional system used to generate Tet/GSK-3β mice exploring whether the biochemical, histopathological and behavioural consequences of increased GSK-3 activity are susceptible to reverting after restoration of normal GSK-3 levels after shutdown of the transgene expressing with doxicicline.[76] It was demonstrated that, in symptomatic mice, doxycycline administration in the drinking water leads to normal GSK-3 activity, normal phospho-tau levels, diminished neuronal death, and suppression of the cognitive deficit, thus further supporting the potential of GSK-3 inhibitors for AD therapeutics. However, permanent GSK-3 inhibition obtained in DN-GSK-3β mice and its conseguences[96] warn us of the likely toxicity of GSK-3 inhibitors.

A good example for GSK-3 inhibitors is lithium.[97] Lithium is a non-competitive inhibitor with respect to the GSK-3 substrates, ATP and the protein to be phosphorylated. Also, it has been suggested that lithium is a competitive inhibitor with respect to magnesium.[98] The use of lithium as a therapy for AD has been tested in several animal models. Thus, inhibition of GSK-3β by lithium inhibits β-amyloid-induced neurodegeneration of cortical and hippo-campal primary cultures.[12] Lithium also blocks the accumulation of Aβ peptides in the brains of mice that overproduce APP.[51]

With respect to tau metabolism, lithium is able to prevent tau phosphor-ylation in several mouse models of tauopathies.[99–103] In most of these models, lithium is able to prevent the development of tau pathology when administered early in disease progression. On the other hand, if lithium is administered at late stages of the disease, it is still able to reduce tau hyperphosphorylation but cannot reverse tau aggregation.[101] Taking into account these data, a first conclusion could be that lithium is able to prevent tau pathology, but that this is not the case when NFTs are already formed. However, in this case lithium would be protective as growing evidence suggests that the formation of the aberrant aggregates present in several neurodegenerative diseases may be a way to neutralize aberrant/toxic proteins generated during the process of neuro-degeneration. In good agreement with this, a learning deficit in a mouse model of FTDP-17 correlated with hyperphosphorylated tau, but not with the pre-sence of NFTs.[104] Thus, it can be proposed that lithium could also be effective even when the disease is in an advanced phase. In any case, the neurological side effects of lithium are well known as well as its risk for overdose-induced toxicity. Recently, it has been demonstrated that GSK-3 inhibition by lithium increased translocation of nuclear factor of activated T cells c3/4 (NFATc3/4) transcription factors to the nucleus, leading to increased Fas ligand (FasL) levels and Fas activation.[105] Interestingly, motor deficits and nuclear factor of activated T-cells (NFAT) nuclear translocation were prevented by cyclosporin A, opening the way to combined therapies.

In addition to lithium, other GSK-3 inhibitors are those that bind to the ATP-binding site. These inhibitors are those like indirubins,[106] maleimides[107] and thiazoles.[102,108] Other GSK-3 inhibitors are those like thiadiazolidinones, which are able to inhibit GSK-3 in a non-competitive manner.[109] More recently, marine compounds have been tested as a source for new GSK-3 inhibitors.[110]

4.7 Conclusions

A growing amount of data supports the hypothesis that GSK-3 *in vivo* may contribute to AD. Thus, GSK-3 deregulation induces severe tau pathology and interferes with the metabolism of β-amyloid, the other hallmark of AD (Figure 4.1). Given the significant role of GSK-3 in a variety of effects linked to mechanisms related to AD, GSK-3 mouse models are important to understand its contribution to neurodegenerative processes and, as a consequence, their contribution to AD. In addition, these models will allow testing of small molecule inhibitors as promising therapeutics tool for AD and related tauopathies. These GSK-3 inhibitors will be probably be tested on the different mouse models mimicking some pathological aspects of Alzheimer's disease as a first step for a possible use in clinic.

Acknowledgements

This work was supported by grants from the Comunidad de Madrid (NEURODEGMODELS-CM), the Spanish Comisión Interministerial de Ciencia y Tecnologia, Fundación Centro Investigación Enfermedades Neurológicas (Fundación CIEN) and the CIBER on Neurodegeneration, and by institutional grants from the Fundación Ramón Areces.

References

1. J. Gotz, F. Chen, J. van Dorpe and R. M. Nitsch, *Science*, 2001, **293**, 1491–1495.
2. J. Lewis, D. W. Dickson, W. L. Lin, L. Chisholm, A. Corral, G. Jones, S. H. Yen, N. Sahara, L. Skipper, D. Yager, C. Eckman, J. Hardy, M. Hutton and E. McGowan, *Science*, 2001, **293**, 1487–1491.
3. M. Rapoport, H. N. Dawson, L. I. Binder, M. P. Vitek and A. Ferreira, *Proc. Natl. Acad. Sci. U. S. A.*, 2002, **99**, 6364–6369.
4. K. Ishiguro, A. Omori, M. Takamatsu, K. Sato, M. Arioka, T. Uchida and K. Imahori, *Neurosci. Lett.*, 1992, **148**, 202–206.
5. K. Ishiguro, A. Shiratsuchi, S. Sato, A. Omori, M. Arioka, S. Kobayashi, T. Uchida and K. Imahori, *FEBS Lett.*, 1993, **325**, 167–172.
6. P. Cohen and S. Frame, *Nat. Rev. Mol. Cell Biol.*, 2001, **2**, 769–776.
7. J. R. Woodgett, *EMBO J.*, 1990, **9**, 2431–2438.
8. A. Takashima, K. Noguchi, G. Michel, M. Mercken, M. Hoshi, K. Ishiguro and K. Imahori, *Neurosci. Lett.*, 1996, **203**, 33–36.
9. J. Busciglio, A. Lorenzo, J. Yeh and B. A. Yankner, *Neuron*, 1995, **14**, 879–888.
10. A. Ferreira, Q. Lu, L. Orecchio and K. S. Kosik, *Mol. Cell. Neurosci.*, 1997, **9**, 220–234.
11. A. Takashima, K. Noguchi, K. Sato, T. Hoshino and K. Imahori, *Proc. Natl. Acad. Sci. U. S. A.*, 1993, **90**, 7789–7793.
12. G. Alvarez, J. R. MunozMontano, J. Satrustegui, J. Avila, E. Bogonez and J. DiazNido, *FEBS Lett.*, 1999, **453**, 260–264.
13. S. Lovestone, C. H. Reynolds, D. Latimer, D. R. Davis, B. H. Anderton, J. M. Gallo, D. Hanger, S. Mulot, B. Marquardt, S. Stabel, J. R. Woodgett and C.C. J. Miller, *Curr. Biol.*, 1994, **4**, 1077–1086.

14. M. Hong, D. C. R. Chen, P. S. Klein and V. M. Y. Lee, *J. Biol. Chem.*, 1997, **272**, 25326–25332.
15. J. R. Munoz-Montano, F. J. Moreno, J. Avila and J. DiazNido, *FEBS Lett.*, 1997, **411**, 183–188.
16. H. Yamaguchi, K. Ishiguro, T. Uchida, A. Takashima, C. A. Lemere and K. Imahori, *Acta Neuropathol.*, 1996, **92**, 232–241.
17. K. Imahori and T. Uchida, *J. Biochem.*, 1997, **121**, 179–188.
18. J. J. Pei, E. Braak, H. Braak, I. GrundkeIqbal, K. Iqbal, B. Winblad and R. F. Cowburn, *J. Neuropathol. Exp. Neurol.*, 1999, **58**, 1010–1019.
19. S. Frame and P. Cohen, *Biochem. J.*, 2001, **359**, 1–16.
20. R. S. Jope and G. V. Johnson, *Trends Biochem. Sci.*, 2004, **29**, 95–102.
21. W. Y. Kim, X. Wang, Y. Wu, B. W. Doble, S. Patel, J. R. Woodgett and W. D. Snider, *Nat. Neurosci.*, 2009, **12**, 1390–1397.
22. L. Hansen, K. C. Arden, S. B. Rasmussen, C. S. Viars, H. Vestergaard, T. Hansen, A. M. Moller, J. R. Woodgett and O. Pedersen, *Diabetologia*, 1997, **40**, 940–946.
23. P. C. Shaw, A. F. Davies, K. F. Lau, M. Garcia-Barcelo, M. M. Waye, S. Lovestone, C. C. Miller and B. H. Anderton, *Genome*, 1998, **41**, 720–727.
24. F. Mukai, K. Ishiguro, Y. Sano and S. C. Fujita, *J. Neurochem.*, 2002, **81**, 1073–1083.
25. K. F. Lau, C. C. Miller, B. H. Anderton and P. C. Shaw, *Genomics*, 1999, **60**, 121–128.
26. C. Russ, S. Lovestone and J. F. Powell, *Mol. Psychiatry*, 2001, **6**, 320–324.
27. K. Hughes, E. Nikolakaki, S. E. Plyte, N. F. Totty and J. R. Woodgett, *EMBO J.*, 1993, **12**, 803–808.
28. B. Bax, P. S. Carter, C. Lewis, A. R. Guy, A. Bridges, R. Tanner, G. Pettman, C. Mannix, A. A. Culbert, M. J. Brown, D. G. Smith and A. D. Reith, *Structure*, 2001, **9**, 1143–1152.
29. L. Kim, J. Liu and A. R. Kimmel, *Cell*, 1999, **99**, 399–408.
30. M. Lesort, R. S. Jope and G. V. Johnson, *J. Neurochem.*, 1999, **72**, 576–584.
31. A. Cole, S. Frame and P. Cohen, *Biochem. J.*, 2004, **377**, 249–255.
32. J. J. Pei, E. Braak, H. Braak, I. Grundke-Iqbal, K. Iqbal, B. Winblad and R. F. Cowburn, *J. Neuropathol. Exp. Neurol.*, 1999, **58**, 1010–1019.
33. Y. Tomidokoro, K. Ishiguro, Y. Harigaya, E. Matsubara, M. Ikeda, J. M. Park, K. Yasutake, T. Kawarabayashi, K. Okamoto and M. Shoji, *Neurosci. Lett.*, 2001, **299**, 169–172.
34. E. ter Haar, J. T. Coll, D. A. Austen, H. M. Hsiao, L. Swenson and J. Jain, *Nat. Struct. Biol.*, 2001, **8**, 593–596.
35. R. Dajani, E. Fraser, S. M. Roe, N. Young, V. Good, T. C. Dale and L. H. Pearl, *Cell*, 2001, **105**, 721–732.
36. L. Xie, E. Helmerhorst, K. Taddei, B. Plewright, W. Van Bronswijk and R. Martins, *J. Neurosci.*, 2002, **22**, RC221.
37. I. Ferrer, T. Gomez-Isla, B. Puig, M. Freixes, E. Ribe, E. Dalfo and J. Avila, *Curr. Alzheimer Res.*, 2005, **2**, 3–18.
38. T. Y. Eom and R. S. Jope, *Mol. Cancer*, 2009, **8**, 14.

39. A. Agarwal-Mawal, H. Y. Qureshi, P. W. Cafferty, Z. Yuan, D. Han, R. Lin and H. K. Paudel, *J. Biol. Chem.*, 2003, **278**, 12722–12728.

40. Z. Yuan, A. Agarwal-Mawal and H. K. Paudel, *J. Biol. Chem.*, 2004, **279**, 26105–26114.

41. Y. Mao, X. Ge, C. L. Frank, J. M. Madison, A. N. Koehler, M. K. Doud, C. Tassa, E. M. Berry, T. Soda, K. K. Singh, T. Biechele, T. L. Petryshen, R. T. Moon, S. J. Haggarty and L. H. Tsai, *Cell*, 2009, **136**, 1017–1031.

42. H. Aberle, A. Bauer, J. Stappert, A. Kispert and R. Kemler, *EMBO J.*, 1997, **16**, 3797–3804.

43. A. Sengupta, Q. L. Wu, I. GrundkeIqbal, K. Iqbal and T. J. Singh, *Mol. Cell Biochem.*, 1997, **167**, 99–105.

44. W. Noble, V. Olm, K. Takata, E. Casey, O. Mary, J. Meyerson, K. Gaynor, J. LaFrancois, L. Wang, T. Kondo, P. Davies, M. Burns, Veeranna, R. Nixon, D. Dickson, Y. Matsuoka, M. Ahlijanian, L. F. Lau and K. Duff, *Neuron*, 2003, **38**, 555–565.

45. I. Nishimura, Y. Yang and B. Lu, *Cell*, 2004, **116**, 671–682.

46. S. Amit, A. Hatzubai, Y. Birman, J. S. Andersen, E. Ben-Shushan, M. Mann, Y. Ben-Neriah and I. Alkalay, *Genes Dev.*, 2002, **16**, 1066–1076.

47. T. J. Singh, N. Haque, I. Grundkeiqbal and K. Iqbal, *FEBS Lett.*, 1995, **358**, 267–272.

48. P. Goni-Oliver, J. J. Lucas, J. Avila and F. Hernandez, *J. Biol. Chem.*, 2007, **282**, 22406–22413.

49. A. E. Aplin, G. M. Gibb, J. S. Jacobsen, J. M. Gallo and B. H. Anderton, *J. Neurochemistry*, 1996, **67**, 699–707.

50. A. E. Aplin, J. S. Jacobsen, B. H. Anderton and J. M. Gallo, *Neuroreport*, 1997, **8**, 639–643.

51. C. J. Phiel, C. A. Wilson, V. M. Lee and P. S. Klein, *Nature*, 2003, **423**, 435–439.

52. J. Ryder, Y. Su, F. Liu, B. Li, Y. Zhou and B. Ni, *Biochem. Biophys. Res. Commun.*, 2003, **312**, 922–929.

53. D. Scheuner, C. Eckman, M. Jensen, X. Song, M. Citron, N. Suzuki, T. D. Bird, J. Hardy, M. Hutton, W. Kukull, E. Larson, E. Levy-Lahad, M. Viitanen, E. Peskind, P. Poorkaj, G. Schellenberg, R. Tanzi, W. Wasco, L. Lannfelt, D. Selkoe and S. Younkin, *Nat. Med.*, 1996, **2**, 864–870.

54. A. L. Brunkan and A. M. Goate, *J. Neurochem.*, 2005, **93**, 769–792.

55. D. J. Selkoe, *J. Biol. Chem.*, 1996, **271**, 18295–18298.

56. J. Hardy and D. J. Selkoe, *Science*, 2002, **297**, 353–356.

57. O. Murayama, T. Tomita, N. Nihonmatsu, M. Murayama, X. Y. Sun, T. Honda, T. Iwatsubo and A. Takashima, *Neurosci. Lett.*, 1999, **265**, 61–63.

58. B. Dermaut, S. Kumar-Singh, S. Engelborghs, J. Theuns, R. Rademakers, J. Saerens, B. A. Pickut, K. Peeters, M. van den Broeck, K. Vennekens, S. Claes, M. Cruts, P. Cras, J. J. Martin, C. Van Broeckhoven and P. P. De Deyn, *Ann. Neurol.*, 2004, **55**, 617–626.

59. C. C. Weihl, G. D. Ghadge, S. G. Kennedy, N. Hay, R. J. Miller and R. P. Roos, *J. Neurosci.*, 1999, **19**, 5360–5369.
60. L. Baki, J. Shioi, P. Wen, Z. Shao, A. Schwarzman, M. Gama-Sosa, R. Neve and N. K. Robakis, *EMBO J.*, 2004, **23**, 2586–2596.
61. J. H. Lee, W. H. Yu, A. Kumar, S. Lee, P. S. Mohan, C. M. Peterhoff, D. M. Wolfe, M. Martinez-Vicente, A. C. Massey, G. Sovak, Y. Uchiyama, D. Westaway, A. M. Cuervo and R. A. Nixon, *Cell*.
62. M. Morishimakawashima, M. Hasegawa, K. Takio, M. Suzuki, H. Yoshida, K. Titani and Y. Ihara, *J. Biol. Chem.*, 1995, **270**, 823–829.
63. J. H. Cho and G. V. Johnson, *J. Biol. Chem.*, 2003, **278**, 187–193.
64. C. Bancher, C. Brunner, H. Lassmann, H. Budka, K. Jellinger, G. Wiche, F. Seitelberger, I. I. Grundke, K. Iqbal and H. M. Wisniewski, *Brain Res.*, 1989, **477**, 90–99.
65. W. Gordon-Krajcer, L. Yang and H. Ksiezak-Reding, *Brain Res.*, 2000, **856**, 163–175.
66. K. Baumann, E. M. Mandelkow, J. Biernat, H. Piwnicaworms and E. Mandelkow, *FEBS Lett.*, 1993, **336**, 417–424.
67. H. K. Paudel, J. Lew, Z. Ali and J. H. Wang, *J. Biol. Chem.*, 1993, **268**, 23512–23518.
68. J. C. Augustinack, J. L. Sanders, L. H. Tsai and B. T. Hyman, *J. Neuropathol. Exp. Neurol.*, 2002, **61**, 557–564.
69. J. J. Pei, I. Grundke-Iqbal, K. Iqbal, N. Bogdanovic, B. Winblad and R. F. Cowburn, *Brain Res.*, 1998, **797**, 267–277.
70. C. Van den Haute, K. Spittaels, J. Van Dorpe, R. Lasrado, K. Vandezande, I. Laenen, H. Geerts and F. Van Leuven, *Neurobiol. Dis.*, 2001, **8**, 32–44.
71. M. Hashiguchi, T. Saito, S. Hisanaga and T. Hashiguchi, *J. Biol. Chem.*, 2002, **277**, 44525–44530.
72. J. Brownlees, N. G. Irving, J. P. Brion, B. J. M. Gibb, U. Wagner, J. Woodgett and C. C. J. Miller, *Neuroreport*, 1997, **8**, 3251–3255.
73. K. P. Hoeflich, J. Luo, E. A. Rubie, M. S. Tsao, O. Jin and J. R. Woodgett, *Nature*, 2000, **406**, 86–90.
74. J. J. Lucas, F. Hernandez, P. Gomez-Ramos, M. A. Moran, R. Hen and J. Avila, *EMBO J.*, 2001, **20**, 27–39.
75. F. Hernandez, J. Borrell, C. Guaza, J. Avila and J. J. Lucas, *J. Neurochem.*, 2002, **83**, 1529–1533.
76. T. Engel, F. Hernandez, J. Avila and J. J. Lucas, *J. Neurosci.*, 2006, **26**, 5083–5090.
77. F. Lim, F. Hernandez, J. J. Lucas, P. Gomez-Ramos, M. A. Moran and J. Avila, *Mol. Cell Neurosci.*, 2001, **18**, 702–714.
78. E. J. McManus, K. Sakamoto, L. J. Armit, L. Ronaldson, N. Shpiro, R. Marquez and D. R. Alessi, *EMBO J.*, 2005, **24**, 1571–1583.
79. T. Y. Eom and R. S. Jope, *Biol. Psychiatry*, 2009, **66**, 494–502.
80. A. Fuster-Matanzo, E. G. de Barreda, H. N. Dawson, M. P. Vitek, J. Avila and F. Hernandez, *FEBS Lett.*, 2009, **583**, 3063–3068.

81. X. P. Hong, C. X. Peng, W. Wei, Q. Tian, Y. H. Liu, X. Q. Yao, Y. Zhang, F. Y. Cao, Q. Wang and J. Z. Wang, *Hippocampus*, 2009.
82. K. Spittaels, C. Van den Haute, J. Van Dorpe, D. Terwel, K. Vandezande, R. Lasrado, K. Bruynseels, M. Irizarry, M. Verhoye, J. Van Lint, J. R. Vandenheede, D. Ashton, M. Mercken, R. Loos, B. Hyman, A. Van der Linden, H. Geerts and F. Van Leuven, *Neuroscience*, 2002, **113**, 797–808.
83. K. Spittaels, C. Van den Haute, J. Van Dorpe, H. Geerts, M. Mercken, K. Bruynseels, R. Lasrado, K. Vandezande, I. Laenen, T. Boon, J. Van Lint, J. Vandenheede, D. Moechars, R. Loos and F. Van Leuven, *J. Biol. Chem.*, 2000, **275**, 41340–41349.
84. K. Spittaels, C. VandenHaute, J. VanDorpe, K. Bruynseels, K. Vandezande, I. Laenen, H. Geerts, M. Mercken, R. Sciot, A. VanLommel, R. Loos and F. VanLeuven, *Am. J. Pathol.*, 1999, **155**, 2153–2165.
85. D. Terwel, D. Muyllaert, I. Dewachter, P. Borghgraef, S. Croes, H. Devijver and F. Van Leuven, *Am. J. Pathol.*, 2008, **172**, 786–798.
86. B. Li, J. Ryder, Y. Su, S. A. Moore Jr., F. Liu, P. Solenberg, K. Brune, N. Fox, B. Ni, R. Liu and Y. Zhou, *Transgenic Res.*, 2004, **13**, 385–396.
87. E. Rockenstein, M. Torrance, A. Adame, M. Mante, P. Bar-on, J. B. Rose, L. Crews and E. Masliah, *J. Neurosci.*, 2007, **27**, 1981–1991.
88. F. Zhang, C. J. Phiel, L. Spece, N. Gurvich and P. S. Klein, *J. Biol. Chem.*, 2003, **278**, 33067–33077.
89. E. Planel, K. Yasutake, S. C. Fujita and K. Ishiguro, *J. Biol. Chem.*, 2001, **276**, 34298–34306.
90. T. Kimura, S. Yamashita, S. Nakao, J. M. Park, M. Murayama, T. Mizoroki, Y. Yoshiike, N. Sahara and A. Takashima, *PloS One*, 2008, **3**, e3540.
91. K. MacAulay, B. W. Doble, S. Patel, T. Hansotia, E. M. Sinclair, D. J. Drucker, A. Nagy and J. R. Woodgett, *Cell Metab.*, 2007, **6**, 329–337.
92. O. Kaidanovich-Beilin, T. V. Lipina, K. Takao, M. van Eede, S. Hattori, C. Laliberte, M. Khan, K. Okamoto, J. W. Chambers, P. J. Fletcher, K. Macaulay, B. W. Doble, M. Henkelman, T. Miyakawa, J. Roder and J. R. Woodgett, *Mol. Brain*, 2009, **2**, 35.
93. N. Sato, L. Meijer, L. Skaltsounis, P. Greengard and A. H. Brivanlou, *Nat. Med.*, 2004, **10**, 55–63.
94. H. K. Bone, T. Damiano, S. Bartlett, A. Perry, J. Letchford, Y. S. Ripoll, A. S. Nelson and M. J. Welham, *Chem. Biol.*, 2009, **16**, 15–27.
95. M. A. Sirerol-Piquer, P. Gomez-Ramos, F. Hernández, M. Perez, M. A. Morán, A. Fuster-Matanzo, J. J. Lucas, J. Avila and J. M. García-Verdugo, *Hippocampus*, 2010, DOI: 10.1002/hipo.20805.
96. R. Gomez-Sintes, F. Hernandez, A. Bortolozzi, F. Artigas, J. Avila, P. Zaratin, J. P. Gotteland and J. J. Lucas, *EMBO J.*, 2007, **26**, 2743–2754.
97. P. S. Klein and D. A. Melton, *Proc. Natl. Acad. Sci. U. S. A.*, 1996, **93**, 8455–8459.

98. W. J. Ryves, R. Dajani, L. Pearl and A. J. Harwood, *Biochem. Biophys. Res. Commun.*, 2002, **290**, 967–972.

99. M. Perez, F. Hernandez, F. Lim, J. Diaz-Nido and J. Avila, *J. Alzheimer Dis.*, 2003, **5**, 301–308.

100. H. Nakashima, T. Ishihara, P. Suguimoto, O. Yokota, E. Oshima, A. Kugo, S. Terada, T. Hamamura, J. Q. Trojanowski, V. M. Lee and S. Kuroda, *Acta Neuropathol.*, 2005, **110**, 547–556.

101. T. Engel, P. Goni-Oliver, J. J. Lucas, J. Avila and F. Hernandez, *J. Neurochemistry*, 2006, **99**, 1445–1455.

102. W. Noble, E. Planel, C. Zehr, V. Olm, J. Meyerson, F. Suleman, K. Gaynor, L. Wang, J. LaFrancois, B. Feinstein, M. Burns, P. Krishnamurthy, Y. Wen, R. Bhat, J. Lewis, D. Dickson and K. Duff, *Proc. Natl. Acad. Sci. U. S. A.*, 2005, **102**, 6990–6995.

103. A. Caccamo, S. Oddo, L. X. Tran and F. M. LaFerla, *Am. J. Pathol.*, 2007, **170**, 1669–1675.

104. K. Santacruz, J. Lewis, T. Spires, J. Paulson, L. Kotilinek, M. Ingelsson, A. Guimaraes, M. DeTure, M. Ramsden, E. McGowan, C. Forster, M. Yue, J. Orne, C. Janus, A. Mariash, M. Kuskowski, B. Hyman, M. Hutton and K. H. Ashe, *Science*, 2005, **309**, 476–481.

105. R. Gomez-Sintes and J. J. Lucas, *J. Clin. Invest.*, 2010, **120**, 2432–2445.

106. S. Leclerc, M. Garnier, R. Hoessel, D. Marko, J. A. Bibb, G. L. Snyder, P. Greengard, J. Biernat, Y. Z. Wu, E. M. Mandelkow, G. Eisenbrand and L. Meijer, *J. Biol. Chem.*, 2001, **276**, 251–260.

107. M. P. Coghlan, A. A. Culbert, D. A. Cross, S. L. Corcoran, J. W. Yates, N. J. Pearce, O. L. Rausch, G. J. Murphy, P. S. Carter, L. Roxbee Cox, D. Mills, M. J. Brown, D. Haigh, R. W. Ward, D. G. Smith, K. J. Murray, A. D. Reith and J. C. Holder, *Chem. Biol.*, 2000, **7**, 793–803.

108. R. Bhat, Y. Xue, S. Berg, S. Hellberg, M. Ormo, Y. Nilsson, A. C. Radesater, E. Jerning, P. O. Markgren, T. Borgegard, M. Nylof, A. Gimenez-Cassina, F. Hernandez, J. J. Lucas, J. Diaz-Nido and J. Avila, *J. Biol. Chem.*, 2003, **278**, 45937–45945.

109. A. Martinez, M. Alonso, A. Castro, C. Perez and F. J. Moreno, *J. Med. Chem.*, 2002, **45**, 1292–1299.

110. K. V. Rao, M. S. Donia, J. Peng, E. Garcia-Palomero, D. Alonso, A. Martinez, M. Medina, S. G. Franzblau, B. L. Tekwani, S. I. Khan, S. Wahyuono, K. L. Willett and M. T. Hamann, *J. Nat. Prod.*, 2006, **69**, 1034–1040.

CHAPTER 5

Invertebrate and Vertebrate Models of Tauopathies

JÜRGEN GÖTZ,[*a] LARS M. ITTNER,[a] NAEMAN N. GÖTZ,[a] HONG LAM[b] AND HANNAH R. NICHOLAS[b]

[a] Alzheimer's & Parkinson's Disease Laboratory, Brain & Mind Research Institute, University of Sydney, Australia; [b] School of Molecular Bioscience, University of Sydney, Australia

5.1 Introduction

Animal models are widely used in neurodegenerative research and the question arises how to best define a model.[1] Wikipedia (http://en.wikipedia.org/wiki/Animal_model, accessed 5 May 2010) emphasizes two key aspects:

(a) 'Animal models serving in research may have an existing, inbred or induced disease or injury that is similar to a human condition'.
(b) 'A modeled disease must be similar in etiology (mechanism of cause) and function to the human equivalent'.

In the field of neurodegenerative disease, key aspects of the human pathology (such as neurofibrillary tangle formation, see below) were not successfully reproduced in animals for some time. Having now overcome this problem, today's challenge lies in understanding what causes disease in the first place.

In modelling Alzheimer's disease (AD), the most prevalent form of all dementias, and frontotemporal dementia (FTD), the most important form with an onset before the age of 65,[2] familial forms of these diseases, as well as histopathological and clinical features in the human patient, provide guidance. In familial AD

RSC Drug Discovery Series No. 6
Animal Models for Neurodegenerative Disease
Edited by Jesús Avila, Jose J. Lucas and Félix Hernández
© Royal Society of Chemistry 2011
Published by the Royal Society of Chemistry, www.rsc.org

(FAD), autosomal dominant mutations have been identified in three genes: *APP*, *presenilin 1 (PSEN1)* and *presenilin 2 (PSEN2)*[3] (Figure 5.1A). The amyloid precursor protein, APP, is a membrane-associated protein from which the peptide amyloid-β (Aβ) is derived by proteolytic cleavage. The presenilins are components of the γ-secretase complex that, together with β-secretase, generates Aβ, while α-secretase activity precludes its formation. In addition to the three FAD genes, several susceptibility genes have been identified in sporadic AD (SAD).[3] Compared to AD, FTD is more heterogeneous as reflected by the mutated genes, the deposited proteins and the clinical features[4] (Figure 5.1B). The first mutations in FTD were in familial FTD with Parkinsonism linked to chromosome 17 (FTDP-17), where they were found in the *MAPT* gene that encodes the microtubule-associated protein tau[4] (Figure 5.1A). This subset of FTD cases is characterized by tau inclusions, a focus of this chapter.

Histopathologically, AD is characterized by both Aβ plaques and tau-containing neurofibrillary lesions (Figure 5.1A). Tau in the lesions is hyperphosphorylated and fibrillar. The tau aggregates are found in cell bodies and apical dendrites as neurofibrillary tangles (NFTs), in distal dendrites as neurophil threads, and in the abnormal neurites that are associated with some amyloid-β plaques. Truncation of tau, in addition to hyperphosphorylation,[5] may play a role in pathogenesis.[6] As stated above, NFTs are also abundant in a significant subset of FTD such as FTDP-17, in which there is an absence of overt plaques. Another subset of FTD with lesions that are not immune-reactive for tau but for ubiquitin, also in the absence of plaques, has been termed FTLD-U or FTDU-17.[4] Here, TDP-43 has been identified in ubiquitin-positive inclusions; these have been also found in sporadic ALS (amyotrophic lateral sclerosis). Similar to tau, TDP-43 in the aggregates is hyperphosphorylated, ubiquitinated and truncated.[7] Both types of protein aggregates can be visualized under the light microscope either by immunohistochemistry, thioflavin-S staining, or by silver impregnation methods such as those developed by Gallyas and Bielschowsky. Massive tau fibrillization as evidenced by Gallyas reactivity is a hallmark feature of the human pathology that the early animal models tried to achieve but in which they failed (see below): although tau formed aggregates as the mice aged, NFTs did not develop.[1]

Another remarkable feature of the human pathology is the characteristic spreading of the tau lesions in the AD brain that is subject to little inter-individual variation, hence providing the basis for distinguishing six stages: the transentorhinal stages I and II representing silent cases; the limbic stages III and IV; and the neocortical stages V and VI[8] (Figure 5.1C). A similar type of staging has subsequently been defined for the Aβ pathology.[9] What causes this spreading is not understood at all, although it implies selective vulnerability of distinct neuronal populations.[10] It is an aspect that so far has not been reproduced in animal models. A further remarkable aspect of the human disease is that of the distinct age of onset and disease duration until death that, for example, discriminates carriers with *PGRN* (progranulin) mutations (leading to TDP-43 deposition) from those with *MAPT* mutations (leading to tau

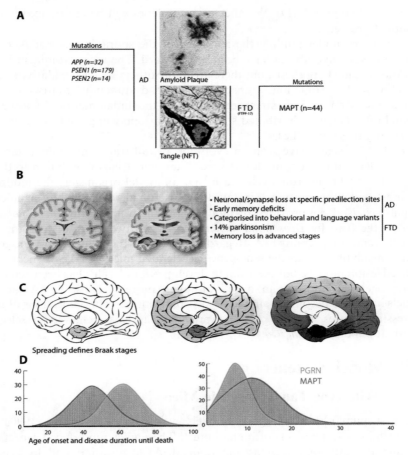

Figure 5.1 Animal models have become indispensable in basic and biomedical research. (**A**) The identification of pathogenic mutations in familial forms of Alzheimer's disease (AD) and frontotemporal dementia (FTD) greatly assisted in establishing transgenic animals that model key aspects of the human disease. In AD, mutations have been identified in the *APP*, *PSEN1* and *PSEN2* genes (n indicating current numbers of mutations), whereas in FTD with Parkinsonism linked to chromosome 17 (FTDP-17), mutations have been identified in the *MAPT* gene encoding tau. AD is characterized by amyloid plaques and neurofibrillary tangles (NFTs), while in FTDP-17, NFTs occur in the absence of overt plaques. The formation of these two lesions has been faithfully reproduced *in vivo*. (**B**) Of the clinical features that characterize AD and FTD, some have been reproduced in model organisms. (**C**) Other aspects such as the distinct spreading of the NFT pathology in AD (visualized by the intensity of magenta for three representative Braak stages) or (**D**) the distinct age of onset and disease duration that discriminates, for example, carriers of mutations in the tau-encoding *MAPT* gene and the progranulin-encoding *PGRN* gene (adapted from ref. 4) have not been reproduced in mice.

deposition)[4] (Figure 5.1D). Whether such subtleties will be reproduced *in vivo* remains to be seen.

In trying to understanding the aetiology of the sporadic cases of AD and FTD, a host of hypotheses have been put forward. The most prominent is the amyloid cascade hypothesis, but there are many others such as oxidative stress/ mitochondrial dysfunction, transport/synaptic dysfunction, mitosis failure, proteasome dysfunction, stress response, or the inflammation hypotheses. Animal models have contributed significantly in supporting the basic concepts underlying these hypotheses.

The last 15 years have seen an evolution of tau transgenic animal models from models with a very modest degree of tau hyperphosphorylation to those with considerable neuronal cell death.[1] Of all model organisms, the mouse is the most widely used species in tau-directed animal research, but significant insight has also been provided by modelling in the roundworm *Caenorhabditis elegans*, the fruit fly *Drosophila melanogaster*, and two types of fish, the sea lamprey and the zebrafish.[11] Regarding rodents, there are non-transgenic models available such as the senescence-accelerated mice (SAM), or chemically induced lesion models, but these fail to develop NFTs.[12] This has been achieved by transgenesis,[13] mostly in mice,[1] although more recently, transgenic rat models have also become available.[14,15] Also, by intracerebral injections of tau-expressing viruses, such as AAV, expression has been achieved in selected neuronal populations *in vivo*.[16]

5.2   Model Systems

5.2.1   Wild-type Tau Transgenic Mice

The first human tau transgenic models were generated in the 1990s by expressing wild-type forms of human tau.[17] These mice reproduced aspects of the human pathology such as somatodendritic localization and hyperphosphorylation of tau. Subsequently, stronger promoters were used to drive expression, resulting in more pronounced phenotypes.[18–20] Although tau formed aggregates in these mice, despite an age-dependent decrease in tau solubility, NFTs did not form until a very old age.[21] Signs of Wallerian degeneration, including axonal breakdown and segmentation of myelin into ellipsoids, were observed. Furthermore, neurogenic muscle atrophy, with groups of small angular muscle fibres, was present in the hindlimb musculature of transgenic mice.[20] Another characteristic was the presence of large numbers of axonal spheroids in the brainstem and spinal cord.[18–20] Similar swellings have been also described in transgenic mice with Aβ plaque formation[22] and in the human AD brain.[22] The role of impaired axonal transport in spheroid formation is discussed elsewhere.[23]

5.2.2   Mutant Tau Transgenic Mice

Following the identification of pathogenic mutations in *MAPT* in FTDP-17, tau filament formation was achieved in both neurons and glia, both by constitutive and doxycycline-regulated expression of mutant human tau.[1,24] The first published mouse strain with NFT formation, JNPL3, expressed human P301L tau under the control of the murine prion protein (PrP) promoter.[25] These mice developed severe motor disturbances by 10 months of age. NFTs were present in the brain and spinal cord, and numbers of motor neurons in the spinal cord were reduced by 50%.[25] The mice developed abnormal tau filaments in neurons, astrocytes and oligodendrocytes, reflecting the promoter used for transgene expression.[26] Using a different human tau isoform, P301L mutant tau was expressed under the control of the mThy1.2 promoter in pR5 mice, which also caused NFT formation, though in the absence of a motor phenotype.[27] Experimental diabetes (as induced by administration of the islet cell toxin, STZ) caused an early-onset and increased NFT formation in the pR5 mice, indicating that diabetes can accelerate onset and increase severity of disease in individuals with a predisposition to develop a tauopathy.[28] Because in humans the P301S mutation causes clinical signs of FTD earlier than in P301L carriers,[29] it was not surprising to find that transgenic mice expressing the P301S mutation developed a particularly early and pronounced pathology.[30]

The tauopathies PSP (progressive supranuclear palsy) and CBD (corticobasal degeneration) are characterized by substantial glial tau pathology. Aspects of this pathology have been reproduced, when G272V mutant tau expression was driven by an autoregulatory transactivator system under the control of the PrP promoter that resulted in high transgene expression in a subset of both neurons and oligodendrocytes. Electron microscopy established that tau filament formation was associated with hyperphosphorylation of tau. Thioflavin-S positive fibrillary inclusions were identified in oligodendrocytes and motor neurons;[31] however, the clinical phenotype of the mice was subtle. In contrast, when human wild-type tau was overexpressed in neurons and glial cells under the control of the mouse Tα1 α-tubulin promoter, a glial pathology was obtained that resembled the astrocytic plaques of CBD and the coiled bodies of both CBD and PSP.[32] A significant age-related neuronal loss was only found at 18 months of age, whereas oligodendrocytes were lost already at six months. The same team employed the 2′,3′-cyclic nucleotide 3′-phosphodiesterase promoter to express human P301L tau exclusively in oligodendrocytes.[33] Interestingly, the structural disruption of myelin and axons preceded the emergence of tau inclusions in oligodendrocytes; likewise, impaired axonal transport preceded the motor deficits in these mice.[33] Together, these studies highlight a role for glial tau in disease.

To determine whether NFTs are integral components of the neurotoxic cascade in AD or whether they represent a protective neuronal response, transgenic mice were generated in which expression of P301L tau could be regulated.[34] A 15-fold overexpression caused a progressive formation of NFTs, a remarkable neuronal loss (70% in the CA1 region) and behavioural impairment, whereas a reduction to 2.5-fold overexpression caused a recovery of memory functions and stabilization of neuron numbers while NFTs continued to accumulate. These data imply that NFTs *per se* are not sufficient to cause cognitive decline or neuronal death in this model.[34] In agreement, cognitive impairment in our P301L tau transgenic mice occurs in the absence of NFT formation.[35,36] An inducible system was also used to express truncated forms of tau that comprised mainly the carboxy-terminal microtubule-binding domain. Of a pro- and an anti-aggregation mutant generated, only the former resulted in neuronal loss.[37] An interesting observation was that human 'pro-aggregation' tau co-aggregated with mouse tau, the possibility of tau forming mixed aggregates being an issue of discrepancy in the field for quite some time.[38] In the pro-aggregation mouse model, again, NFTs persisted when tau levels were reduced with doxycycline.[37] In related work in flies, overexpression of both wild-type and R406W mutant tau caused premature death and neurodegeneration, with mutant tau leading to an earlier age of onset.[39] Again, the tau-associated pathology occurred in the absence of NFT formation suggesting that they may not be required for tau to execute its toxic effects.

In many tau transgenic models, memory functions have been addressed as have motor functions. Another clinical feature of a significant subset of FTD, parkinsonism, has also been modelled in mice.[40–43] We established an early-onset model based on the identification of the K369I mutation of tau in a patient with Pick's disease, an extreme form of FTD.[44] The tau lesions in Pick's disease (and also in the brain of the K369I human carrier) show a remarkable feature: they are Bielschowsky-positive and Gallyas-negative; in addition tau in the lesions is hyperphosphorylated at many epitopes but Ser262/Ser356 (12E8). We succeeded in reproducing this distinct pathology in the K3 mouse strain that expresses K369I mutant tau.[41] As the K3 mice express the transgene also in the substantia nigra pars compacta, they develop early-onset parkinsonism (i.e. resting tremor, bradykinesia, postural instability and gait anomalies). The mice show an increased cataleptic response to haloperidol and an early, but not late response to L-Dopa (levodopa), which models the limited efficacy of L-Dopa treatment in FTD patients. In the K3 mice, there is a selectively impaired axonal transport of distinct cargos such as mitochondria or TH (tyrosine hydroxylase)-containing vesicles.[41] This is because a component of the kinesin motor machinery, the scaffold/adapter protein JIP1, is trapped in the soma by elevated tau, preventing it from executing its physiological function. In a follow-up study the pathological interaction between tau and JIP1 was also revealed in AD brain.[42] Interestingly, this pathological interaction requires phosphorylation of tau. Since JIP1 is involved in regulating cargo binding to kinesin motors, these findings may, at least in part, explain how hyperphosphorylated forms of tau mediate impaired axonal transport in AD and FTD before tau starts to aggregate.[42] This finding presents inhibition of the hyperphosphorylation of tau as a promising therapeutic target for the development of disease-modifying drugs.[45]

The transmission, secretion and spreading of tau is another interesting aspect of tau that has emerged recently from animal studies. Using a stereotactic injection approach,[46] a donor brain extract derived from six-months-old NFT-forming P301S mice[30] was injected into brains of three-months-old ALZ17 mice,[20,47] a human wild-type tau transgenic strain, which develops a tau-associated amyotrophy but fails to develop NFTs.[46] The study found that the donor extract induced NFTs in the recipient brain, a feature termed 'transmission'. Secondly, NFT induction was not confined to the injection site, a feature referred to by the investigators as 'spreading'. While this did not reproduce the spreading that defines the Braak stages, the experiment provides some insight into how spreading of a pathology may be achieved in the first place (Figure 5.1C). Finally, insoluble rather than soluble tau was found to be responsible for these effects.[46] In related work, extracellular tau aggregates were shown to transmit a misfolded state from the outside of a cell to the inside, a feature reminiscent of the infectivity of prions.[48]

Tau secretion has further been reported in the sea lamprey by injecting tau-encoding plasmids into the giant neurons of the fish.[49] Interneuronal transmission has also been found for α-synuclein, a protein with a structure very similar to tau.[50] Another interesting finding obtained in animals is that both the formation of amyloid plaques and NFTs can be a rapid, occurring within 24 hours.[51,52]

Many of the treatment strategies that are currently being tested in clinical trials have their foundation in animal work. This includes the vaccination approach, which has been pioneered by researchers of the biotech company Elan.[53–55] Oddo and co-workers demonstrated that a single intracerebral injection of anti-Aβ antibodies into the hippocampus of triple transgenic mice not only reduced Aβ levels, but also cleared an early tau pathology.[56] As the brain is generally conceived to be immune-privileged, the efficacy of the Aβ-targeted immunization approach came as a surprise. As far as tau is concerned, it seems that efficacy would be even more difficult to achieve, as unlike Aβ, tau is an intracellular protein. However, immunization of tau transgenic mice with the PHF1 phospho-tau-peptide (containing the phospho-sites Ser396/Ser404) caused a reduction of aggregated tau levels, a shift from the insoluble to the soluble pool of tau, and a slowing of the progression of an NFT-related motor phenotype.[57]

Modulating tau phosphorylation and cleavage is another therapeutic strategy. Several candidate tau kinases have been targeted *in vivo*. A major problem is that most kinases are promiscuous.[58–60] The relatively non-specific protein kinase inhibitor K252a prevented motor deficits in the P301L tau transgenic JNPL mice and reduced levels of aggregated tau; however, NFT numbers were not reduced, suggesting that the main cytotoxic effects of tau are not exerted by NFTs but by lower order aggregates.[61] This is in line with studies where a reduction in the expression of transgenic human P301L tau led to a recovery of memory function and stabilization of neuron numbers, despite the continued accumulation of NFTs.[34] Another strategy is the targeting of phosphatases.[45] Here, PP2A is particularly promising, as it may be possible to activate one of its many regulatory subunits that determine substrate (tau) specificity instead of modulating general catalytic activity.[62–64] Together this presents a strategy that modulates either kinases or phosphatases as an attractive therapeutic approach.[45]

5.2.3   Invertebrate and Non-rodent Models

With rising costs associated with rodent husbandry and an increased awareness of animal ethics issues, it is anticipated that modelling in invertebrates and non-rodent species will become increasingly important. Already now, many tau transgenic models are available in species as diverse as the fruit fly *D. melanogaster*, the nematode *C. elegans*, the sea lamprey *Petromyzon marinus* and the zebrafish *Danio rerio*.[65,66]

Drosophila offers the advantage of low cost, small size and short lifespan, making it useful for drug testing and modifier screens. When wild-type and FTDP-17 mutant forms of human tau were expressed in *Drosophila*, the flies exhibited key features of the human disorder including adult-onset, progressive neurodegeneration, early death, enhanced toxicity of mutant tau, accumulation of abnormal tau and anatomical selectivity. However, neurodegeneration occurred without NFT formation.[39] Interestingly, in AD, approximately 15% of the neuronal loss cannot be explained by NFTs.[67,68] When expression of wild-type human tau was combined with that of the *Drosophila* GSK-3 homolog Shaggy, this resulted in a neurofibrillary pathology with tau filaments.[69] In a related study, expression of wild-type human tau caused impaired axonal transport with vesicle aggregation and associated loss of locomotor function, in the absence of neuronal loss.[70] Interestingly, co-expression of a constitutively active form of GSK-3β enhanced both axonal transport and locomotor phenotype, and two GSK-3β inhibitors, lithium and AR-A014418, reversed these changes, supporting previous studies in mice.[71] Both studies suggest that GSK-3 is an attractive drug target.

Directed expression of wild-type human and *Drosophila* tau to adult mushroom body neurons (centres of olfactory learning and memory) strongly impaired associative olfactory learning and memory, while olfactory conditioning relevant osmotactic and mechanosensory responses remained intact.[72] For comparison, studies in P301L tau transgenic mice revealed an impairment of conditioned taste aversion memory, a specific form of olfactory learning in mice.[35] In another *Drosophila* model, wild-type human tau was expressed in larval motor neurons. This caused synaptic dysfunction, possibly due to a reduction in numbers of functional mitochondria.[73] Data from wild-type and mutant tau-expressing *Drosophila* also support a role for cell cycle activation downstream of tau phosphorylation and upstream of apoptosis, a key player being TOR (target of rapamycin) kinase.[74] Evidence for a role of cell cycle regulated genes in neurodegeneration has also been obtained in wild-type human tau transgenic mice[75] and in tau-expressing human neuroblastoma cells.[76] The DNA-damage-activated Checkpoint kinase 2 (Chk2) has been identified as a kinase in *Drosophila* that phosphorylates tau at Ser262 (12E8 epitope), with a possible role in tau toxicity.[77]

Drosophila has been used extensively in modifier screens. In a *Drosophila* model of tauopathy, protein kinases and phosphatases comprised the major classes of modifiers and several candidate kinases were shown to enhance tau toxicity *in vivo*.[78] Overall, increased tau phosphorylation correlated with an increase in toxicity. One exception was MARK/PAR-1, which behaved as a genetic suppressor.[78] MARK is a kinase that phosphorylates tau at the 12E8 epitope Ser262/Ser356 which is in the microtubule-binding domain.[79,80] In a related study, this kinase was shown to act as an important enhancer of tau toxicity by triggering a temporally ordered phosphorylation process. It was shown that MARK/PAR-1 is a physiological tau kinase that plays a central role in regulating tau toxicity in *Drosophila*, without promoting NFT formation.[80] Mutating Ser262/Ser356 by alanine substitutions abolished tau toxicity. When human *tau* was expressed in the *Drosophila* eye along with *GSK-3β/ Shaggy* or *Cdk5*, toxicity was enhanced by either of the two kinases.[81]

The protein kinase inhibitor rapamycin has been reported to reduce toxicity in *Drosophila* expressing wild-type or mutant forms of tau, probably by reducing the amount of insoluble tau.[82] Rapamycin induces autophagy through inhibition of the protein kinase mammalian target of rapamycin (mTOR).[83] It had a similar beneficial effect in a *Drosophila* model of Huntington's disease.[84,85] It remains to be seen whether these effects of rapamycin are related to TOR-dependent abnormal cell cycle activation.[74] Targeting kinases for AD therapy is not easy. For example, lithium administration was found to inhibit GSK-3β activity and thus reduce tau phosphorylation, but the treatment did not ameliorate tau-induced toxicity.[81]

Several aspects of the human tau pathology have been modelled in the nematode *C. elegans*.[86] This roundworm has a number of useful features:

1. It is easy to culture as it feeds on bacteria grown on agar plates.
2. Within three days it develops from an egg to an adult worm, with about 300 progeny originating from one self-fertilized hermaphrodite.
3. Its small size allows for assays in microtitre format, studying hundreds of animals in a single well.
4. The worm is transparent, which is ideal for the use of fluorescence markers *in vivo*.
5. It is a complex multicellular animal: an adult hermaphrodite has exactly 959 somatic cells that form different organs, including 302 neurons forming the nervous system.
6. It has a short life span of only 2–3 weeks.
7. Genetic modifications, such as transgenic over-expression or RNAi-mediated gene knockdown, are quite easy compared with other *in vivo* systems.

Furthermore, most human disease genes and pathways are present in *C. elegans*,[87] such as the single tau-like protein called PTL-1 (protein with tau-like repeats).[88,89] Expression in *C. elegans* of FTD mutant, but not wild-type tau, resulted in neurodegeneration with an accumulation of hyperphosphorylated tau and associated uncoordinated locomotion (Unc).[90–92] In a genome-wide RNA interference screen, of more than 16 000 genes tested, 75 were found to enhance the transgene-induced behavioural phenotype.[93] Forty-six of these had sequence similarities to known human genes and fell into a number of broad classes, including kinases (such as MARK and GSK-3β), phosphatases (PP2A), chaperones (CHIP) and proteases—all of which have been implicated in tauopathies.

Different from *Drosophila,* the sea lamprey *P. marinus* is not used routinely as an experimental organism. Its set of six giant neurons in the hindbrain, the anterior bulbar cells, is readily accessible for manipulation. As discussed, this system provides support for tau secretion;[49] it further reproduced the stereotypic sequence of degenerative changes in AD, with the earliest changes occurring in distal dendrites.[94]

Only recently, zebrafish has been established as a model which reproduced tau hyperphosphorylation, NFT formation, neuronal and behavioural disturbances as well as cell death. Of the many inhibitors of the kinase GSK-3β tested in the fish model, AR-534 turned out to reduce tau hyperphosphorylation *in vivo*.[95] Methylene blue is an compound approved by the US Food and Drug Administration (FDA) that has recently been shown in a highly publicized phase 2 clinical trial to slow cognitive decline in AD patients. Incubation of P301L tau transgenic zebrafish larvae with methylene blue rescued neither abnormal phosphorylation, neuronal cell loss nor neurite outgrowth and swimming defects. The authors demonstrated that lack of efficacy was not due to methylene blue not being biologically active in zebrafish.[96]

5.2.4   The Aβ–Tau Axis

Transgenic mice have been used to address the cross-talk between tau and Aβ (Figure 5.1A). While in previous years, plaques and NFTs were assumed to be the key players in AD pathology, more recently, the focus has shifted from the end-stage lesions to the precursors, soluble oligomeric forms of Aβ and tau, as the major (but not exclusive) culprits. When P301L mutant tau transgenic JNPL3 mice were crossed with APP mutant Tg2576 mice, this caused a seven-fold NFT induction, while the Aβ pathology was not altered.[97] Similarly, intracerebral injections of fibrillar preparations of Aβ42 into P301L tau mutant pR5 mice caused a five-fold induction of NFTs.[98] The pathology was induced in the amygdala and is reflected by an impairment in amygdala-dependent tasks.[35,36] Injections into wild-type tau transgenic ALZ17 mice that lack an NFT pathology[20] failed to induce NFTs, suggesting that the toxic effect of Aβ is dependent on a pre-existing tau pathology.[98]

Subsequently, 3xtg-AD mice were generated that combine an NFT and plaque pathology. Their formation was found to be preceded by synaptic and LTP deficits.[99] The clinical features were correlated with intraneuronal Aβ formation,[100] and blocking of Aβ was shown to delay onset and progression of the tau pathology.[101] Injecting extracts from Aβ plaque-forming APP23 mice into P301L tau mutant JNPL3/B6 mice induced a tau pathology in areas with a neuronal projection to the injection site.[102] When the injections were performed in APP23/JNPL3/B6 mice with both a plaque and NFT pathology, a more pronounced tau pathology was induced in areas with a high Aβ plaque load.[102] Levels of tau were reduced in 3xtg-AD mice upon treatment with memantine, an *N*-methyl-D-aspartate (NMDA) receptor antagonist approved for the treatment of moderate to severe AD.[103] A combined plaque and NFT pathology was also achieved in ^{triple}AD mice in which there was found an enhanced phosphorylation of tau at Ser422 mediated by Aβ.[104] More importantly, Aβ and tau were found to synergistically impair the oxidative phosphorylation system.[105] Notably, deregulation of complex I was tau-dependent, whereas deregulation of complex IV was Aβ-dependent. Together, these studies suggest that Aβ and tau contribute synergistically to neurodegeneration in AD.[11] Interestingly, this concept can be extended to amyloidogenic peptides and proteins other than Aβ as shown by crossing P301S tau transgenic mice with mice expressing ADan, a model of familial Danish dementia (FDD).[106]

While the findings described above support the amyloid cascade hypothesis in mice, there is also a role for tau. Previously, Aβ toxicity in primary neuronal cultures was shown to depend on the presence of tau, as wild-type neurons degenerated when incubated with Aβ42 and primary neurons derived from tau knock-out mice turned out to be resistant.[107] This was translated *in vivo* by breeding APP mutant J20 mice onto a tau knock-out background.[108] Most if not all APP mutant strains with a robust Aβ plaque pathology are characterized by premature death of unknown cause.[109–111] Roberson and colleagues reported the remarkable finding that tau reduction ameliorated the premature lethality of J20 mice, increased resistance to excitotoxin-induced seizures, and prevented behavioural deficits in the Morris water maze.[108] This finding highlights the importance of tau and suggests that, as Aβ-directed therapies are likely to be ineffective in tauopathies without a plaque pathology, targeting tau might be a sensible approach.[112]

Acknowledgements

This work has been supported by the University of Sydney, the National Health & Medical Research Council (NHMRC), the Australian Research Council (ARC), and the J.O. & J.R. Wicking Trust. Postgraduate scholarship support has been provided by the Wenkart Foundation, GlaxoSmithKline and Alzheimer's Australia.

References

1. J. Gotz and L. M. Ittner, *Nat. Rev. Neurosci.*, 2008, **9**(7), 532.
2. C. Ballatore, V. M. Lee and J. Q. Trojanowski, *Nat. Rev. Neurosci.*, 2007, **8**(9), 663.
3. L. Bertram and R. E. Tanzi, *Nat. Rev. Neurosci.*, 2008, **9**(10), 768.
4. M. Cruts and C. Van Broeckhoven, *Trends Genet.*, 2008.
5. F. Chen, D. David, A. Ferrari and J. Gotz, *Curr. Drug Targets*, 2004, **5**(6), 503.
6. P. M. Horowitz, K. R. Patterson, A. L. Guillozet-Bongaarts, M. R. Reynolds, C. A. Carroll, S. T. Weintraub, D. A. Bennett, V. Cryns, R. W. Berry and L. I. Binder, *J. Neurosci.*, 2004, **24**(36), 7895.

7. M. Neumann, D. M. Sampathu, L. K. Kwong, A. C. Truax, M. C. Micsenyi, T. T. Chou, J. Bruce, T. Schuck, M. Grossman, C. M. Clark, L. F. McCluskey, B. L. Miller, E. Masliah, I. R. Mackenzie, H. Feldman, W. Feiden, H. A. Kretzschmar, J. Q. Trojanowski and V. M. Lee, *Science*, 2006, **314**(5796), 130.

8. H. Braak and E. Braak, *Acta Neuropathol.*, 1991, **82**(4), 239.

9. D. R. Thal, U. Rub, M. Orantes and H. Braak, *Neurology*, 2002, **58**(12), 1791.

10. J. Gotz, N. Schonrock, B. Vissel and L. M. Ittner, *J. Alzheimers Dis.*, 2009, **18**(2), 243.

11. J. Gotz, N. Deters, A. Doldissen, L. Bokhari, Y. Ke, A. Wiesner, N. Schonrock and L. M. Ittner, *Brain Pathol.*, 2007, **17**, 91.

12. D. Van Dam and P. P. De Deyn, *Nat. Rev. Drug. Discov.*, 2006, **5**(11), 956.

13. L. M. Ittner and J. Gotz, *Nat. Protoc.*, 2007, **2**(5), 1206.

14. N. Zilka, P. Filipcik, P. Koson, L. Fialova, R. Skrabana, M. Zilkova, G. Rolkova, E. Kontsekova and M. Novak, *FEBS Lett.*, 2006, **580**(15), 3582.

15. P. Koson, N. Zilka, A. Kovac, B. Kovacech, M. Korenova, P. Filipcik and M. Novak, *Eur. J. Neurosci.*, 2008, **28**(2), 239.

16. T. Jaworski, I. Dewachter, B. Lechat, S. Croes, A. Termont, D. Demedts, P. Borghgraef, H. Devijver, R. K. Filipkowski, L. Kaczmarek, S. Kugler and F. Van Leuven, *PLoS One*, 2009, **4**(10), e7280.

17. J. Gotz, A. Probst, M. G. Spillantini, T. Schafer, R. Jakes, K. Burki and M. Goedert, *EMBO J.*, 1995, **14**(7), 1304.

18. T. Ishihara, M. Hong, B. Zhang, Y. Nakagawa, M. K. Lee, J. Q. Trojanowski and V. M. Lee, *Neuron*, 1999, **24**(3), 751.

19. K. Spittaels, C. Van den Haute, J. Van Dorpe, K. Bruynseels, K. Vandezande, I. Laenen, H. Geerts, M. Mercken, R. Sciot, A. Van Lommel, R. Loos and F. Van Leuven, *Am. J. Pathol.*, 1999, **155**(6), 2153.

20. A. Probst, J. Gotz, K. H. Wiederhold, M. Tolnay, C. Mistl, A. L. Jaton, M. Hong, T. Ishihara, V. M. Lee, J. Q. Trojanowski, R. Jakes, R. A. Crowther, M. G. Spillantini, K. Burki and M. Goedert, *Acta Neuropathol.*, 2000, **99**(5), 469.

21. T. Ishihara, B. Zhang, M. Higuchi, Y. Yoshiyama, J. Q. Trojanowski and V. M. Lee, *Am. J. Pathol.*, 2001, **158**(2), 555.

22. G. B. Stokin, C. Lillo, T. L. Falzone, R. G. Brusch, E. Rockenstein, S. L. Mount, R. Raman, P. Davies, E. Masliah, D. S. Williams and L. S. Goldstein, *Science*, 2005, **307**(5713), 1282.

23. J. Gotz, L. M. Ittner and S. Kins, *J. Neurochem.*, 2006, **98**(4), 993.

24. J. Gotz and N. N. Gotz, *ASN Neuro*, 2009, 1(4), pii: e00019.

25. J. Lewis, E. McGowan, J. Rockwood, H. Melrose, P. Nacharaju, M. Van Slegtenhorst, K. Gwinn-Hardy, P. M. Murphy, M. Baker, X. Yu, K. Duff, J. Hardy, A. Corral, W. L. Lin, S. H. Yen, D. W. Dickson, P. Davies and M. Hutton, *Nat. Genet.*, 2000, **25**(4), 402.

26. W. L. Lin, J. Lewis, S. H. Yen, M. Hutton and D. W. Dickson, *Am. J. Pathol.*, 2003, **162**(1), 213.

27. J. Gotz, F. Chen, R. Barmettler and R. M. Nitsch, *J. Biol. Chem.*, 2001, **276**(1), 529.

28. Y. D. Ke, F. Delerue, A. Gladbach, J. Gotz and L. M. Ittner, *PLoS One*, 2009, **4**(11), e7917.

29. O. Bugiani, J. R. Murrell, G. Giaccone, M. Hasegawa, G. Ghigo, M. Tabaton, M. Morbin, A. Primavera, F. Carella, C. Solaro, M. Grisoli, M. Savoiardo, M. G. Spillantini, F. Tagliavini, M. Goedert and B. Ghetti, *J. Neuropathol. Exp. Neurol.*, 1999, **58**(6), 667.

30. B. Allen, E. Ingram, M. Takao, M. J. Smith, R. Jakes, K. Virdee, H. Yoshida, M. Holzer, M. Craxton, P. C. Emson, C. Atzori, A. Migheli, R. A. Crowther, B. Ghetti, M. G. Spillantini and M. Goedert, *J. Neurosci.*, 2002, **22**(21), 9340.

31. J. Gotz, M. Tolnay, R. Barmettler, F. Chen, A. Probst and R. M. Nitsch, *Eur. J. Neurosci.*, 2001, **13**(11), 2131.

32. M. Higuchi, T. Ishihara, B. Zhang, M. Hong, A. Andreadis, J. Trojanowski and V. M. Lee, *Neuron*, 2002, **35**(3), 433.

33. M. Higuchi, B. Zhang, M. S. Forman, Y. Yoshiyama, J. Q. Trojanowski and V. M. Lee, *J. Neurosci.*, 2005, **25**(41), 9434.

34. K. Santacruz, J. Lewis, T. Spires, J. Paulson, L. Kotilinek, M. Ingelsson, A. Guimaraes, M. DeTure, M. Ramsden, E. McGowan, C. Forster, M. Yue, J. Orne, C. Janus, A. Mariash, M. Kuskowski, B. Hyman, M. Hutton and K. H. Ashe, *Science*, 2005, **309**(5733), 476.

35. L. Pennanen, H. Welzl, P. D'Adamo, R. M. Nitsch and J. Gotz, *Neurobiol. Dis.*, 2004, **15**(3), 500.

36. L. Pennanen, D. P. Wolfer, R. M. Nitsch and J. Gotz, *Genes Brain Behav.*, 2006, **5**(5), 369.

37. M. M. Mocanu, A. Nissen, K. Eckermann, I. Khlistunova, J. Biernat, D. Drexler, O. Petrova, K. Schonig, H. Bujard, E. Mandelkow, L. Zhou, G. Rune and E. M. Mandelkow, *J. Neurosci.*, 2008, **28**(3), 737.

38. C. Andorfer, Y. Kress, M. Espinoza, R. de Silva, K. L. Tucker, Y. A. Bardem, K. Duff and P. Davies, *J. Neurochem.*, 2003, **86**(3), 582.

39. C. W. Wittmann, M. F. Wszolek, J. M. Shulman, P. M. Salvaterra, J. Lewis, M. Hutton and M. B. Feany, *Science*, 2001, **293**(5530), 711.

40. H. N. Dawson, V. Cantillana, L. Chen and M. P. Vitek, *J. Neurosci.*, 2007, **27**(34), 9155.

41. L. M. Ittner, T. Fath, Y. D. Ke, M. Bi, J. van Eersel, K. M. Li, P. Gunning and J. Gotz, *Proc. Natl. Acad. Sci. U. S. A.*, 2008, **105**(41), 15997.

42. L. M. Ittner, Y. D. Ke and J. Gotz, *J. Biol. Chem.*, 2009, **284**(31), 20909.

43. M. Takenokuchi, K. Kadoyama, S. Chiba, M. Sumida, S. Matsuyama, K. Saigo and T. Taniguchi, *Neurosci. Lett.*, 2010, **473**(3), 182.

44. M. Neumann, W. Schulz-Schaeffer, R. A. Crowther, M. J. Smith, M. G. Spillantini, M. Goedert and H. A. Kretzschmar, *Ann. Neurol.*, 2001, **50**(4), 503.

45. K. Iqbal and I. Grundke-Iqbal, *J. Cell. Mol. Med.*, 2008, **12**(1), 38.

46. F. Clavaguera, T. Bolmont, R. A. Crowther, D. Abramowski, S. Frank, A. Probst, G. Fraser, A. K. Stalder, M. Beibel, M. Staufenbiel, M. Jucker, M. Goedert and M. Tolnay, *Nat. Cell. Biol.*, 2009, **11**(7), 909.
47. J. Gotz and R. M. Nitsch, *Neuroreport*, 2001, **12**(9), 2007.
48. B. Frost, R. L. Jacks and M. I. Diamond, *J. Biol. Chem.*, 2009, **284**(19), 12845.
49. W. Kim, S. Lee, C. Jung, A. Ahmed, G. Lee and G. F. Hall, *J. Alzheimers Dis.*, 2010, **19**(2), 647.
50. P. Desplats, H. J. Lee, E. J. Bae, C. Patrick, E. Rockenstein, L. Crews, B. Spencer, E. Masliah and S. J. Lee, *Proc. Natl. Acad. Sci. U. S. A.*, 2009, **106**, 13010.
51. M. Meyer-Luehmann, T. L. Spires-Jones, C. Prada, M. Garcia-Alloza, A. de Calignon, A. Rozkalne, J. Koenigsknecht-Talboo, D. M. Holtzman, B. J. Bacskai and B. T. Hyman, *Nature*, 2008, **451**(7179), 720.
52. A. de Calignon, L. M. Fox, R. Pitstick, G. A. Carlson, B. J. Bacskai, T. L. Spires-Jones and B. T. Hyman, *Nature*, 2010, **464**(7292), 1201.
53. D. Schenk, R. Barbour, W. Dunn, G. Gordon, H. Grajeda, T. Guido, K. Hu, J. Huang, K. Johnson-Wood, K. Khan, D. Kholodenko, M. Lee, Z. Liao, I. Lieberburg, R. Motter, L. Mutter, F. Soriano, G. Shopp, N. Vasquez, C. Vandevert, S. Walker, M. Wogulis, T. Yednock, D. Games and P. Seubert, *Nature*, 1999, **400**(6740), 173.
54. F. Bard, C. Cannon, R. Barbour, R. L. Burke, D. Games, H. Grajeda, T. Guido, K. Hu, J. Huang, K. Johnson-Wood, K. Khan, D. Kholodenko, M. Lee, I. Lieberburg, R. Motter, M. Nguyen, F. Soriano, N. Vasquez, K. Weiss, B. Welch, P. Seubert, D. Schenk and T. Yednock, *Nat. Med.*, 2000, **6**(8), 916.
55. E. M. Sigurdsson, H. Scholtzova, P. D. Mehta, B. Frangione and T. Wisniewski, *Am. J. Pathol.*, 2001, **159**(2), 439.
56. S. Oddo, L. Billings, J. P. Kesslak, D. H. Cribbs and F. M. LaFerla, *Neuron*, 2004, **43**(3), 321.
57. A. A. Asuni, A. Boutajangout, D. Quartermain and E. M. Sigurdsson, *J. Neurosci.*, 2007, **27**(34), 9115.
58. F. Hernandez, F. Lim, J. J. Lucas, C. Perez-Martin, F. Moreno and J. Avila, *Mini Rev. Med. Chem.*, 2002, **2**(1), 51.
59. G. Li, A. Faibushevich, B. J. Turunen, S. O. Yoon, G. Georg, M. L. Michaelis and R. T. Dobrowsky, *J. Neurochem.*, 2003, **84**(2), 347.
60. C. J. Phiel, C. A. Wilson, V. M. Lee and P. S. Klein, *Nature*, 2003, **423**(6938), 435.
61. S. Le Corre, H. W. Klafki, N. Plesnila, G. Hubinger, A. Obermeier, H. Sahagun, B. Monse, P. Seneci, J. Lewis, J. Eriksen, C. Zehr, M. Yue, E. McGowan, D. W. Dickson, M. Hutton and H. M. Roder, *Proc. Natl. Acad. Sci. U. S. A.*, 2006, **103**(25), 9673.
62. S. Kins, A. Crameri, D. R. Evans, B. A. Hemmings, R. M. Nitsch and J. Gotz, *J. Biol. Chem.*, 2001, **276**(41), 38193.
63. A. Schild, L. M. Ittner and J. Gotz, *Biochem. Biophys. Res. Commun.*, 2006, **343**(4), 1171.

64. N. Deters, L. M. Ittner and J. Gotz, *Biochem. Biophys. Res. Commun.*, 2009, **379**(2), 400.
65. J. Gotz, J. R. Streffer, D. David, A. Schild, F. Hoerndli, L. Pennanen, P. Kurosinski and F. Chen, *Mol. Psychiatry*, 2004, **9**, 664.
66. V. M. Lee, T. K. Kenyon and J. Q. Trojanowski, *Biochim. Biophys. Acta*, 2005, **1739**(2–3), 251.
67. T. Gomez-Isla, R. Hollister, H. West, S. Mui, J. H. Growdon, R. C. Petersen, J. E. Parisi and B. T. Hyman, *Ann. Neurol.*, 1997, **41**(1), 17.
68. P. Giannakopoulos, F. R. Herrmann, T. Bussiere, C. Bouras, E. Kovari, D. P. Perl, J. H. Morrison, G. Gold and P. R. Hof, *Neurology*, 2003, **60**(9), 1495.
69. G. R. Jackson, M. Wiedau-Pazos, T.-K. Sang, N. Wagle, C. A. Brown, S. Massachi and D. H Geschwind, *Neuron*, 2002, **34**(4), 509.
70. A. Mudher, D. Shepherd, T. A. Newman, P. Mildren, J. P. Jukes, A. Squire, A. Mears, J. A. Drummond, S. Berg, D. MacKay, A. A. Asuni, R. Bhat and S. Lovestone, *Mol. Psychiatry*, 2004, **9**(5), 522.
71. K. Spittaels, C. Van Den Haute, J. Van Dorpe, H. Geerts, M. Mercken, K. Bruynseels, R. Lasrado, K. Vandezande, I. Laenen, T. Boon, J. Van Lint, J. Vandenheede, D. Moechars, R. Loos and F. Van Leuven, *J. Biol. Chem.*, 2000, **275**(52), 41340.
72. A. Mershin, E. Pavlopoulos, O. Fitch, B. C. Braden, D. V. Nanopoulos and E. M. Skoulakis, *Learn. Mem.*, 2004, **11**(3), 277.
73. F. C. Chee, A. Mudher, M. F. Cuttle, T. A. Newman, D. MacKay, S. Lovestone and D. Shepherd, *Neurobiol. Dis.*, 2005, **20**(3), 918.
74. V. Khurana, Y. Lu, M. L. Steinhilb, S. Oldham, J. M. Shulman and M. B. Feany, *Curr Biol.*, 2006, **16**(3), 230.
75. C. Andorfer, C. M. Acker, Y. Kress, P. R. Hof, K. Duff and P. Davies, *J. Neurosci.*, 2005, **25**(22), 5446.
76. F. J. Hoerndli, S. Pelech, A. Papassotiropoulos and J. Götz, *Eur. J. Neurosci.*, 2007, **26**(1), 60.
77. K. Iijima-Ando, L. Zhao, A. Gatt, C. Shenton and K. Iijima, *Hum. Mol. Genet.*, 2010, **19**(10), 1930.
78. J. M. Shulman and M. B. Feany, *Genetics*, 2003, **165**(3), 1233.
79. J. C. Augustinack, A. Schneider, E. M. Mandelkow and B. T. Hyman, *Acta Neuropathol.*, 2002, **103**(1), 26.
80. I. Nishimura, Y. Yang and B. Lu, *Cell*, 2004, **116**(5), 671.
81. K. W. Chau, W. Y. Chan, P. C. Shaw and H. Y. Chan, *Biochem. Biophys. Res. Commun.*, 2006, **346**(1), 150.
82. Z. Berger, B. Ravikumar, F. M. Menzies, L. G. Oroz, B. R. Underwood, M. N. Pangalos, I. Schmitt, U. Wullner, B. O. Evert, C. J. O'Kane and D. C. Rubinsztein, *Hum. Mol. Genet.*, 2006, **15**(3), 433.
83. T. Noda and Y. Ohsumi, *J. Biol. Chem.*, 1998, **273**(7), 3963.
84. B. Ravikumar, R. Duden and D. C. Rubinsztein, *Hum. Mol. Genet.*, 2002, **11**(9), 1107.
85. B. Ravikumar, C. Vacher, Z. Berger, J. E. Davies, S. Luo, L. G. Oroz, F. Scaravilli, D. F. Easton, R. Duden, C. J. O'Kane and D. C. Rubinsztein, *Nat. Genet.*, 2004, **36**(6), 585.

86. M. Morcos and H. Hutter, *J. Alzheimers Dis.*, 2009, **16**(4), 897.
87. T. Kaletta and M. O. Hengartner, *Nat. Rev. Drug Discov.*, 2006, **5**(5), 387.
88. J. B. McDermott, S. Aamodt and E. Aamodt, *Biochemistry*, 1996, **35**(29), 9415.
89. M. Goedert, C. P. Baur, J. Ahringer, R. Jakes, M. Hasegawa, M. G. Spillantini, M. J. Smith and F. Hill, *J. Cell. Sci.*, 1996, **109**(Pt 11), 2661.
90. B. C. Kraemer, B. Zhang, J. B. Leverenz, J. H. Thomas, J. Q. Trojanowski and G. D. Schellenberg, *Proc. Natl. Acad. Sci. U. S. A.*, 2003, **100**(17), 9980.
91. T. Miyasaka, Z. Ding, K. Gengyo-Ando, M. Oue, H. Yamaguchi, S. Mitani and Y. Ihara, *Neurobiol. Dis.*, 2005, **20**(2), 372.
92. R. Brandt, A. Gergou, I. Wacker, T. Fath and H. Hutter, *Neurobiol. Aging*, 2009, **30**(1), 22.
93. B. C. Kraemer, J. K. Burgess, J. H. Chen, J. H. Thomas and G. D. Schellenberg, *Hum. Mol. Genet.*, 2006, **15**(9), 1483.
94. G. F. Hall, V. M. Lee, G. Lee and J. Yao, *Am. J. Pathol.*, 2001, **158**(1), 235.
95. D. Paquet, R. Bhat, A. Sydow, E. M. Mandelkow, S. Berg, S. Hellberg, J. Falting, M. Distel, R. W. Koster, B. Schmid and C. Haass, *J. Clin. Invest.*, 2009, **119**(5), 1382.
96. F. van Bebber, D. Paquet, A. Hruscha, B. Schmid and C. Haass, *Neurobiol. Dis.*, 2010, **39**(3), 265.
97. J. Lewis, D. W. Dickson, W.-L. Lin, L. Chisholm, A. Corral, G. Jones, S.-H. Yen, N. Sahara, L. Skipper, D. Yager, C. Eckman, J. Hardy, M. Hutton and E. McGowan, *Science*, 2001, **293**(5534), 1487.
98. J. Gotz, F. Chen, J. van Dorpe and R. M. Nitsch, *Science*, 2001, **293**(5534), 1491.
99. S. Oddo, A. Caccamo, J. D. Shepherd, M. P. Murphy, T. E. Golde, R. Kayed, R. Metherate, M. P. Mattson, Y. Akbari and F. M. LaFerla, *Neuron*, 2003, **39**(3), 409.
100. L. M. Billings, S. Oddo, K. N. Green, J. L. McGaugh and F. M. LaFerla, *Neuron*, 2005, **45**(5), 675.
101. S. Oddo, A. Caccamo, B. Tseng, D. Cheng, V. Vasilevko, D. H. Cribbs and F. M. LaFerla, *J. Neurosci.*, 2008, **28**(47), 12163.
102. T. Bolmont, F. Clavaguera, M. Meyer-Luehmann, M. C. Herzig, R. Radde, M. Staufenbiel, J. Lewis, M. Hutton, M. Tolnay and M. Jucker, *Am. J. Pathol.*, 2007, **171**(6), 2012.
103. H. Martinez-Coria, K. N. Green, L. M. Billings, M. Kitazawa, M. Albrecht, G. Rammes, C. G. Parsons, S. Gupta, P. Banerjee and F. M. LaFerla, *Am. J. Pathol.*, 2010, **176**(2), 870.
104. F. Grueninger, B. Bohrmann, C. Czech, T. M. Ballard, J. R. Frey, C. Weidensteiner, M. von Kienlin and L. Ozmen, *Neurobiol. Dis.*, 2010, **37**(2), 294.
105. V. Rhein, X. Song, A. Wiesner, L. M. Ittner, G. Baysang, F. Meier, L. Ozmen, H. Bluethmann, S. Drose, U. Brandt, E. Savaskan, C. Czech, J. Gotz and A. Eckert, *Proc. Natl. Acad. Sci. U. S. A.*, 2009, **106**(47), 20057.

106. J. Coomaraswamy, E. Kilger, H. Wolfing, C. Schafer, S. A. Kaeser, B. M. Wegenast-Braun, J. K. Hefendehl, H. Wolburg, M. Mazzella, J. Ghiso, M. Goedert, H. Akiyama, F. Garcia-Sierra, D. P. Wolfer, P. M. Mathews and M. Jucker, *Proc. Natl. Acad. Sci. U. S. A.*, 2010, **107**(17), 7969.
107. M. Rapoport, H. N. Dawson, L. I. Binder, M. P. Vitek and A. Ferreira, *Proc. Natl. Acad. Sci. U. S. A.*, 2002, **99**(9), 6364.
108. E. D. Roberson, K. Scearce-Levie, J. J. Palop, F. Yan, I. H. Cheng, T. Wu, H. Gerstein, G. Q. Yu and L. Mucke, *Science*, 2007, **316**(5825), 750.
109. J. J. Palop, J. Chin, E. D. Roberson, J. Wang, M. T. Bien-Thwin, N. Ly, J. Yoo, K. O. Ho, G. Q. Yu, A. Kreitzer, S. Finkbeiner, J. L. Noebels and L. Mucke, *Neuron*, 2007, **55**(5), 697.
110. R. Minkeviciene, S. Rheims, M. B. Dobszay, M. Zilberter, J. Hartikainen, L. Fulop, B. Penke, Y. Zilberter, T. Harkany, A. Pitkanen and H. Tanila, *J. Neurosci.*, 2009, **29**(11), 3453.
111. J. J. Palop and L. Mucke, *Arch. Neurol.*, 2009, **66**(4), 435.
112. J. Gotz, L. M. Ittner and N. Schonrock, *Med. J. Aust.*, 2006, **185**(7), 381.

CHAPTER 6
Animal Models of Parkinson's Disease

HARDY J. RIDEOUT[a] AND LEONIDAS STEFANIS[*a,b]

[a] Department of Neurology, Columbia University, New York, USA;
[b] Division of Basic Neurosciences, Biomedical Research Foundation of the Academy of Athens, Athens, Greece

6.1 Introduction

Parkinson's disease (PD) is a common progressive, neurodegenerative disorder, characterized predominantly by motor symptoms such as tremor, rigidity and bradykinesia. Only about 10% of PD cases demonstrate Mendelian inheritance of the disease, whereas the rest are sporadic.

The neuropathological hallmark of the disorder is loss of dopamine neurons in the substantia nigra pars compacta (SNpc), reduction of striatal dopamine terminals and levels, and the presence of cytoplasmic inclusions termed Lewy bodies (LBs) and neuritic inclusions termed Lewy neurites (LNs). Such inclusions are present predominantly in the nigra and other brainstem nuclei, but can also be found in autonomic, limbic and cortical regions. Furthermore, the neuronal degeneration is quite extensive, and involves not only the nigra but also other brainstem nuclei such as the locus coeruleus, and more widespread neuronal networks and neurotransmitter systems, both in the central and the autonomic nervous system. A corollary of this is that, although PD manifests predominantly as a dopamine deficiency-dependent motoric deficit, it is also characterized by non-dopamine related, non-motor symptoms such as olfactory impairment, depression, autonomic dysfunction, and cognitive and sleep disturbances.

RSC Drug Discovery Series No. 6
Animal Models for Neurodegenerative Disease
Edited by Jesús Avila, Jose J. Lucas and Félix Hernández
© Royal Society of Chemistry 2011
Published by the Royal Society of Chemistry, www.rsc.org

Ideally, animal models of the disease should recapitulate such clinical and neuropathological features. Here, we address mainly four types of animal models of PD: those based on neurotoxins that preferentially target the dopamine system, those based on manipulation of alpha-synuclein, those based on other genetic causes of PD, and finally some additional models, that, although not based on specific genetic defects found in PD patients, provide exceptional insights into PD pathogenesis and pathophysiology.

6.2   Genetic Models

6.2.1   Animal Models Based on Manipulation of Alpha-synuclein

Alpha-synuclein (ASYN) is a small presynaptic neuronal protein which is heavily implicated as a pathogenetic factor in PD. In 1997, a missense point mutation leading to an A to T substitution of residue 53 in *SNCA*, the gene encoding for ASYN, was identified in affected members of families with autosomally dominant inherited PD.[1] Two other missense point mutations, A30P and E46K, were subsequently found in other populations.[2,3] Furthermore, duplications or triplications of the wild-type *SNCA* gene were identified in other families with autosomal dominant PD.[4] It should be noted that all such cases with Mendelian inheritance of *SNCA* mutations or locus multiplications are exceedingly rare. The link of ASYN to PD is not only genetic. ASYN constitutes the major component of LBs and LNs that characterize sporadic PD.[5,6] Cementing an epipathogenic link with PD, Genome-Wide Association Studies (GWAS) have recently shown that ASYN is one of the very few genes that show a significant correlation with sporadic PD.[7] Therefore, ASYN is not only linked to PD through rare familial cases, but, more importantly, also through its genetic and neuropathological association with the much more common sporadic disease. The predominant theory for the underlying basis for this link between ASYN and PD is that ASYN has the propensity to misfold and aggregate and that something along this aggregation pathway, but not necessarily the full-blown LB, is neurotoxic. The bulk of the evidence suggests that intermediate soluble oligomeric forms are toxic, although the exact offending species are unknown.[8]

Based on such genetic defects, with autosomal dominant inheritance and the clear indication from the multiplication cases that an excess of even the wild-type protein can lead to PD, attempts based on ASYN to model PD in animals have used overexpression of the wild-type protein or its mutants in various model systems. Before discussing these efforts, however, it is worth noting that the lack of ASYN in null mice leads to no overt phenotype but to subtle electrophysiological abnormalities, reflecting the likely role of this protein in control of neurotransmitter release.[9–11] Something that is not widely appreciated is that such mice also demonstrate, for unclear reasons, a slightly reduced number of nigral dopamine neurons,[12–14] even though there is no excess programmed cell death or increased vulnerability to neuronal apoptosis.[15] Furthermore, a gradual loss of striatal dopamine terminals and levels occurs in these mice,[16] and such mice are especially vulnerable to synaptic toxicity.[17] Although the relevance of such observations to PD is unclear as there is no indication of a loss of function of ASYN in PD (except for the idea that aggregated ASYN may precipitate within inclusions and lose its function at the level of the presynaptic membrane), this raises the issue of potential harmful effects of loss of ASYN in the central nervous system (CNS).

6.2.1.1 Genetic Mammalian Models

Masliah *et al.*[18] were the first to demonstrate that overexpression of wild-type ASYN in the mouse CNS (involving basically the neocortex and the hippocampus, with some lower expression in the substantia nigra) through a platelet-derived growth factor (PDGF) B promoter led to accumulation of cytoplasmic and nuclear ASYN, sometimes in the form of granular, but not fibrillar, inclusions. At later time points, at 12 months of age, the mice showed loss of striatal dopamine (without overt cell loss) and motor dysfunction.[18] Expression of the A53T mutant form led to enhanced neurotoxicity, although inclusion formation was less marked, suggesting that inclusions *per se* were not toxic.[19] Sharon *et al.*[20,21] also generated PDGF-B transgenic mice, but concentrated solely on the formation of soluble oligomers under the influence of fatty acids.

A number of groups, including the Masliah lab, have created mice expressing wild-type or mutant ASYN under the control of the Thy-1 promoter.[17,22–27] This leads to more widespread expression in the CNS, including higher expression in the nigra and expression in the brainstem and spinal cord. Accumulation of proteinase K-resistant ASYN aggregates, but not frank fibrillar inclusions, was identified in some, but not all, cases. Remarkably, neuronal damage, manifested by synaptic and axonal degeneration mainly in the brainstem and spinal cord, was seen even in the absence of frank aggregation.[27] An early behavioural phenotype was seen, including alterations in sensorimotor and cognitive tasks.[22,25,27–29] In certain instances, under comparable expression levels, PD-related mutant forms caused more aggregation, gliosis, neuronal degeneration and motoric dysfunction compared to the wild-type protein; interestingly, expression of the mouse protein did not lead to any abormalities.[17]

There has been scant evidence of a link of dopamine dysfunction to the observed behavioural phenotypes, as a basic anatomical and behavioural analysis of such mice at the time when they manifest clear behavioural abnormalities did not reveal alterations in the dopaminergic system. It has been suggested, instead, that the noradrenergic dysfunction that has been observed in such mice may be more linked to these early behavioural abnormalities.[28] It should also be noted that, by expressing at relatively low levels an aggregation-prone artificial mutant form of ASYN, Zhou *et al.*[30] demonstrated significant neuronal degeneration the CNS, but again with little damage to the dopaminergic system, using the Thy1 promoter.

Transgenic mice expressing various forms of ASYN under the prion protein (PrP) promoter have also been created;[31–34] this leads to high levels of expression throughout the CNS. Although the picture is not homogeneous, it appears that under this promoter only the mutant, but not the wild-type forms, of the protein are toxic; a major determinant is also the expression level of the transgene. Interestingly, in one case, lack of endogenous ASYN enhanced the observed pathology, arguing for a protective effect of endogenous mouse ASYN, as noted earlier.[10,32] The neurotoxicity in these models manifests mainly with gliosis and neuritic degeneration,[31,33,34] but can also include frank neuronal death that has some features of apoptosis; mitochondrial damage is especially prominent.[35] In these PrP–ASYN transgenic models, somewhat fibrillar, but not LB-like inclusions are seen in some[31,34] but not in other[32,33] cases. The dopamine system appears to be little affected, although in one line the observed hyperactivity was linked to alterations in dopaminergic neuro-transmission both at the pre- and post-synaptic level.[36] We have examined the Giasson mice[31] and have found the dopaminergic system to be intact, even at 14 months of age. However, we have detected a progressive degeneration of the noradrenergic system in the CNS of these mice, raising the possibility that this system may be more vulnerable to ASYN-mediated toxicity.[37]

A number of groups have attempted to overexpress ASYN through the tyrosine hydroxylase (TH) promoter in order to achieve expression within dopaminergic neurons. It should be noted that this would also be expected to lead to ASYN expression within noradrenergic neurons, although the effects of the transgene expression on this system have not been studied. Expression of the full length wild-type or mutant protein has not led to any significant abnormalities, despite variable ASYN accumulation.[38–41] However, a double mutant form (A53T plus A30P) did show, despite the lack of obvious aggre-gation, dopaminergic terminal degeneration without nigral neuron loss.[42] In addition, two groups have created transgenic mice expressing *C*-terminal truncated forms of wild-type or mutant ASYN under the control of the TH promoter.[43,44] *C*-terminal truncated forms were used because of their pro-pensity to fibrillize more than the wild-type protein. In one study,[44] this led to early developmental, non-progressive dopaminergic neuron death without frank aggregation. In the other,[43] dopaminergic terminal degeneration but not neuron death was observed, together with somewhat fibrillar inclusions. Of note, dopaminergic terminal degeneration appeared at a few months of life, but did not progress further.

Some studies have used inducible expression systems to overexpress ASYN in a defined temporal or spatial fashion within the CNS. This has the advantage that, once a phenotype is seen, expression of the transgene can be suppressed and the reversibility of the phenotype can be assessed. Nuber *et al.*[45] generated a transgenic mouse expressing wild-type ASYN in a tetracycline-regulated fashion under the control of the calmodulin kinase (CaM) and the PrP promoter. The former mouse expressed moderate amounts of ASYN in olfactory bulb, cortex, basal ganglia and substantia nigra, which were partially reduced with treatment with doxycycline. Such expression resulted in neuritic and cell soma degeneration, mostly in the form of 'dark neurons'. Substantia nigra dopaminergic neurons also appeared to degenerate as 'dark neurons', but a slight reduction in numbers of nigral TH-positive neurons was not statistically significant and no reduction in striatal dopamine levels was seen. The authors argued that this may be due to compensatory mechanisms at the level of sprouting or dopamine metabolism. Overexpressed ASYN accumulated in neurites and cell bodies, sometimes in intracellular 'patches', but fibrillar inclusions were not seen. The mice also demonstrated progressive motoric and memory deficits; however, their relationship to dopaminergic damage was unclear. Interestingly, the behavioural phenotype was arrested, but not reversed, with tetracycline treatment, suggesting that continuous presence of the transgene was necessary for the progression of observed deficits. Daher et al.[46] have used the Cre recombinase technology to express truncated 1-119 A53T ASYN in a Cre-dependent fashion with the TH promoter either within widespread CNS regions, with the nestin promoter, or in cathecholaminergic neurons. In the latter case, loss of striatal dopamine, but not nigral dopamine neurons, could be demonstrated in homozygous mice.

Overall, a major limitation of genetic mouse models of ASYN has been the lack of significant dopaminergic dysfunction. Even when this occurs, as in the Tofaris et al.[43] or the Richfield et al.[42] models, it appears to be non-progressive and to be unaccompanied by dopaminergic neuronal death. In fact, a consistent feature that has been observed is that, when neuronal degeneration occurs, it affects predominantly axonal processes and not cell bodies. In addition, there is a lack of bona fide LBs; even when inclusions occur, they lack the radial fibrillar arrangement of LBs. Another feature that emerges is the lack of association between inclusion formation and neurodegeneration, as the latter may occur without the former and *vice versa*. No doubt the inability to achieve high levels of expression of ASYN within the dopaminergic system contributes significantly to the lack of an overt dopaminergic phenotype in these models. A potential intrinsic ability of mouse dopaminergic neurons to withstand such insults, perhaps related to endogenous ASYN, may also play a role.

Despite these limitations, such transgenic mice have already led to significant insights into potential pathological functions of ASYN and have led to the testing of experimental therapies. For example, our studies in such mice led to the identification of ASYN species co-eluting with and inhibiting the proteasome;[47] others showed that soluble oligomeric ASYN species derived from areas where inclusions occur carry neurotoxic properties,[48] or that ASYN is phosphorylated at various residues within inclusions and this may influence its aggregation potential,[49] or that ASYN-mediated neuritic degeneration is associated with an early microglial activation.[50] On the experimental therapy front, examples include the use of statins and autophagy inducers as potential therapies against such ASYN-related neurodegeneration.[51,52] Transgenic mice from the Masliah lab have also been used to good advantage to support the theory of 'propagation', as they appear to 'transfer' pathological human ASYN to transplanted immature cells.[53]

6.2.1.2   Viral Mammalian Models

A number of labs have also created rodent models of excess ASYN based on viral transduction within the substantia nigra. Kirik *et al.*[54] were the first to use this approach successfully, as they expressed wild-type and A53T ASYN in the rat nigra using adeno-associated vector (AAV), leading to ASYN inclusions, gradual striatal dopaminergic depletion and nigral neuron death. A behavioural phenotype could be seen in the most severely affected rats. Importantly, Kirik's group has successfully extended this approach to non-human primates.[55] Others, such as the Aebischer lab, have used mostly viral transduction *via* lentivirus, with similar effects.[56,57] It should be noted, however, that the Aebischer lab has also recently turned to AAV transduction, as this appears to achieve more reliably nigral transduction.[58] Experimental approaches using nigral viral transduction of ASYN have led to significant insights into ASYN pathological functions; for example:

- phosphorylation at S129 was identified as a likely protective mechanism, preventing aggregation and neurodegeneration;[58,59]
- Hsp104 overexpression was found to prevent both aggregation and neurodegeneration;[57]
- microglial cells appeared to exhibit a biphasic response depending on the presence of neuronal death.[60]

A major advantage of these viral models is that actual dopaminergic nigral neuron death occurs. Furthermore, different variants of ASYN can be created relatively quickly and used in the same system in order to test pathogenetic mechanisms; in the case of transgenic mouse models, the creation of a new mouse line may take years. The use of rats over mice is another advantage as the former have a more robust motor phenotype upon dopaminergic neurodegeneration. The ability to extend this approach to non-human primates is another plus. Disadvantages include the fact that the widespread use of this approach is hampered by the need for significant specialized expertise to generate the required viruses and to inject them appropriately in the brain. Furthermore, viral transduction does not represent a stable genetic model and it only mimics part of the PD pathological and phenotypic spectrum, which includes many non-nigral components, as mentioned in section 6.1.

6.2.1.3   Invertebrate Models

Invertebrate models are especially useful due to the rapidity with which they can be generated and the potential they offer for high-throughput genetic screens. Invertebrates do not express endogenous ASYN, a fact that may be advantageous or a handicap, depending on the questions addressed. The lack of endogenous ASYN may be a reason why ASYN overexpression in invertebrates is more toxic than in vertebrate systems. In fly ASYN models, ASYN variants can be expressed in specific regions of the fly CNS, including dopaminergic neurons. This leads to rather characteristic LB-like inclusions, neuronal degeneration and a motor behavioural phenotype in a climbing assay.[61] There has been some controversy as to whether true dopaminergic neuron degeneration occurs in this model; it appears that this is due to the different techniques used to assess the dopaminergic phenotype.[62] In most cases, loss of TH staining can be demonstrated, but the question remains whether this is *bona fide* neuronal death or merely loss of phenotype. The protective effects of chaperones in this system were nicely demonstrated using molecular or pharmacological approaches[63,64] before being verified in some mammalian systems. The role of ASYN phosphorylation was addressed in other studies and, although the results on ASYN fibrillization appeared compatible with those in mammalian systems, effects of such modifications on ASYN neurotoxicity were discordant, raising concerns about the validity of this model.[58,59,65] On the other hand, a variant of ASYN that does not oligomerize or fibrillize does not cause neurodegeneration, providing conclusive evidence that ASYN aggregation is required for its toxic effects in this system.[66] A role for oxidative stress in ASYN-mediated toxicity in flies has been demonstrated,[67–69] but whether this applies to mammalian systems is unknown.

Models based on *C. elegans* ASYN also represent a powerful tool, with the added advantage that dopaminergic neurons can be directly visualized and their loss in transgenic lines expressing ASYN under a P_{dat-1} promoter could be demonstrated.[70,71] Such nematodes have been used to great advantage by the Lindquist lab, in collaboration with the Caldwell lab, to transfer insights generated by the yeast model to the organismal level (see, for example, refs. 72,73).

6.2.2 LRRK2

LRRK2 is the other major gene linked to autosomal dominant PD. Despite the advances in our understanding of LRRK2 biology and the mechanism of neurotoxicity as a result of cell-based studies, there are, compared to α-synuclein, relatively few murine models of LRRK2 neurodegeneration at this stage. As such, with a few notable exceptions, the data available so far do not reveal a clear mechanism of mutant LRRK2 pathogenesis at the organism level, though they do suggest LRRK2 plays a role in neurotransmitter release.

LRRK2 is a large multi-domain protein with multiple signalling regions that place it in several protein families. For example, the central signalling core consisting of a small GTPase-like domain (ROC), an intermediate domain (COR; *C*-terminal of ROC), followed by a Ser/Thr kinase domain, places LRRK2 in the ROCO family of proteins.[74] On the other hand, the basic domain structure of the protein with a kinase domain flanked by multiple protein–protein interaction domains, together with a high sequence identity, places LRRK2 in the RIP kinase protein family.[75] Mutations clearly associated with the development of PD span most of the core signalling domains (*e.g.* R1441C/G/H in the ROC domain, and G1029S or I2010T in the kinase domain).

Because LRRK2-associated PD is a dominantly inherited disorder, most models described to date have taken the approach of overexpressing mutant forms of the human cDNA, or their equivalent mutations in the mouse cDNA. Studies examining the physiological role of normal LRRK2, in contrast, rely on models in which the endogenous locus is deleted. While no overt phenotype, particularly in the nervous system, has yet been described in mice lacking the gene encoding LRRK2 (*e.g.* refs. 76,77); it was recently reported that in the kidney, which expresses large amounts of LRRK2 relative to brain, deletion of LRRK2 results in the dramatic accumulation and aggregation of α-synuclein, coupled with disruptions of the lysosomal protein degradation pathway.[77] Additionally, and further suggesting the possibility of an interaction between LRRK2 and α-synuclein, mice lacking LRRK2 are resistant to many of the neuropathological changes induced by transgenic over-expression of A53T mutant α-synuclein[78] under an inducible CaMKII promoter expressing α-synuclein widely throughout the cortex and striatum.

The first *in vivo* model of mutant LRRK2 in mice, reported by Li and colleagues in 2009 used the bacterial artificial chromosomes approach.[80] Bacterial artificial chromosomes (BAC) are F-element based circular genomic clones containing 180–200 kb of genomic DNA. BAC clones often span entire loci including not only the open reading frame (ORF) of a particular gene (with intact exonic and intronic structure), but also the endogenous promoter and other upstream and downstream expression regulatory elements. For a more detailed review of BAC technology, the reader is referred to the excellent review on this area by Heintz.[79]

The main advantage of this approach, compared to traditional transgenic overexpression, is the fact that its native promoter regulates expression of the 'transgene'. Although copy number integration can influence expression levels, unlike a gene-targeting approach (discussed below) in which the endogenous alleles are modified, the regional and temporal pattern of BAC-driven expression remains similar to the endogenous gene. In this initial report, the authors created BAC transgenic mice expressing human R1441G mutant LRRK2. The mice displayed an age-dependent decline in motor activity, as assessed by measurements of rearing in an open field test, which was partially rescued by administration of L-Dopa (levodopa).[80] Dopaminergic function was also altered in these mice. Striatal dopamine (DA) levels following a nomifensine challenge were decreased in R1441G-BAC mice, suggesting a defect in dopamine release. Somewhat surprising was the absence of dopamine neuronal loss in the R1441-BAC mice, although there was some evidence of neuropathological changes (labelling of neurites with antibodies recognizing phosphorylated tau and a slight reduction in TH-positive dendrites in the substantia nigra pars reticulata, SNpr). While some issues concerning comparable expression levels of the transgene, species differences and incomplete penetrance of mutant LRRK2 may have contributed to the lack of dopamine neuron death in this model, it nevertheless provided an early clue to the emerging role of LRRK2 in the regulation of neurotransmitter release.

Following the initial demonstration, in an *in vivo* setting, of the potential role of LRRK2 in modulating neurotransmitter (in particular DA) release, Tong and colleagues[81] utilized a targeted knock-in approach to examine the effects of R1441C mutant LRRK2 on DA neurotransmission in a much more rigorous fashion. As with BAC transgenic mice, R1441C knock-in mice have normal steady-state levels of striatal dopamine and its metabolites suggesting that mutant LRRK2 does not disrupt DA turnover.[81] In contrast to the BAC transgenic mice reported by Li *et al.*,[80] R1441C knock-in mice showed no abnormal baseline changes across a battery of behavioural assays,[81] but did exhibit reduced amphetamine-induced hyperactivity compared to control animals. Electrophysiological study of catecholamine release from chromaffin cells derived from R1441C knock-in mice revealed a deficit not only in the frequency of vesicular evoked DA release, but also in the amount of transmitter released from each vesicle (*i.e.* quantal size), suggesting that mutant LRRK2 can have multiple deleterious effects on dopamine release.[81] The altered DA transmission was postulated to be the result of disrupted function of D2 dopamine receptors, as R1441C knock-in mice also showed reduced inhibition of locomotor activity following pharmacological challenge with quinpirole, a D2 receptor agonist.[81]

Altered DA physiology was also reported in BAC mice expressing a different mutant form of LRRK2 associated with PD, the G1029S substitution at a conserved residue within the kinase domain. In contrast to mice expressing R1441C or R1441G mutant LRRK2, G1029S mutant LRRK2 mice showed a progressive loss in steady-state DA and DA metabolite levels.[82] In fact, in this report, mice overexpressing wild-type BAC LRRK2 showed a slight *enhancement* of baseline DA levels compared to control mice. This was accompanied by a slight hyperactivity in wild-type mice compared to controls, which was returned to baseline levels in G2019S mutant LRRK2 BAC mice. Notwithstanding the differences between this report and previous reports on steady-state DA levels in wild-type mice, the reduction in DA levels in G1029S mice, but not mice harbouring a mutation at the R1441 residue, could be attributed to functional consequences of the different mutations on LRRK2 kinase activity. Multiple groups have consistently found that the G1029S mutation causes a very robust increase in intrinsic LRRK2 kinase activity (assessed by autophosphorylation) compared to the other PD-linked mutations, including R1441C (see ref. 83). Although much more work remains to be completed *directly* comparing the effects of the different LRRK2 mutations on DA release, the possibility exists that LRRK2 kinase activity may be linked to its apparent regulation of neurotransmitter turnover.

In addition to the protection afforded by deletion of LRRK2 in mice against the neuropathological alterations induced by A53T mutant α-synuclein overexpression noted above, the work of Lin and colleagues demonstrated that the inducible overexpression of G2019S mutant LRRK2 throughout the cortex and striatum led to an age-dependent abnormal hyperactivity phenotype in the open field test.[78] This appeared to be strictly a neuronal circuitry abnormality in that no observable neuronal loss was reported in these aged mice. However, when A53T α-synuclein and G2019S LRRK2 overexpressing mice were crossed, the presence of mutant LRRK2 accelerated and worsened the effects of mutant α-synuclein both in terms of neuronal loss as well as astroglial and microglial activation.[78] These findings, coupled with the observations that the expression of both α-synuclein and LRRK2 in the developing mouse appears to be correlated[84] and that deletion of α-synuclein prevents neuronal apoptotic death induced by mutant LRRK2 (Jorgensen, Rideout and Dauer, unpublished observations), raise the intriguing possibility of a genetic interaction between two dominantly inherited genes linked to the pathogenesis of PD.

Finally the pathogenesis—as well as the normal physiological role—of LRRK2 has also been investigated using the powerful invertebrate model systems of *Drosophila melanogaster* and *Caenorhabditis elegans*. For example, *C. elegans* mutants lacking the LRRK2 homologue, lrk-1, show deficits in synaptic vesicle trafficking,[85] consistent with the apparent involvement of LRRK2 in neurotransmitter release. Similarly, in *Drosophila*, loss of the endogenous LRRK2 homologue, LRRK, resulted in impaired locomotive activity and apparent DA neuron dysfunction, both of which were restored by exogenous expression of LRRK.[86] As noted above, both model systems provide numerous advantages owing largely to their relative ease of genetic manipulation and identification of DA neuronal circuits. For example, in *Drosophila*, six clusters of DA neurons, expressing tyrosine hydroxylase (TH), are present in the adult brain, whereas *C. elegans* contains eight phenotypically defined DA neurons.

Recently, Yao and colleagues created *C. elegans* transgenic lines expressing human wild-type, R1441C or G2019S mutant LRRK2 under the control of the endogenous *C. elegans* DAT promoter (P*dat*). Worms expressing R1441C or G2019S mutant LRRK2 showed a progressive age-dependent degeneration of the CEP cluster of DA neurons.[87] The authors additionally reported that worms expressing the wild-type form of human LRRK2 also displayed a loss of DA neurons in this assay, but it was less profound than that caused by expression of mutant LRRK2. To link this toxicity to the GTPase and kinase activities of LRRK2, double mutant forms of LRRK2 were expressed that contained an additional mutation (K1347A) that prevents GTP binding to the ROC GTPase domain of LRRK2 and inhibits its intrinsic kinase activity. This mutation blocked the degeneration of DA neurons caused by the PD-linked mutants R1441C and G2019S,[87] confirming previous cellular studies demonstrating the importance of both signalling domains for LRRK2-induced neurotoxicity.[88–90] In contrast, the group of Wolozin and colleagues[91] showed that widespread overexpression of human wild-type LRRK2 caused a modest enhancement of worm lifespan compared to non-transgenic controls or worms expressing G2019S mutant LRRK2. This difference may be due to the directed expression of the LRRK2 transgene in DA neurons in the former study, and possibly due to altered DA metabolism or vesicular packaging resulting in increased ROS production (see below).

Saha *et al.*[91] additionally reported that wild-type LRRK2 was protective against toxicity induced by exposure of the worms to the mitochondrial toxins rotenone or paraquat, linking LRRK2 to oxidative stress. Strengthening this link, at least in the *C. elegans* model system, is the observation that functional deletion of the homologue lrk-1 also blocked toxicity induced by rotenone or paraquat.[87] This finding, however, may be unique to *C. elegans*. Mice lacking the gene encoding murine LRRK2 showed no loss of sensitivity to the dopaminergic toxin MPTP.[92] It is possible, though, that the partial loss-of-function deletion of *C. elegans* lrk-1 retained some residual activity that altered sensitivity to oxidative insults. This aspect of LRRK2 biology awaits further study in multiple model systems to clarify its physiological role and determine if it is linked to its capacity to induce neuronal death when carrying PD-linked mutations.

Because of the dominant inheritance of LRRK2 mutations, the fact that the pathogenic mutations identified thus far are amino acid substitutions as opposed to improper truncations or deletions, it is generally thought that most of the PD-linked mutations in LRRK2 appear to be associated with a toxic gain-of-function, rather than a loss of function. In contrast to the dopaminergic-mediated locomotive defects in *Drosophila* lacking the endogenous LRRK allele, no overt dopaminergic phenotype was observed when an equivalent mutation was inserted into the ROC domain at R1069C, mimicking the human PD-linked R1441C mutation.[86] However, overexpression of human wild-type or G1029S mutant LRRK2 in *Drosophila*, either in retina[93] or selectively in DA neurons,[93,94] caused degeneration in the relevant cell population. This was accompanied by locomotive impairment when expressed in DA neurons.[93,94] The discrepancy between the findings of the latter two reports compared to the initial report could be due to the different mutation analysed (though Venderova and colleagues assessed not only the G2019S mutation but also mutations in other domains[94]) or due to species differences between the human and *Drosophila* homologues. For example, human LRRK2 contains a significantly larger *N*-terminal region compared to *Drosophila* LRRK, which may mediate additional cell death signalling.

6.2.3   Models of Recessive Parkinson's Disease

Three genes have been identified thus far that are inherited in an autosomal recessive manner: *parkin*, *pink1* and *dj1*. Attempts to model the pathophysiology of these proteins initially yielded mixed results. However, recent work— particularly in the invertebrate system *Drosophila melanogaster*—has identified several potential points of convergence of these proteins that may have implications for PD. Dozens of mutations in *parkin* (consisting of both point mutations as well as truncations or premature stop signals), an E3 ubiquitin ligase, have been identified thus far, usually resulting in a loss-of-function of its ligase activity. Similarly, mutations in PINK1 (PTEN induced kinase I), a mitochondrially localized kinase, and DJ-1, a redox-sensitive chaperone peroxidase, lead to loss-of-function in both proteins.

Most murine models of recessive PD have relied on deletion of the endogenous alleles. Disappointingly, knock-out of the *parkin* gene does not result in SN dopaminergic neuronal loss or gross motor abnormalities;[95,96] nor are *parkin* null mice more sensitive to the dopaminergic/mitochondrial toxins MPTP[97] or 6-OHDA.[98] In fact, triple knock-out mice, lacking the genes encoding Parkin, PINK1 and DJ-1 show normal numbers of SNpc DA neurons even in aged mice.[99] However, a recent report demonstrated a loss of spontaneous locomotor activity in aged mice lacking the gene encoding PINK1, as well as a reduction in general mitochondrial function.[100] In addition, aging mice lacking the DJ-1 gene show some behavioural abnormalities, yet no evidence of nigrostriatal dopaminergic damage.[101] In contrast, a report by Lu and colleagues elegantly demonstrated profound DA neuron loss accompanied by reduced striatal DA and metabolite levels, behavioural deficits, and accumulation of proteinase K resistant α-synuclein in mice expressing mutant *parkin* with a premature stop signal from a BAC construct selectively in SNpc DA neurons.[102] This raises the possibility that some Parkin mutants may actually confer a toxic gain of function, something that has also been suggested by our own cellular studies.[103]

The bulk of the data obtained from both mammalian and invertebrate models of each of the recessive PD-linked genes seem to suggest not only a protective role for these proteins, as opposed to a role crucial for DA neuron survival, but also that there is significant interaction between these proteins. Overexpression of wild-type Parkin can protect against the toxic effects of mutant α-synuclein in mouse models[104] as well as the *Drosophila* model.[105] Similarly, expression of wild-type PINK1 can rescue the behavioural defects (loss of climbing activity) in *Drosophila* expressing mutant α-synuclein.[106] Similar to the mitochondrial phenotype reported in aged PINK1 deficient mice, studies in *Drosophila* have demonstrated a convergence of Parkin and PINK1 at the level of mitochondria. Two seminal works in 2006 demonstrated that exogenous expression of Parkin reversed the phenotypes (mitochondrial morphological defects and dysfunction) associated with loss-of-function mutants, or deletion, of PINK1.[107,108] The functional localization of Parkin to *Drosophila* mitochondria is dependent upon PINK1,[109,110] perhaps through its phosphorylation of Parkin,[109] and together both PINK1 and Parkin can regulate mitochondrial morphology and thus its function through the mitochondrial fission pathway.[111]

6.3 Toxin Models

A variety of toxin-based models of PD have been developed, as expected with different targets and mixed results. The bulk of these models employ agents that induce some form of mitochondrial dysfunction (*e.g.* the MPTP model), whereas other agents target other cellular processes such as the proteasome degradation system. Each approach has distinct advantages and successfully reproduces aspects of PD-neurodegeneration. However, like the genetic approaches described above, none fully and consistently reproduces the hallmark pathological components of PD, the death of dopaminergic neurons of the ventral midbrain, glial cell activation, and the formation of ubiquitinated inclusions in surviving neurons.

6.3.1 Mitochondrial Toxins

The role of mitochondrial dysfunction in the pathogenesis of PD has been extensively studied since an initial report of parkinsonian symptoms in a group of drug users accidentally exposed to the mitochondrial toxin 1-methyl-4-phenyl-1,2,3,4-tetrahydropyridine (MPTP). MPTP readily crosses the blood–brain–barrier (BBB), where it is metabolized, primarily in astrocytes, into the active metabolite 1-methyl-4-phenylpyridinium (MPP+) *via* the activity of monoamine oxidase. MPP+ is transported out of astrocytes *via* the organic cation transporter-3[112] where it binds with high affinity, and is taken up by the DAT to accumulate in high concentrations within DA neurons. Interestingly, there appears to be a unique vulnerability of DA neurons to this, as well as other, insults. Although the DA neurons of the ventral tegmental area express DAT at even higher levels than substantia nigra (pars compacta) DA neurons, the neurons of the VTA are comparatively resistant to the neurotoxic effects of MPP+.[113] MPP+ can also be sequestered within noradrendergic or serotonergic neurons through actions of their respective transporters; however, these neuronal populations are also relatively spared in comparison to the dopaminergic neurons of the SNpc.

There are a variety of administration paradigms of MPTP producing variable time courses of neuronal death, as well as differing mechanisms of cell death. MPTP insults can be:

- *acute*, administered every few hours during the course of a single day;
- *semi-chronic*, where the toxin is administered once per day for a total of five days;
- or more recently, *chronic*, where a low dose is continuously infused through an implanted micropump.

Regardless of the dosing schedule, the target of the converted MPP+ is Complex I of the mitochondrial respiratory chain. It is, in part, because of the information garnered from the MPTP model of PD that led to the discovery of mitochondrial dysfunction in PD. For example, respiratory chain enzymatic activity is decreased in fibroblasts obtained from patients with PD (*e.g.* ref. 114), and cultured neuroblastoma cell lines re-populated with mitochondria isolated from PD patients show altered mitochondrial physiology and with prolonged culture, develop ubiquitinated cytoplasmic inclusions.[115]

In non-human primate models in particular, MPTP induces an incredibly faithful reproduction of human PD in terms of behavioural responses and neuropathological features. The exception being the lack of formation of LB-like inclusions, even following long intervals after the exposure to the toxin.[116] This is a consistent finding in all acute or semi-chronic regimens of MPTP administration in both murine models as well as non-human primate models (*e.g.* ref. 117). It is because of the remarkable similarity in behavioural symptoms and neuropathological changes between monkeys exposed to MPTP and human PD that this model is used extensively to assess potential therapeutic strategies. MPTP in monkeys induces a very similar nigrostriatal dopaminergic insult as is seen in PD,[118] and monkeys similarly respond well to treatment with L-DOPA, even to the extent of suffering dyskinesias after long-term treatment. In fact, the monkey model of MPTP has been utilized as a powerful tool to investigate this aspect of PD therapeutics (*e.g.* refs. 119,120).

Despite the widespread use of this toxin to model SN DA neuron death, there are a number of weaknesses that accompany this approach. First, there is a wide variability in both species and even strains with respect to sensitivity to MPTP. For this reason, use of MPTP is largely restricted to the C57/BL6 strain of mouse. Another weakness of the (acute) MPTP model is the rapidity of neuronal death following administration of the toxin, in contrast to the gradual progressive death of SN dopaminergic neurons in the human disease.

As an alternative, several groups have developed a more chronic model of administering either MPTP or its metabolite, MPP+.[121,122] In the first report of chronic administration of MPTP, Fornai and colleagues[121] utilized an implanted mini-osmotic pump to deliver different doses of the toxin gradually over a period of 28 days. Similar to acute administration of MPTP, the chronic regimen resulted in loss of TH-positive cell bodies in the SNpc as well as terminals in the striatum. In contrast to the acute or semi-chronic approaches, chronic infusion of MPTP caused a persistent drop in all three activities of the proteasome, lasting up to two weeks following termination of MPTP infusion. Additionally, mice receiving chronic MPTP infusions displayed a dose-dependent accumulation of ubiquitin and α-synuclein positive neuronal inclusions.[121] Importantly, and also distinguishing this approach from the acute or semi-chronic regimen, mice receiving MPTP infusions showed a profound measurable behavioural deficit in the open field test and rearing assays. While very encouraging as a progressive chronic model, more closely approximating the human disease condition, and displaying most of the neuropathological and basic behavioural phenotypes characteristic of PD, more sophisticated behavioural assessment need to be completed as well as replication of these findings in multiple independent laboratories before this approach can be applied more broadly.

With the dual goals of overcoming the species restrictiveness of MPTP as well as establishing a more protracted SNpc lesion, Yazdani and colleagues[122] developed a chronic infusion approach using rats administered with the MPP+ metabolite unilaterally into the lateral ventricle over a period of four weeks. With this approach, the investigators reported a progressive loss of dopaminergic neurons of the SNpc, continuing even 14 days after the end of the MPP+ infusion.[122] Additionally, although LB-like neuronal inclusions were not observed within nigral dopamine neurons themselves, they were present within intrinsic striatal neuronal cell bodies close to the site of infusion. While other groups have employed a direct injection of MPP+ into either the SN or striatum of Sprague Dawley rats, the chronic infusion approach, like the similar report using MPTP in mice, needs to be explored further as the initial reports seem to hold some promise for modelling multiple aspects of PD neuropathology in a chronic progressive state.

One of the most promising toxin-based models to be developed in recent years was the systemic rotenone model. However, its initial promise has been tempered somewhat due to inconsistencies in reproducing the original findings in other labs. In 2000, the Greenamyre group[123] reported that rats given chronic systemic infusions of the pesticide rotenone develop a progressive selective loss of dopamine neurons in the SN, the formation of ubiquitin and α-synuclein positive inclusions, and behavioural abnormalities. Analysis of TH immunostaining revealed a progressive, exposure-dependent, dying back profile of SNpc DA neuron damage, while DA neurons of the ventral tegmental area (VTA) were relatively spared. Rotenone-treated rats, like the chronic MPTP/MPP+ models, also showed LB-like inclusion formation (α-synuclein and ubiquitin positive) and motor disturbances. Preceding loss of striatal TH staining is evidence of microglial activation, both in the SN as well as the striatum.[124] Rats exposed to chronic rotenone also display progressive gastrointestinal dysfunction, similar to human PD, including reduced motility, loss of myenteric neurons, and the accumulation and aggregation of α-synuclein in inclusions reminiscent of enteric LBs.[125] The systemic rotenone model has been subjected to rigorous evaluation by multiple independent groups, yielding mixed results. While some groups have reproduced most of the major findings of the initial report (*e.g.* refs. 126,127), others have failed to do so, and even in the reports replicating and extending the original findings, considerable animal-to-animal variability exists in the response to chronic rotenone exposure,[128] making this model's utility as a potential drug-screening tool unlikely.

This variability can potentially be exploited to uncover potential modifier genes or other genetic differences that can interact with environmental toxins to modulate PD pathogenesis.[129] To address this concern in particular, the Greenamyre group reported a modified administration protocol, in which lower doses of rotenone in a modified injection vehicle were injected intraperitoneally over a period of up to 60 days. This approach greatly reduced inter-animal variability and mortality, and yielded a considerably more consistent dopaminergic lesion and phenotype.[130] Importantly, this model underscores the significance of mitochondrial dysfunction in PD, particularly when taken in context with genetic models of familial PD (described above).

6.3.2   6-Hydroxydopamine

Lesioning the nigrostriatal dopaminergic pathway by 6-hydroxydopamine (6-OHDA) is one of the earliest developed toxin-based approaches to model PD neurodegeneration. 6-OHDA is a noradrenaline analogue which is readily taken up by catecholaminergic neurons through the dopamine (DAT) or norepinephrine transporter. It produces cell death which, depending upon the injection site, can be quite rapid with non-apoptotic features, or more protracted with some apoptotic characteristics. Since 6-OHDA cannot pass the BBB, it has to be administered by stereotactic injection directly into the brain. The typical paradigm involves a unilateral stereotactic injection into the striatum, the medial forebrain bundle (MFB), or directly into the SN of rats. Its mechanism of toxicity has been well established, and is mediated by the combination of the formation of reactive oxygen species (ROS) and reactive quinones.

When injected into the SN or MFB, a rapid loss of dopamine neurons occurs within 24 hours,[131] whereas when 6-OHDA is administered into the striatum (*i.e.* in a more distal site relative to dopamine neuron cell bodies), a more protracted cell death occurs.[132,133] It is this target site that allows a more detailed study of the specific molecular mechanisms of apoptotic dopaminergic neuronal death. There are several clear advantages associated with this model. First, because of its structural similarities with DA, coupled with its activation of ROSs and cytosolic quinones, 6-OHDA may be the best approach to model the formation of DA-quinones, an oxidative by-product of DA metabolism that has been implicated in the pathogenesis of PD.[134,135] Secondly, the unilateral administration of 6-OHDA allows clear and objective behavioural assessment of both dopaminergic denervation as well as potential antiparkinsonian therapeutic strategies. In contrast, a general weakness of this—as well as the other commonly used toxin models—is the lack of formation of LB-like inclusions in surviving neurons. To the extent that their formation is integral to the pathogenesis of PD, 6-OHDA is a comparatively poor approach to model this aspect of PD. There is also some evidence that 6-OHDA may induce more of a phenotypic loss (*e.g.* of tyrosine hydroxlase expression) rather than DA neuron death.[136]

6.3.3 Proteasome Inhibition

A key pathological hallmark of neurodegeneration typical of PD is the formation of ubiquitin and α-synuclein proteinaceous inclusions known as Lewy Bodies (LBs). It is generally suspected that the presence of polyubiquitinated proteins within such inclusions reflects a diminished capacity of the neuronal ubiquitin-dependent proteasome system to degrade substrate proteins, or alternatively an attempt by the neuron to sequester potentially toxic proteins within cytosolic inclusions.

This feature of PD has been modeled extensively in cell culture systems (*e.g.* refs. 137–139). Measurement of proteasome activity in PD brain using artificial substrates as well as the levels of certain proteasome subunits has revealed decreased activity compared to control brain.[140,141] In 2004, McNaught and colleagues reported an *in vivo* model of systemic proteasome inhibition which resulted in selective dopaminergic cell death, marked behavioural abnormalities accompanied by biochemical alterations in DA turnover, and the formation of fibrillar α-synuclein and ubiquitin positive neuronal inclusions, thus replicating all of the key neuropathological features of PD.[142] However, this model has been met with considerable disagreement concerning its apparent lack of reproducibility by other groups (including our own[143–146]).

More consistent results have been obtained with a modification of the approach using direct injection of a proteasome inhibitor into the nigra or striatum. Miwa and colleagues administered increasing doses of the proteasome inhibitor lactacystin unilaterally into the striatum, resulting in retrograde nigral DA neuron degeneration together with α-synuclein positive inclusions.[147] Likewise, unilateral injection of lactacystin into the SN of rats induces a progressive loss of DA neuron cell bodies as well as striatal projections, the formation of α-synuclein positive inclusions, and behavioural abnormalities;[148,149] however, the inclusions did not occur exclusively within dopaminergic neurons,[149] suggesting that these neurons may not be as selectively affected by proteasome inhibition as originally believed. Despite the controversies regarding the systemic administration of proteasome inhibitors as a model of PD, the approach in general provides an attractive model specifically of inclusion body formation, a pathological feature that is lacking in other *in vivo* models of PD.

6.4   Other PD models

An especially interesting PD mammalian model is that of the VMAT2 hypomorph mice. VMAT2 is the vesicular monoamine transporter responsible for packaging of monoamines within vesicles. In its absence, monoamines accumulate in the cytosol, with potential detrimental consequences in part mediated through oxidative stress. VMAT2 nulls are lethal, but hypomorphs, which express about 5% of normal VMAT2, demonstrate a remarkable phenotype, with progressive loss of DA terminals and cell bodies in the SNpc, as well as α-synuclein accumulation.[150] In fact, such mice also demonstrate a host of non-motor symptoms reminiscent of those present in PD, such as disturbed sleep, mood, olfaction and autonomic function, which may be due to dysfunction of non-dopamine monoamine systems, and thus may provide a good model for both motor and non-motor manifestations of PD.[151] However, the link to PD pathogenesis is unclear as there are no known genetic defects in VMAT2 in PD patients.

Given the strength of the evidence linking mitochondrial dysfunction to the pathogenesis of not only sporadic forms of PD but certain familial forms as well, the development of the so called MitoPark mouse represents an important complementary addition to the currently available models of mitochondrial dysfunction, which are currently only toxin-based. Ekstrand and colleagues selectively deleted the mitochondrial transcription factor A (Tfam) in DA neurons by crossing *loxP*-flanked Tfam mice with mice expressing Cre recombinase under the control of the endogenous DAT promoter.[152] Tfam is a crucial transcription factor that is necessary for the regulation of mitochondrial DNA copy as well as transcription of mitochondrially encoded genes. Its deletion results in severe respiratory chain deficiencies. Mice lacking the *Tfam* gene selectively in DA neurons develop adult-onset dopaminergic behavioural abnormalities including decreased general locomotion, reduced frequency of rearing, and even signs of tremor. These behavioural deficits are the result of dopaminergic dysfunction, as they improve following administration of L-DOPA. Accordingly, DA terminal loss occurs in the striatum followed by loss of DA neuron cell bodies in the SNpc.[152] Interestingly, surviving neurons, including neurons of the VTA, contained widespread inclusion formation, which did not contain α-synuclein. Frequently, the inclusions were adjacent to or partially overlapped with mitochondria, suggesting that the dysfunctional mitochondria may be undergoing macrophagic elimination,[152] although this question was not addressed by the authors. Additionally, the dopaminergic phenotype of the DAT-*Tfam* null mice is not dependent on the presence of α-synuclein, since crossing these mice with α-synuclein null mice did not affect the development of the phenotype. As initially observed in *Drosophila*, given the involvement of some of the autosomal recessive genes with mitochondrial function, an interesting extension of this report will be to determine if loss-of-function mutations in any of these genes (*e.g.* Parkin or PINK1) worsens the phenotype of the MitoPark mouse.

Finally, to model proteasome dysfunction genetically, Bedford and colleagues generated mice in which the Rpt2 subunit of the 19S regulatory proteasome was deleted in specific brain regions, including TH-expressing catecholaminergic neurons. The result of this deletion is the inhibition of 26S/ubiquitin-dependent proteasomal degradation, leaving 20S/ubiquitin-independent degradation unaffected.[153] While DA neuron-specific deletion of this subunit resulted in lethality usually around the fourth week of age, with almost complete loss of nigral TH-positive neurons by the third week, deletion throughout the cortex resulted in a more gradual neuronal loss associated with the formation of ubiquitinated α-synuclein containing inclusions in surviving neurons. This approach has a distinct advantage over the pharmacological inhibitor-based model of proteasome dysfunction in that it can be selectively targeted not only to specific regions and cellular phenotypes (a feature that can also be considered a weakness), but also to specific proteasomal subunits, allowing the precise link between proteasomal dysfunction and inclusion formation to be teased out.

References

1. M. H. Polymeropoulos, C. Lavedan, E. Leroy, S. E. Ide, A. Dehejia, A. Dutra, B. Pike, H. Root, J. Rubenstein, R. Boyer, E. S. Stenroos, S. Chandrasekharappa, A. Athanassiadou, T. Papapetropoulos, W. G. Johnson, A. M. Lazzarini, R. C. Duvoisin, G. Di Iorio, L. I. Golbe and R. L. Nussbaum, *Science*, 1997, **276**, 2045–2047.
2. R. Kruger, W. Kuhn, T. Muller, D. Woitalla, M. Graeber, S. Kosel, H. Przuntek, J. T. Epplen, L. Schols and O. Riess, *Nat. Genet.*, 1998, **18**, 106–108.
3. J. J. Zarranz, J. Alegre, J. C. Gomez-Esteban, E. Lezcano, R. Ros, I. Ampuero, L. Vidal, J. Hoenicka, O. Rodriguez, B. Atares, V. Llorens, E. Gomez Tortosa, T. del Ser, D. G. Munoz and J. G. de Yebenes, *Ann. Neurol.*, 2004, **55**, 164–173.
4. A. B. Singleton, M. Farrer, J. Johnson, A. Singleton, S. Hague, J. Kachergus, M. Hulihan, T. Peuralinna, A. Dutra, R. Nussbaum,

S. Lincoln, A. Crawley, M. Hanson, D. Maraganore, C. Adler, M. R. Cookson, M. Muenter, M. Baptista, D. Miller, J. Blancato, J. Hardy and K. Gwinn-Hardy, *Science*, 2003, **302**, 841.

5. H. Braak, K. Del Tredici, U. Rub, R. A. de Vos, E. N. Jansen Steur and E. Braak, *Neurobiol. Aging*, 2003, **24**, 197–211.

6. M. G. Spillantini, R. A. Crowther, R. Jakes, M. Hasegawa and M. Goedert, *Proc. Natl. Acad. Sci. U. S. A.*, 1998, **95**, 6469–6473.

7. J. Simon-Sanchez, C. Schulte, J. M. Bras, M. Sharma, J. R. Gibbs, D. Berg, C. Paisan-Ruiz, P. Lichtner, S. W. Scholz, D. G. Hernandez, R. Kruger, M. Federoff, C. Klein, A. Goate, J. Perlmutter, M. Bonin, M. A. Nalls, T. Illig, C. Gieger, H. Houlden, M. Steffens, M. S. Okun, B. A. Racette, M. R. Cookson, K. D. Foote, H. H. Fernandez, B. J. Traynor, S. Schreiber, S. Arepalli, R. Zonozi, K. Gwinn, M. van der Brug, G. Lopez, S. J. Chanock, A. Schatzkin, Y. Park, A. Hollenbeck, J. Gao, X. Huang, N. W. Wood, D. Lorenz, G. Deuschl, H. Chen, O. Riess, J. A. Hardy, A. B. Singleton and T. Gasser, *Nat. Genet.*, 2009, **41**, 1308–1312.

8. K. Vekrellis, H. J. Rideout and L. Stefanis, *Mol. Neurobiol.*, 2004, **30**, 1–21.

9. A. Abeliovich, Y. Schmitz, I. Farinas, D. Choi-Lundberg, W. H. Ho, P. E. Castillo, N. Shinsky, J. M. Verdugo, M. Armanini, A. Ryan, M. Hynes, H. Phillips, D. Sulzer and A. Rosenthal, *Neuron*, 2000, **25**, 239–252.

10. D. E. Cabin, K. Shimazu, D. Murphy, N. B. Cole, W. Gottschalk, K. L. McIlwain, B. Orrison, A. Chen, C. E. Ellis, R. Paylor, B. Lu and R. L. Nussbaum, *J. Neurosci.*, 2002, **22**, 8797–8807.

11. S. Chandra, F. Fornai, H. B. Kwon, U. Yazdani, D. Atasoy, X. Liu, R. E. Hammer, G. Battaglia, D. C. German, P. E. Castillo and T. C. Sudhof, *Proc. Natl. Acad. Sci. U. S. A.*, 2004, **101**, 14966–14971.

12. V. L. Buchman and N. Ninkina, *Neurotox. Res.*, 2008, **14**, 329–341.

13. W. Dauer, N. Kholodilov, M. Vila, A. C. Trillat, R. Goodchild, K. E. Larsen, R. Staal, K. Tieu, Y. Schmitz, C. A. Yuan, M. Rocha, V. Jackson-Lewis, S. Hersch, D. Sulzer, S. Przedborski, R. Burke and R. Hen, *Proc. Natl. Acad. Sci. U. S. A.*, 2002, **99**, 14524–14529.

14. D. C. Robertson, O. Schmidt, N. Ninkina, P. A. Jones, J. Sharkey and V. L. Buchman, *J. Neurochem.*, 2004, **89**, 1126–1136.

15. L. Stefanis, Q. Wang, T. Oo, I. Lang-Rollin, R. E. Burke and W. T. Dauer, *Eur. J. Neurosci.*, 2004, **20**, 1969–1972.

16. A. Al-Wandi, N. Ninkina, S. Millership, S. J. Williamson, P. A. Jones and V. L. Buchman, *Neurobiol. Aging*, 2010, **31**, 796–804.

17. S. Chandra, G. Gallardo, R. Fernandez-Chacon, O. M. Schluter and T. C. Sudhof, *Cell*, 2005, **123**, 383–396.

18. E. Masliah, E. Rockenstein, I. Veinbergs, M. Mallory, M. Hashimoto, A. Takeda, Y. Sagara, A. Sisk and L. Mucke, *Science*, 2000, **287**, 1265–1269.

19. M. Hashimoto, E. Rockenstein, L. Crews and E. Masliah, *Neuromolecular Med.*, 2003, **4**, 21–36.

20. R. Sharon, I. Bar-Joseph, M. P. Frosch, D. M. Walsh, J. A. Hamilton and D. J. Selkoe, *Neuron*, 2003, **37**, 583–595.
21. R. Sharon, M. S. Goldberg, I. Bar-Josef, R. A. Betensky, J. Shen and D. J. Selkoe, *Proc. Natl. Acad. Sci. U. S. A.*, 2001, **98**, 9110–9115.
22. S. M. Fleming, J. Salcedo, P. O. Fernagut, E. Rockenstein, E. Masliah, M. S. Levine and M. F. Chesselet, *J. Neurosci.*, 2004, **24**, 9434–9440.
23. P. J. Kahle, M. Neumann, L. Ozmen and C. Haass, *Ann. N. Y. Acad. Sci.*, 2000, **920**, 33–41.
24. P. J. Kahle, M. Neumann, L. Ozmen, V. Muller, S. Odoy, N. Okamoto, H. Jacobsen, T. Iwatsubo, J. Q. Trojanowski, H. Takahashi, K. Wakabayashi, N. Bogdanovic, P. Riederer, H. A. Kretzschmar and C. Haass, *Am. J. Pathol.*, 2001, **159**, 2215–2225.
25. M. Neumann, P. J. Kahle, B. I. Giasson, L. Ozmen, E. Borroni, W. Spooren, V. Muller, S. Odoy, H. Fujiwara, M. Hasegawa, T. Iwatsubo, J. Q. Trojanowski, H. A. Kretzschmar and C. Haass, *J. Clin. Invest.*, 2002, **110**, 1429–1439.
26. E. Rockenstein, M. Mallory, M. Hashimoto, D. Song, C. W. Shults, I. Lang and E. Masliah, *J. Neurosci. Res.*, 2002, **68**, 568–578.
27. H. van der Putten, K. H. Wiederhold, A. Probst, S. Barbieri, C. Mistl, S. Danner, S. Kauffmann, K. Hofele, W. P. Spooren, M. A. Ruegg, S. Lin, P. Caroni, B. Sommer, M. Tolnay and G. Bilbe, *J. Neurosci.*, 2000, **20**, 6021–6029.
28. M. F. Chesselet, *Exp. Neurol.*, 2008, **209**, 22–27.
29. C. Freichel, M. Neumann, T. Ballard, V. Muller, M. Woolley, L. Ozmen, E. Borroni, H. A. Kretzschmar, C. Haass, W. Spooren and P. J. Kahle, *Neurobiol. Aging*, 2007, **28**, 1421–1435.
30. W. Zhou, J. B. Milder and C. R. Freed, *J. Biol. Chem.*, 2008, **283**, 9863–9870.
31. B. I. Giasson, J. E. Duda, S. M. Quinn, B. Zhang, J. Q. Trojanowski and V. M. Lee, *Neuron*, 2002, **34**, 521–533.
32. S. Gispert, D. Del Turco, L. Garrett, A. Chen, D. J. Bernard, J. Hamm-Clement, H. W. Korf, T. Deller, H. Braak, G. Auburger and R. L. Nussbaum, *Mol. Cell. Neurosci.*, 2003, **24**, 419–429.
33. T. Gomez-Isla, M. C. Irizarry, A. Mariash, B. Cheung, O. Soto, S. Schrump, J. Sondel, L. Kotilinek, J. Day, M. A. Schwarzschild, J. H. Cha, K. Newell, D. W. Miller, K. Ueda, A. B. Young, B. T. Hyman and K. H. Ashe, *Neurobiol. Aging*, 2003, **24**, 245–258.
34. M. K. Lee, W. Stirling, Y. Xu, X. Xu, D. Qui, A. S. Mandir, T. M. Dawson, N. G. Copeland, N. A. Jenkins and D. L. Price, *Proc. Natl. Acad. Sci. U. S. A.*, 2002, **99**, 8968–8973.
35. L. J. Martin, Y. Pan, A. C. Price, W. Sterling, N. G. Copeland, N. A. Jenkins, D. L. Price and M. K. Lee, *J. Neurosci.*, 2006, **26**, 41–50.
36. E. L. Unger, D. J. Eve, X. A. Perez, D. K. Reichenbach, Y. Xu, M. K. Lee and A. M. Andrews, *Neurobiol. Dis.*, 2006, **21**, 431–443.
37. E. Sotiriou, D. K. Vassilatis, M. Vila and L. Stefanis, *Neurobiol. Aging*, 2009.

38. A. B. Manning-Bog, A. L. McCormack, M. G. Purisai, L. M. Bolin and D. A. Di Monte, *J. Neurosci.*, 2003, **23**, 3095–3099.
39. Y. Matsuoka, M. Vila, S. Lincoln, A. McCormack, M. Picciano, J. LaFrancois, X. Yu, D. Dickson, W. J. Langston, E. McGowan, M. Farrer, J. Hardy, K. Duff, S. Przedborski and D. A. Di Monte, *Neurobiol. Dis.*, 2001, **8**, 535–539.
40. S. Rathke-Hartlieb, P. J. Kahle, M. Neumann, L. Ozmen, S. Haid, M. Okochi, C. Haass and J. B. Schulz, *J. Neurochem.*, 2001, **77**, 1181–1184.
41. M. Wakamatsu, A. Ishii, Y. Ukai, J. Sakagami, S. Iwata, M. Ono, K. Matsumoto, A. Nakamura, N. Tada, K. Kobayashi, T. Iwatsubo and M. Yoshimoto, *J. Neurosci. Res.*, 2007, **85**, 1819–1825.
42. E. K. Richfield, M. J. Thiruchelvam, D. A. Cory-Slechta, C. Wuertzer, R. R. Gainetdinov, M. G. Caron, D. A. Di Monte and H. J. Federoff, *Exp. Neurol.*, 2002, **175**, 35–48.
43. G. K. Tofaris, P. Garcia Reitbock, T. Humby, S. L. Lambourne, M. O'Connell, B. Ghetti, H. Gossage, P. C. Emson, L. S. Wilkinson, M. Goedert and M. G. Spillantini, *J. Neurosci.*, 2006, **26**, 3942–3950.
44. M. Wakamatsu, A. Ishii, S. Iwata, J. Sakagami, Y. Ukai, M. Ono, D. Kanbe, S. Muramatsu, K. Kobayashi, T. Iwatsubo and M. Yoshimoto, *Neurobiol. Aging*, 2008, **29**, 574–585.
45. S. Nuber, E. Petrasch-Parwez, B. Winner, J. Winkler, S. von Horsten, T. Schmidt, J. Boy, M. Kuhn, H. P. Nguyen, P. Teismann, J. B. Schulz, M. Neumann, B. J. Pichler, G. Reischl, C. Holzmann, I. Schmitt, A. Bornemann, W. Kuhn, F. Zimmermann, A. Servadio and O. Riess, *J. Neurosci.*, 2008, **28**, 2471–2484.
46. J. P. Daher, M. Ying, R. Banerjee, R. S. McDonald, M. D. Hahn, L. Yang, M. Flint Beal, B. Thomas, V. L. Dawson, T. M. Dawson and D. J. Moore, *Mol. Neurodegener.*, 2009, **4**, 34.
47. E. Emmanouilidou, L. Stefanis and K. Vekrellis, *Neurobiol. Aging*, 2010, **31**, 953–968.
48. E. Tsika, M. Moysidou, J. Guo, M. Cushman, P. Gannon, R. Sandaltzopoulos, B. I. Giasson, D. Krainc, H. Ischiropoulos and J. R. Mazzulli, *J. Neurosci.*, 2010, **30**, 3409–3418.
49. K. E. Paleologou, A. Oueslati, G. Shakked, C. C. Rospigliosi, H. Y. Kim, G. R. Lamberto, C. O. Fernandez, A. Schmid, F. Chegini, W. P. Gai, D. Chiappe, M. Moniatte, B. L. Schneider, P. Aebischer, D. Eliezer, M. Zweckstetter, E. Masliah and H. A. Lashuel, *J. Neurosci.*, 2010, **30**, 3184–3198.
50. X. Su, H. J. Federoff and K. A. Maguire-Zeiss, *Neurotox. Res.*, 2009, **16**, 238–254.
51. A. O. Koob, K. Ubhi, J. F. Paulsson, J. Kelly, E. Rockenstein, M. Mante, A. Adame and E. Masliah, *Exp. Neurol.*, 2010, **221**, 267–274.
52. B. Spencer, R. Potkar, M. Trejo, E. Rockenstein, C. Patrick, R. Gindi, A. Adame, T. Wyss-Coray and E. Masliah, *J. Neurosci.*, 2009, **29**, 13578–13588.

53. P. Desplats, H. J. Lee, E. J. Bae, C. Patrick, E. Rockenstein, L. Crews, B. Spencer, E. Masliah and S. J. Lee, *Proc. Natl. Acad. Sci. U. S. A.*, 2009, **106**, 13010–13015.
54. D. Kirik, C. Rosenblad, C. Burger, C. Lundberg, T. E. Johansen, N. Muzyczka, R. J. Mandel and A. Bjorklund, *J. Neurosci.*, 2002, **22**, 2780–2791.
55. D. Kirik, L. E. Annett, C. Burger, N. Muzyczka, R. J. Mandel and A. Bjorklund, *Proc. Natl. Acad. Sci. U. S. A.*, 2003, **100**, 2884–2889.
56. C. Lo Bianco, J. L. Ridet, B. L. Schneider, N. Deglon and P. Aebischer, *Proc. Natl. Acad. Sci. U. S. A.*, 2002, **99**, 10813–10818.
57. C. Lo Bianco, J. Shorter, E. Regulier, H. Lashuel, T. Iwatsubo, S. Lindquist and P. Aebischer, *J. Clin. Invest.*, 2008, **118**, 3087–3097.
58. S. Azeredo da Silveira, B. L. Schneider, C. Cifuentes-Diaz, D. Sage, T. Abbas-Terki, T. Iwatsubo, M. Unser and P. Aebischer, *Hum. Mol. Genet.*, 2009, **18**, 872–887.
59. O. S. Gorbatyuk, S. Li, L. F. Sullivan, W. Chen, G. Kondrikova, F. P. Manfredsson, R. J. Mandel and N. Muzyczka, *Proc. Natl. Acad. Sci. U. S. A.*, 2008, **105**, 763–768.
60. V. Sanchez-Guajardo, F. Febbraro, D. Kirik and M. Romero-Ramos, *PLoS One*, 2010, **5**, e8784.
61. M. B. Feany and W. W. Bender, *Nature*, 2000, **404**, 394–398.
62. J. A. Botella, F. Bayersdorfer, F. Gmeiner and S. Schneuwly, *Neuromolecular Med.*, 2009, **11**, 268–280.
63. P. K. Auluck, H. Y. Chan, J. Q. Trojanowski, V. M. Lee and N. M. Bonini, *Science*, 2002, **295**, 865–868.
64. P. K. Auluck, M. C. Meulener and N. M. Bonini, *J. Biol. Chem.*, 2005, **280**, 2873–2878.
65. L. Chen and M. B. Feany, *Nat. Neurosci.*, 2005, **8**, 657–663.
66. M. Periquet, T. Fulga, L. Myllykangas, M. G. Schlossmacher and M. B. Feany, *J. Neurosci.*, 2007, **27**, 3338–3346.
67. J. A. Botella, F. Bayersdorfer and S. Schneuwly, *Neurobiol. Dis.*, 2008, **30**, 65–73.
68. K. Trinh, K. Moore, P. D. Wes, P. J. Muchowski, J. Dey, L. Andrews and L. J. Pallanck, *J. Neurosci.*, 2008, **28**, 465–472.
69. R. Wassef, R. Haenold, A. Hansel, N. Brot, S. H. Heinemann and T. Hoshi, *J. Neurosci.*, 2007, **27**, 12808–12816.
70. T. Kuwahara, A. Koyama, K. Gengyo-Ando, M. Masuda, H. Kowa, M. Tsunoda, S. Mitani and T. Iwatsubo, *J. Biol. Chem.*, 2006, **281**, 334–340.
71. M. Lakso, S. Vartiainen, A. M. Moilanen, J. Sirvio, J. H. Thomas, R. Nass, R. D. Blakely and G. Wong, *J. Neurochem.*, 2003, **86**, 165–172.
72. A. A. Cooper, A. D. Gitler, A. Cashikar, C. M. Haynes, K. J. Hill, B. Bhullar, K. Liu, K. Xu, K. E. Strathearn, F. Liu, S. Cao, K. A. Caldwell, G. A. Caldwell, G. Marsischky, R. D. Kolodner, J. Labaer, J. C. Rochet, N. M. Bonini and S. Lindquist, *Science*, 2006, **313**, 324–328.

73. J. A. Kritzer, S. Hamamichi, J. M. McCaffery, S. Santagata, T. A. Naumann, K. A. Caldwell, G. A. Caldwell and S. Lindquist, *Nat. Chem. Biol.*, 2009, **5**, 655–663.

74. I. Marin, W. N. van Egmond and P. J. van Haastert, *FASEB J.*, 2008, **22**, 3103–3110.

75. E. Meylan and J. Tschopp, *Trends Biochem. Sci.*, 2005, **30**, 151–159.

76. E. Andres-Mateos, C. Perier, L. Zhang, B. Blanchard-Fillion, T. M. Greco, B. Thomas, H. S. Ko, M. Sasaki, H. Ischiropoulos, S. Przedborski, T. M. Dawson and V. L. Dawson, *Proc. Natl. Acad. Sci. U. S. A.*, 2007, **104**, 14807–14812.

77. Y. Tong, H. Yamaguchi, E. Giaime, S. Boyle, R. Kopan, R. J. Kelleher, 3rd and J. Shen, *Proc. Natl. Acad. Sci. U. S. A.*, 2010, **107**, 9879–9884.

78. X. Lin, L. Parisiadou, X. L. Gu, L. Wang, H. Shim, L. Sun, C. Xie, C. X. Long, W. J. Yang, J. Ding, Z. Z. Chen, P. E. Gallant, J. H. Tao-Cheng, G. Rudow, J. C. Troncoso, Z. Liu, Z. Li and H. Cai, *Neuron*, 2009, **64**, 807–827.

79. N. Heintz, *Nat. Rev. Neurosci.*, 2001, **2**, 861–870.

80. Y. Li, W. Liu, T. F. Oo, L. Wang, Y. Tang, V. Jackson-Lewis, C. Zhou, K. Geghman, M. Bogdanov, S. Przedborski, M. F. Beal, R. E. Burke and C. Li, *Nat. Neurosci.*, 2009, **12**, 826–828.

81. Y. Tong, A. Pisani, G. Martella, M. Karouani, H. Yamaguchi, E. N. Pothos and J. Shen, *Proc. Natl. Acad. Sci. U. S. A.*, 2009, **106**, 14622–14627.

82. X. Li, J. C. Patel, J. Wang, M. V. Avshalumov, C. Nicholson, J. D. Buxbaum, G. A. Elder, M. E. Rice and Z. Yue, *J. Neurosci.*, 2010, **30**, 1788–1797.

83. E. Greggio and M. R. Cookson, *ASN Neuro*, 2009, **1**.

84. M. Westerlund, C. Ran, A. Borgkvist, F. H. Sterky, E. Lindqvist, K. Lundstromer, K. Pernold, S. Brene, P. Kallunki, G. Fisone, L. Olson and D. Galter, *Mol. Cell. Neurosci.*, 2008, **39**, 586–591.

85. A. Sakaguchi-Nakashima, J. Y. Meir, Y. Jin, K. Matsumoto and N. Hisamoto, *Curr. Biol.*, 2007, **17**, 592–598.

86. S. B. Lee, W. Kim, S. Lee and J. Chung, *Biochem. Biophys. Res. Commun.*, 2007, **358**, 534–539.

87. C. Yao, R. El Khoury, W. Wang, T. A. Byrd, E. A. Pehek, C. Thacker, X. Zhu, M. A. Smith, A. L. Wilson-Delfosse and S. G. Chen, *Neurobiol. Dis.*, 2010.

88. E. Greggio, S. Jain, A. Kingsbury, R. Bandopadhyay, P. Lewis, A. Kaganovich, M. P. van der Brug, A. Beilina, J. Blackinton, K. J. Thomas, R. Ahmad, D. W. Miller, S. Kesavapany, A. Singleton, A. Lees, R. J. Harvey, K. Harvey and M. R. Cookson, *Neurobiol. Dis.*, 2006, **23**, 329–341.

89. P. A. Lewis, E. Greggio, A. Beilina, S. Jain, A. Baker and M. R. Cookson, *Biochem. Biophys. Res. Commun.*, 2007, **357**, 668–671.

90. Y. Li, L. Dunn, E. Greggio, B. Krumm, G. S. Jackson, M. R. Cookson, P. A. Lewis and J. Deng, *Biochim. Biophys. Acta*, 2009, **1792**, 1194–1197.

91. S. Saha, M. D. Guillily, A. Ferree, J. Lanceta, D. Chan, J. Ghosh, C. H. Hsu, L. Segal, K. Raghavan, K. Matsumoto, N. Hisamoto, T. Kuwahara, T. Iwatsubo, L. Moore, L. Goldstein, M. Cookson and B. Wolozin, *J. Neurosci.*, 2009, **29**, 9210–9218.

92. E. Andres-Mateos, R. Mejias, M. Sasaki, X. Li, B. M. Lin, S. Biskup, L. Zhang, R. Banerjee, B. Thomas, L. Yang, G. Liu, M. F. Beal, D. L. Huso, T. M. Dawson and V. L. Dawson, *J. Neurosci.*, 2009, **29**, 15846–15850.

93. Z. Liu, X. Wang, Y. Yu, X. Li, T. Wang, H. Jiang, Q. Ren, Y. Jiao, A. Sawa, T. Moran, C. A. Ross, C. Montell and W. W. Smith, *Proc. Natl. Acad. Sci. U. S. A.*, 2008, **105**, 2693–2698.

94. K. Venderova, G. Kabbach, E. Abdel-Messih, Y. Zhang, R. J. Parks, Y. Imai, S. Gehrke, J. Ngsee, M. J. Lavoie, R. S. Slack, Y. Rao, Z. Zhang, B. Lu, M. E. Haque and D. S. Park, *Hum. Mol. Genet.*, 2009, **18**, 4390–4404.

95. M. S. Goldberg, S. M. Fleming, J. J. Palacino, C. Cepeda, H. A. Lam, A. Bhatnagar, E. G. Meloni, N. Wu, L. C. Ackerson, G. J. Klapstein, M. Gajendiran, B. L. Roth, M. F. Chesselet, N. T. Maidment, M. S. Levine and J. Shen, *J. Biol. Chem.*, 2003, **278**, 43628–43635.

96. R. Von Coelln, B. Thomas, J. M. Savitt, K. L. Lim, M. Sasaki, E. J. Hess, V. L. Dawson and T. M. Dawson, *Proc. Natl. Acad. Sci. U. S. A.*, 2004, **101**, 10744–10749.

97. B. Thomas, R. von Coelln, A. S. Mandir, D. B. Trinkaus, M. H. Farah, K. Leong Lim, N. Y. Calingasan, M. Flint Beal, V. L. Dawson and T. M. Dawson, *Neurobiol. Dis.*, 2007, **26**, 312–322.

98. F. A. Perez, W. R. Curtis and R. D. Palmiter, *BMC Neurosci.*, 2005, **6**, 71.

99. T. Kitada, Y. Tong, C. A. Gautier and J. Shen, *J. Neurochem.*, 2009, **111**, 696–702.

100. S. Gispert, F. Ricciardi, A. Kurz, M. Azizov, H. H. Hoepken, D. Becker, W. Voos, K. Leuner, W. E. Muller, A. P. Kudin, W. S. Kunz, A. Zimmermann, J. Roeper, D. Wenzel, M. Jendrach, M. Garcia-Arencibia, J. Fernandez-Ruiz, L. Huber, H. Rohrer, M. Barrera, A. S. Reichert, U. Rub, A. Chen, R. L. Nussbaum and G. Auburger, *PLoS One*, 2009, **4**, e5777.

101. J. S. Chandran, X. Lin, A. Zapata, A. Hoke, M. Shimoji, S. O. Moore, M. P. Galloway, F. M. Laird, P. C. Wong, D. L. Price, K. R. Bailey, J. N. Crawley, T. Shippenberg and H. Cai, *Neurobiol. Dis.*, 2008, **29**, 505–514.

102. X. H. Lu, S. M. Fleming, B. Meurers, L. C. Ackerson, F. Mortazavi, V. Lo, D. Hernandez, D. Sulzer, G. R. Jackson, N. T. Maidment, M. F. Chesselet and X. W. Yang, *J. Neurosci.*, 2009, **29**, 1962–1976.

103. E. Kyratzi, M. Pavlaki, D. Kontostavlaki, H. J. Rideout and L. Stefanis, *J. Neurochem.*, 2007, **102**, 1292–1303.

104. C. Lo Bianco, B. L. Schneider, M. Bauer, A. Sajadi, A. Brice, T. Iwatsubo and P. Aebischer, *Proc. Natl. Acad. Sci. U. S. A.*, 2004, **101**, 17510–17515.

105. A. F. Haywood and B. E. Staveley, *BMC Neurosci.*, 2004, **5**, 14.

106. A. M. Todd and B. E. Staveley, *Genome*, 2008, **51**, 1040–1046.
107. I. E. Clark, M. W. Dodson, C. Jiang, J. H. Cao, J. R. Huh, J. H. Seol, S. J. Yoo, B. A. Hay and M. Guo, *Nature*, 2006, **441**, 1162–1166.
108. J. Park, S. B. Lee, S. Lee, Y. Kim, S. Song, S. Kim, E. Bae, J. Kim, M. Shong, J. M. Kim and J. Chung, *Nature*, 2006, **441**, 1157–1161.
109. Y. Kim, J. Park, S. Kim, S. Song, S. K. Kwon, S. H. Lee, T. Kitada, J. M. Kim and J. Chung, *Biochem. Biophys. Res. Commun.*, 2008, **377**, 975–980.
110. E. Ziviani, R. N. Tao and A. J. Whitworth, *Proc. Natl. Acad. Sci. U. S. A.*, 2010, **107**, 5018–5023.
111. H. Deng, M. W. Dodson, H. Huang and M. Guo, *Proc. Natl. Acad. Sci. U. S. A.*, 2008, **105**, 14503–14508.
112. M. Cui, R. Aras, W. V. Christian, P. M. Rappold, M. Hatwar, J. Panza, V. Jackson-Lewis, J. A. Javitch, N. Ballatori, S. Przedborski and K. Tieu, *Proc. Natl. Acad. Sci. U. S. A.*, 2009, **106**, 8043–8048.
113. N. A. Seniuk, W. G. Tatton and C. E. Greenwood, *Brain Res.*, 1990, **527**, 7–20.
114. A. H. Schapira, J. M. Cooper, D. Dexter, P. Jenner, J. B. Clark and C. D. Marsden, *Lancet*, 1989, **1**, 1269.
115. P. A. Trimmer, M. K. Borland, P. M. Keeney, J. P. Bennett, Jr. and W. D. Parker Jr., *J. Neurochem.*, 2004, **88**, 800–812.
116. G. Halliday, M. T. Herrero, K. Murphy, H. McCann, F. Ros-Bernal, C. Barcia, H. Mori, F. J. Blesa and J. A. Obeso, *Mov. Disord.*, 2009, **24**, 1519–1523.
117. M. Shimoji, L. Zhang, A. S. Mandir, V. L. Dawson and T. M. Dawson, *Brain Res. Mol. Brain Res.*, 2005, **134**, 103–108.
118. J. Bove, D. Prou, C. Perier and S. Przedborski, *NeuroRx*, 2005, **2**, 484–494.
119. O. Berton, C. Guigoni, Q. Li, B. H. Bioulac, I. Aubert, C. E. Gross, R. J. Dileone, E. J. Nestler and E. Bezard, *Biol. Psychiatry*, 2009, **66**, 554–561.
120. E. Bezard, S. Ferry, U. Mach, H. Stark, L. Leriche, T. Boraud, C. Gross and P. Sokoloff, *Nat. Med.*, 2003, **9**, 762–767.
121. F. Fornai, O. M. Schluter, P. Lenzi, M. Gesi, R. Ruffoli, M. Ferrucci, G. Lazzeri, C. L. Busceti, F. Pontarelli, G. Battaglia, A. Pellegrini, F. Nicoletti, S. Ruggieri, A. Paparelli and T. C. Sudhof, *Proc. Natl. Acad. Sci. U. S. A.*, 2005, **102**, 3413–3418.
122. U. Yazdani, D. C. German, C. L. Liang, L. Manzino, P. K. Sonsalla and G. D. Zeevalk, *Exp. Neurol.*, 2006, **200**, 172–183.
123. R. Betarbet, T. B. Sherer, G. MacKenzie, M. Garcia-Osuna, A. V. Panov and J. T. Greenamyre, *Nat. Neurosci.*, 2000, **3**, 1301–1306.
124. T. B. Sherer, R. Betarbet, J. H. Kim and J. T. Greenamyre, *Neurosci. Lett*, 2003, **341**, 87–90.
125. R. E. Drolet, J. R. Cannon, L. Montero and J. T. Greenamyre, *Neurobiol. Dis.*, 2009, **36**, 96–102.
126. M. Alam and W. J. Schmidt, *Behav. Brain Res.*, 2002, **136**, 317–324.

127. V. Bashkatova, M. Alam, A. Vanin and W. J. Schmidt, *Exp. Neurol.*, 2004, **186**, 235–241.
128. C. Zhu, P. Vourc'h, P. O. Fernagut, S. M. Fleming, S. Lacan, C. D. Dicarlo, R. L. Seaman and M. F. Chesselet, *J. Comp. Neurol.*, 2004, **478**, 418–426.
129. C. Perier, J. Bove, M. Vila and S. Przedborski, *Trends Neurosci.*, 2003, **26**, 345–346.
130. J. R. Cannon, V. Tapias, H. M. Na, A. S. Honick, R. E. Drolet and J. T. Greenamyre, *Neurobiol. Dis.*, 2009, **34**, 279–290.
131. B. S. Jeon, V. Jackson-Lewis and R. E. Burke, *Neurodegeneration*, 1995, **4**, 131–137.
132. S. Przedborski, M. Levivier, H. Jiang, M. Ferreira, V. Jackson-Lewis, D. Donaldson and D. M. Togasaki, *Neuroscience*, 1995, **67**, 631–647.
133. H. Sauer and W. H. Oertel, *Neuroscience*, 1994, **59**, 401–415.
134. D. Sulzer, *Trends Neurosci.*, 2007, **30**, 244–250.
135. D. Sulzer and L. Zecca, *Neurotox. Res.*, 2000, **1**, 181–195.
136. K. E. Bowenkamp, D. David, P. L. Lapchak, M. A. Henry, A. C. Granholm, B. J. Hoffer and T. J. Mahalik, *Exp. Brain Res.*, 1996, **111**, 1–7.
137. K. S. McNaught, C. Mytilineou, R. Jnobaptiste, J. Yabut, P. Shashidharan, P. Jennert and C. W. Olanow, *J. Neurochem.*, 2002, **81**, 301–306.
138. H. J. Rideout, I. C. Lang-Rollin, M. Savalle and L. Stefanis, *J. Neurochem.*, 2005, **93**, 1304–1313.
139. H. J. Rideout, K. E. Larsen, D. Sulzer and L. Stefanis, *J. Neurochem.*, 2001, **78**, 899–908.
140. K. S. McNaught and P. Jenner, *Neurosci. Lett.*, 2001, **297**, 191–194.
141. K. S. McNaught, C. W. Olanow, B. Halliwell, O. Isacson and P. Jenner, *Nat. Rev. Neurosci.*, 2001, **2**, 589–594.
142. K. S. McNaught, D. P. Perl, A. L. Brownell and C. W. Olanow, *Ann. Neurol.*, 2004, **56**, 149–162.
143. J. Bove, C. Zhou, V. Jackson-Lewis, J. Taylor, Y. Chu, H. J. Rideout, D. C. Wu, J. H. Kordower, L. Petrucelli and S. Przedborski, *Ann. Neurol.*, 2006, **60**, 260–264.
144. A. Hawlitschka, S. J. Haas, O. Schmitt, D. G. Weiss and A. Wree, *Brain Res.*, 2007, **1173**, 137–144.
145. J. H. Kordower, N. M. Kanaan, Y. Chu, R. Suresh Babu, J. Stansell, 3rd, B. T. Terpstra, C. E. Sortwell, K. Steece-Collier and T. J. Collier, *Ann. Neurol.*, 2006, **60**, 264–268.
146. B. N. Mathur, M. D. Neely, M. Dyllick-Brenzinger, A. Tandon and A. Y. Deutch, *Brain Res.*, 2007, **1168**, 83–89.
147. H. Miwa, T. Kubo, A. Suzuki, K. Nishi and T. Kondo, *Neurosci. Lett*, 2005, **380**, 93–98.
148. C. Niu, J. Mei, Q. Pan and X. Fu, *Stereotact. Funct. Neurosurg.*, 2009, **87**, 69–81.
149. A. C. Vernon, S. M. Johansson and M. M. Modo, *BMC Neurosci.*, 2010, **11**, 1.

150. W. M. Caudle, J. R. Richardson, M. Z. Wang, T. N. Taylor, T. S. Guillot, A. L. McCormack, R. E. Colebrooke, D. A. Di Monte, P. C. Emson and G. W. Miller, *J. Neurosci.*, 2007, **27**, 8138–8148.
151. T. N. Taylor, W. M. Caudle, K. R. Shepherd, A. Noorian, C. R. Jackson, P. M. Iuvone, D. Weinshenker, J. G. Greene and G. W. Miller, *J. Neurosci.*, 2009, **29**, 8103–8113.
152. M. I. Ekstrand, M. Terzioglu, D. Galter, S. Zhu, C. Hofstetter, E. Lindqvist, S. Thams, A. Bergstrand, F. S. Hansson, A. Trifunovic, B. Hoffer, S. Cullheim, A. H. Mohammed, L. Olson and N. G. Larsson, *Proc. Natl. Acad. Sci. U. S. A.*, 2007, **104**, 1325–1330.
153. L. Bedford, D. Hay, A. Devoy, S. Paine, D. G. Powe, R. Seth, T. Gray, I. Topham, K. Fone, N. Rezvani, M. Mee, T. Soane, R. Layfield, P. W. Sheppard, T. Ebendal, D. Usoskin, J. Lowe and R. J. Mayer, *J. Neurosci.*, 2008, **28**, 8189–8198.

CHAPTER 7

Animal Models and the Pathogenesis of Parkinson's Disease

JOSÉ G. CASTAÑO,*a TERESA IGLESIASb AND
JUSTO G. DE YÉBENESc

a Departamento de Bioquímica, Instituto de Investigaciones Biomédicas
'Alberto Sols' (UAM-CSIC), Centro de Investigación Biomédica en Red
sobre Enfermedades Neurodegenerativas (CIBERNED) and Hospital La Paz
Health Research Institute (IdiPaz), Facultad de Medicina UAM, Madrid,
Spain; b Instituto de Investigaciones Biomédicas 'Alberto Sols' (CSIC-UAM)
and CIBERNED, Madrid, Spain; c Departamento de Neurologia. Hospital
Ramón y Cajal and CIBERNED, Madrid, Spain

7.1 Introduction

Parkinson's disease (PD) was initially described as a clinical entity of unknown
aetiology, characterized by motor symptoms (bradykinesia or akinesia, resting
tremor, rigidity and postural abnormalities), with onset in adulthood and a
progressive course. For a long time, behavioural, perceptive and cognitive
disturbances were considered to be absent. During the 20th century it was
recognized that patients with PD have a severe loss of nigrostriatal dopamine
neurons and a drop-out of dopamine (DA) and its metabolites in striatum.
Brain pathology of PD patients shows the presence of eosinophilic intracyto-
plasmic inclusions in neurons; these inclusions are immunoreactive to

RSC Drug Discovery Series No. 6
Animal Models for Neurodegenerative Disease
Edited by Jesús Avila, Jose J. Lucas and Félix Hernández
© Royal Society of Chemistry 2011
Published by the Royal Society of Chemistry, www.rsc.org

α-synuclein (Snca) and ubiquitin, the so-called Lewy bodies, as well as Snca immunoreactive neuritic pathology.[1]

The cause of PD is still elusive, with the exception of those cases produced by some infectious agents, tumours, trauma, toxins, iatrogenic and vascular diseases of the brain, that are collectively designated as 'secondary parkinsonisms'.[2,3] Several mutations of a continuously increasing number of genes and loci account for a fraction from 10 to 25% depending on the series of cases with familiar forms of PD.[4,5] Environmental agents surely play a role in PD; nowadays PD can be considered a complex multifactorial disease with, very likely, concurrent pathogenic mechanisms which share final common pathways.

Clinical and neuropathological studies performed in the past 25 years have revealed that PD is a dynamic process involving different brains areas in its early stages and as a consequence of disease progression, resulting in a wider spectrum of clinical abnormalities. Most experts agree that patients with PD have a premotor phase characterized by anosmia, constipation, hypotension, changes in personality and depression due to the abnormal function of the dopamine system regulating novelty seeking behaviour, and rapid-eye-movement (REM) sleep behaviour disorder.[6] Later the pathological involvement includes structures such as the olfactory bulb, the dorsal nucleus of the X cranial nerve, the locus coeruleus and the nucleus raphe magnus. The typical asymmetric motor phase of the disease follows with involvement of the substantia nigra (Sn) and depletion of DA in the striatum. The disease then progresses to late stages with symptoms including abnormal behaviours such as punding, hallucinations, and cognitive and perceptive deficits. These late stage symptoms may be related to the extension of the pathology to the limbic system, amigdala, hippocampus, cortical multimodal association areas and cortical primary receptive areas.[7] Aggregated Snca and Lewy bodies are also present in other diseases, known globally as synucleinopathies, which include dementia with Lewy bodies (DLB), multiple system atrophy (MSA), pure autonomic failure, Hallervorden–Spatz disease, REM sleep behaviour disorder, *etc.*[8]

To find a model that reproduces such a variety of clinical, temporal and pathological variables of PD is a difficult task. We believe that the most important requirements for a valid disease model of PD include the following items:

1. Slow but progressive development of clinical deficits in adult animals.
2. The core symptom is akinesia but other motor, behavioural and cognitive symptoms may be present.
3. Severe pathological involvement of the nigrostriatal dopaminergic pathway with frequent additional lesion of brainstem nuclei, hippocampus and neocortex; without important lesions in the optic pathways, the cerebellum or the spinal cord.
4. Aggregation of α-synuclein and formation of Lewy inclusion bodies reproducing the hallmark of PD pathology.

There are excellent reviews (including Chapter 6 in this book by Rideout and Stefanis) on toxic,[9,10] genetic models of PD,[11–13] and general reviews on the pathophysiology and biochemistry of PD[14,15] including cellular and animal models. We review here the different attempts to generate an animal model of PD according to the major causes, or factors, which may be involved in neurodegeneration.

Neurons are 'designed' to survive—with almost no potential for renovation—the entire lifespan of a given species. Neurons, as well as other cells in our organism, have robust systems to cope with different stress conditions—oxidative, inflammatory, trophic, *etc.* The response to stress factors is not a unique function of neurons; other cells in the nervous system also participate in the stress-response process. In fact, it would be better to speak of 'brain degeneration' because the entire organ, like other organs or systems in our organism, responds and contributes to the degeneration process. As a consequence, both cell autonomous (neuronal) and non-cell autonomous (glial cells and vasculature) mechanisms may contribute significantly to compensate or accelerate the process of brain degeneration.

Any disease process can be understood as the result of some common pathological pathways[16] whose particular clinical manifestation are determined individually by the interplay of genetic and epigenetic mechanism in response to environment and manifested in an organ- and tissue-specific manner. The interplay of many factors is implicated in brain degeneration—aging, genetics, environmental stress response, protein aggregation and mitochondrial dysfunction. Each one of these factors alone or concurrently promotes brain degeneration. The major factor in brain degeneration is age; the increased longevity of the human population is the major factor contributing to the increased incidence and prevalence of brain degenerative diseases.

The process of aging is not well understood. Certainly, a key factor in aging is the accumulation of structural and functional damage of cell components as a result of sustained stress during the lifespan of an organism. Nevertheless, it is conceivable that aging is another phase of development of an organism; a genetic programme of aging may exist and can be accelerated, as demonstrated by the progeria syndromes.[17] This developmental aging programme could be a re-programming of cells for active growth and de-differentiation similar to what happens in cancer cells but without cell division.[18]

Genetic contribution to disease is based on the fact that familiar forms of diseases can be attributed to a 'single' malfunctioning gene (in a reduced percentage of cases) in an appropriate genetic background or due to the influence of multifactorial gene expression profiles and interactions.[19] Epigenetic mechanisms are likely to contribute to the tissue and organ specificity of complex disease manifestations as neurodegeneration, cancer, diabetes, atherosclerosis, *etc.* The environmental stress response conditioned by the genetic make-up of an individual is a major cause of any disease, including brain degeneration.

Protein aggregation is a hallmark of brain degenerative diseases; the aggregated proteins are expressed almost ubiquitously, but the aggregation within brain cells can be considered as cause or consequence of the brain degenerative process. Accordingly, protein homeostasis (proteostasis) in the cell is a critical factor in brain degeneration.

Finally, mitochondrial dysfunction can be considered both as a consequence and a trigger of the process of brain degeneration. Except for protein aggregation, the other factors implicated in brain degeneration are common to many 'single' or 'multi' organ chronic diseases. Therefore, it is not surprising that the main recommendations for the prevention of brain degenerative disorders and other chronic diseases are similar, *i.e.* healthy nutritional habits, and moderate mental and physical exercise.

This review is organized into five main sections, each one dedicated to the animal models of PD based on genetics, environmental stress factors, protein aggregation (proteostasis), development and mitochondrial dysfunction. It takes the human model as a reference for the different animal models that have been studied.

7.2   Genetics of Parkinson's Disease

An increasing number of loci and gene mutations have been linked to PD as expected for a complex disease, but have not always been clearly validated.[4] Genome-Wide Association Studies (GWAS) have also provided some links with common single nucleotide polymorphisms (SNPs). SNPs in the loci of *SNCA*, *LRRK2* and the new *PARK16* locus (1q32) are shared risk loci and *BST1* (bone marrow stromal cell antigen 1) and *MAPT* (tau) are loci showing differences between Europeans and Japanese populations.[20,21]

We review the data concerning the human model of genetic PD due to different gene mutations and the corresponding animal models that have been generated trying to mimic the PD pathology observed in humans carrying those mutations. We do not cover those mutations found in human genes related to PD (OMIM #168600) that have not yet been studied in animal models including *PARK9/ATP13A2*, *PARK11/GIGFY2*, *PARK15/FBX07* and the direct relationship between Gaucher's disease alleles and PD.

7.2.1   Alpha-Synuclein: *PARK1/4*

7.2.1.1   *Human Model of Synuclein PARK1/4 Mutations*

Surprisingly enough, α-synuclein (Snca) was not discovered initially in the brains of patients with PD but isolated from the plaques of Alzheimer's disease patients[22] and named the non-amyloid component. This fact together with the recent GWAS showing microtubule-associated protein tau (MAPT) as a risk allele for PD point to some possible common mechanisms in the two most frequent neurodegenerative diseases, *i.e.* Alzheimer's disease and Parkinson's disease.

PD patients with mutations in the *SNCA* gene were initially recognized as the cause of parkinsonism in a large Italo-American family investigated by Larry

Golbe for many years. At the beginning, the clustering of several patients in a small place in USA was interpreted as a probe for the environmental pathogenesis of PD. Afterwards, Golbe realized that in addition to the affected family members who were in North America and had emigrated there in the 18th century, there was another family branch living in Contursi, a small town in Campania, Italy. The two branches of the family had not been in contact for more than two centuries. Then it became apparent that the cause of the disease was genetic. The gene responsible for the Contursi family was mapped to chromosome 4 and the protein encoded identified as Snca.[23]

The mutation found in the Contursi family is the A53T missense mutation, which is also present in about a dozen families from Greece and the eastern Mediterranean. The clinical spectrum of the disease in these patients is variable, but typical PD is the most frequent phenotype. The pathology is characterized by typical PD findings consistent with neuronal loss and Lewy body inclusions in the Sn, locus coeruleus, dorsal nucleus of the X cranial nerve, nucleus raphe magnus, nucleus of Meynert and other brain structures.

Some patients with SNCA mutations have been found to present additional tau and amyloid-β (Aβ) pathology. In addition to the A53Tmutation, other mutations of SNCA have been found in patients with PD. A family from Germany had typical PD associated with a A30P mutation;[24] the first pathological appraisal of these patients showed similar lesions to those described above for the A53T patients.[25] Another large family from Spain shows a mixture of phenotypes from typical PD to the more extensive Lewy body disease. These patients have the E46K mutation.[26] The pathology in this family shows widespread distribution of neuronal atrophy and Lewy bodies in brainstem, diencephalon, allocortex and neocortex.

In addition to point mutations, *SNCA* locus duplication[27,28] and triplication[29] have been described in certain familial forms of PD and hypomethylation of *SNCA* intron 1 is found in sporadic PD patients.[30] These results indicate that increased expression of Snca is also directly involved in the pathogenesis of Lewy body diseases and neurodegeneration. In general, the clinical phenotype is more severe and the disease start earlier in patients with triplications than in those with duplications, and in subjects with the E46K mutation than with other missense mutations of the *SNCA* gene. The role of Snca in PD pathology has also been pointed out recently with the postmortem analysis of the brain of PD patients who had been treated with grafts of embryonic cells and who surprisingly developed PD pathology with Lewy bodies pathology in the grafted neurons[31–34] concomitant with extensive microgliosis.

7.2.1.2   Animal Models of SNCA

Transgenic overexpression of human wild-type Snca or its missense mutants have been used to mimic the human pathology. Expression of human Snca wild-type, A30P and A53T in *Drosophila* produces selective loss of dopaminergic neuronal markers in an age-dependent process, Lewy-like pathology and locomotor deficits that responds to DA treatment[35] but may not cause a reduction in the total number of DA neurons.[36] In *Caenorhabditis elegans*, expression of human Snca wild-type or A53T mutant under the control of a pan neural or DA promoter results in the loss of DA neurons,[37] decreased motor response of the worms to food[38,39] and a 25% increase in their lifespan.[40] The DA phenotype can be rescued by treatment of the worms with probucol, a potent antioxidant and activator of respiratory complex II.[41]

The *SNCA* gene is not essential for mouse development as demonstrated by knock-out (KO) targeting of *SNCA*[42] or spontaneous deletion of the *SNCA* locus.[43] Snca KO mice display some deficits in dopamine release.[42] Transgenic mice overexpressing Snca wild-type, A30P and A53T have been obtained using different promoters to drive its expression. Thy1-driven wild-type and A53T transgenic mice show aggregation of Snca affecting brainstem neurons and motoneurons.[44] Similar results are obtained with Thy-1 driven A30P over-expression.[45] Prion protein (PrP) promoter driven wild-type and A53T trans-genic mice show complex motor impairment and Snca aggregation only by expression of the mutant Snca.[46,47] Another report using the same promoter described no aggregation with motor impairment in both A53T and wild-type transgenic mice.[48] Other authors also using the same promoter reported that Snca A53T overexpression affects mainly the norepinephrine system.[49] Snca aggregates and fibrils are not present in the wild-type and A30P transgenic mice using PrP promoter for expression.[47,50] Platelet-derived growth factor (PDGF) driven Snca wild-type and A53T transgenic mice demonstrated accumulated Snca and inclusions.[51,52] Tyrosine hydroxylase (TH) promoter driving expression of any of the three forms of Snca (wild-type, A30P and A53T) showed neither formation of aggregates nor degeneration of DA neurons.[53] PrP-driven Snca A53T on Snca null mice background showed exacerbated motoneuron pathology.[54]

Recently, it has been reported that bacterial artificial chromosome (BAC) mediated insertion of an entire human *SNCA* gene (wild-type, A53T, A30P) in *SNCA* null mice background produces abnormalities in the enteric nervous system in the case of the mutant Sncas (but not wild-type) with formation of Snca aggregates.[55] Conditional overexpression of Snca wild-type and A53T mutant leads to massive degeneration of neurons in the hippocampus during postnatal development but has no effect if the transgene is turned on in the adult mice.[56] Several models of overexpression of Snca have been generated by viral transfer in the Sn of adult animals,[57] leading to loss of DA neurons when wild-type and mutant Snca are overexpressed by lentiviral vectors in the rat[58] and by adeno-associated vector (AAV) in non-primate humans.[59,60] Viral transfer results in Snca expression both in neurons and glial cells, and the pathology observed can be attributed to both to cell and non-cell autonomous mechanisms.

Factors modifying Snca toxicity have also been examined in animal models. *Drosophila* models of Snca show specific protection of DA neuronal loss by overexpression of Hsp70[61] or histone deacetylase 6[62] and by inhibitors of sirtuin2.[63] In contrast, no beneficial effect of Hsp70 overexpression has been recently reported in A53T transgenic mice.[64] Snca belongs to a protein family that includes two other members: β- and γ-synuclein with high sequence similarity to each other.[65] Transgenic co-expression of β-synuclein in A53T transgenic mice prevents Snca aggregation and fibril formation, motor deficits and neurodegeneration[66] probably as a result of a reduction of Snca protein expression in the double transgenic mice.[67] In contrast to β-synuclein transgenic mice that do not develop neuropathological changes, mice with Thy-1 driven γ-synuclein transgenic expression seem to develop aggregation of γ-synuclein and extensive dose-dependent neurodegeneration in the entire nervous system and specially in the motor neurons of the spinal cord that leads to animal death.[68]

The sensitivity of *SNCA* null and transgenic mice to the toxic 1-methyl-4-phenyl-1, 2, 3,6-tetrahydropyridine (MPTP) has also been studied. Null Snca mice show protection to MPTP with a decrease of DA neuronal loss in the Sn,[69,70] but MPTP is equally effective in the peripheral catecholaminergic system.[71] Similarly, overexpression of miRNA-7, which by binding to the 3'-UTR of Snca mRNA downregulates Snca protein expression, protects against MPTP.[72] In contrast, Snca overexpression (wild-type, A30P and A53T) has no significant effect on MPTP toxicity.[53,73]

Due to the fact that Snca truncated at the *C*-terminus is more prone *in vitro* to aggregation than the full-length protein, different transgenic mice have been obtained with such protein constructs. TH-driven transgenic mice expressing Snca 1-120 show inclusions in the Sn and olfactory tract with decreased striatal DA.[74] Similarly, TH-driven Snca 1-130 transgenic mice[75] show fewer DA neurons compared with transgenic mice expressing full-length Snca, a defect occurring during embryogenesis. Conditional Snca 1-119 mice directing expression to Sn show also an age-dependent decline of DA levels in the striatum.[76]

Finally, several attempts at transgenesis with Snca have been used to mimic another synucleinopathy, MSA. Proteolipid protein promoter (PLP) driven Snca transgenic expression promotes Snca aggregation and inclusion formation in oligodendrogial cells.[77] Myelin basic protein (MBP) promoter driven Snca transgenic mice show also Snca inclusions and degeneration of the neocortex with loss of dopaminergic fibres in the striatum.[78] Treatment of these transgenic mice with an inhibitor of the respiratory complex II, 3-nitropropionic acid[79] exacerbates the pathology and is accompanied by a specific decrease in the levels of GDNF.[80] Transgenic mice with a 2',3'-cyclic nucleotide 3'-phospho-diesterase promoter driving Snca expression show also formation of Snca inclusions and neurodegeneration.[81]

7.2.2 DJ-1/PARK7

7.2.2.1 Human Model of DJ-1/PARK7 Mutations

Mutations in DJ-1 are linked with autosomal recessive early-onset PD. A number of pathogenic mutations have been identified in DJ-1 and include exonic deletions, truncations, homozygous (L166P and M26I) and heterozygous (A104T and D149A) missense mutations; a rare polymorphism (R98Q) has also been identified that is not associated with PD.[82,83]

7.2.2.2 Animal Models of DJ-1/PARK7

DJ-1/PARK7 belongs to the ThiJ/PfpI protein superfamily and is present from bacteria to human. The function of DJ-1 is not clearly established and the only clear consensus is its role in response to oxidative stress. DJ-1 KO has been considered to mimic the human loss of function mutations and produced in different eukaryotic organisms—baker's yeast,[84] *C. elegans*,[85] *Drosophila*[86–91] and mouse.[92–96] The DJ-1 KO organisms show increased sensitivity to oxidative stress produced by different stressors (H_2O_2, MPTP). DJ-1 KO flies show male sterility, shortened lifespan and reduced climbing ability, like PINK-1 KO (see below), and overexpression of DJ-1 seems to rescue PINK1 defect.[97] DJ-1 null mice exhibit no significant reduction in the number of DA neurons and striatum projections, and only a decrease of DA neurons in the ventral tegumental area (VTA) has been described;[98] prolonged follow-up of these null mice have shown some behavioural deficits. Double mutant mice, DJ-1 KO and transgenic for A53T Snca, do not show increased neuronal vulnerability of neurons respect to A53T Snca transgenic mice.[99] Even triple KO of PARKIN/DJ-1/PINK1 does not show reduction in the number of DA neurons.[100] The pro-survival function of DJ-1 on adult DA neurons of Sn (but not VTA) has been demonstrated in double knock-out mice (RET and DJ-1) produced by DA transporter (DAT)-Cre mediated recombination.[101]

7.2.3 PARKIN/PARK2 and PINK1/PARK6

7.2.3.1 Human Model of PARKIN/PARK2 and PINK1/PARK6 Mutations

Several studies published in Japan reported the presence of a type of parkinsonism which was initially considered limited to that country and was called 'autosomal recessive parkinsonism of early onset'. This disease was characterized by autosomal recessive inheritance, early onset, good response to L-DOPA (levodopa) and early development of dyskinesia and fluctuations. Pathological studies of these patients revealed an important degree of atrophy of the Sn, in general without Lewy bodies, although these neuronal inclusions have been found in heterozygous mutation carriers.[102] In 1998, the defective gene was identified and named *PARKIN*, being a member of the E3 ubiquitin ligase family.[103] The pattern of mutations found is very variable, from deletion of exons to point mutations, insertion and deletions, including changes in the open reading frame and internal stop codons. In general, there was a correlation between the kind of mutations and the clinical phenotype, such that subjects with homozygous deletions of one or more exons have a juvenile onset while those with combined or single heterozygotic mutation develop parkinsonism in adulthood and even in presenile years.[102]

Mutations of *PARK2* are the most frequent cause of recessive parkinsonism worldwide. A relative frequency of the different genes mutated in recessive early-onset PD was reported in the Chinese population by Guo *et al.*[104] They reported that out of 29 unrelated families with early onset recessive parkinsonism, 14 have mutations of *PARKIN*, two have mutations of *PINK1* and one has mutation of *DJ-1*. In most cases patients with mutations of *PARKIN*, *PINK1* and *DJ-1* do not have the smell disturbances so frequent in idiopathic PD. The neuropathology of PD cases with mutations in *PINK1* and *DJ-1* mutations is relatively unknown.

7.2.3.2 Animal Models of PARKIN/PARK2 and PINK1/PARK6 Mutations

Parkin is a RING-finger E3 ligase able to autoubiquitylate and ubquitylate several protein substrates, leading to monoubiquitylation or K63, K48 poly-ubiquitylation.[105] Pink1 has an *N*-terminal mitochondrial targeting motif and belongs to the Ser/Thr protein kinase family.[106] Pink1 has been described to be localized in the mitochondria and with its catalytic domain facing the cytoplasm.[107] Pink1 has also been described to be present in the cytoplasm.[108–111]

PARKIN and *PINK1* genes are conserved in *Drosophila*. *PARKIN* KO in *Drosophila* produces male sterility, severe muscle deficits affecting both flight and climbing capabilities of the flies due to reduced muscle fibre number and muscle mitochondrial abnormalities, and also some decrease in DA by degeneration of a subset of DA neurons.[112,113] The phenotype of *PINK1* null flies is very similar to the *PARKIN* null.[114,115] Furthermore, *PINK1* null phenotype can be rescued by *PARKIN* overexpression while *PARKIN* null flies can not be rescued by *PINK1*, indicating that *PINK1* is upstream of *PARKIN*.[114,115] Parkin also relates to neurotoxicity caused by panneural expression of Parkin-associated endothelin receptor-like receptor Pael-R in *Drosophila* that produces age-dependent selective degeneration of DA neurons, and can be rescued by parkin expression that mediates the ubiquitylation and degradation of Pael-R.[116] The interplay between Parkin and Pink1 and its trafficking is related to the regulation of mitochondrial fusion/fission[117,118] and mitoautophagy of damaged mitochondria[119–122] being Pink1 required for Parkin localization at the mitochondria.

There are several mouse models of *PARKIN*[123–126] and *PINK1*[127–129] deficient mice. These mice show no significant abnormalities in DA neurons or striatal DA while they have some deficits in DA transmission, a poor phenotype compared with *Drosophila*. In contrast with this situation, expression of mutant and truncated forms of Parkin either in *Drosophila* by expression of Parkin with the mutation T240R or the *C*-terminal deleted Parkin mutant (Q311X) but not wild-type Parkin;[130] or Parkin R275W but not Parkin wild-type or G328E mutant[131] show progressive DA degeneration and locomotor deficits. The expression of Parkin Q311X driven by a DAT promoter in BAC transgenic mice[132] produces a progressive hypokinetic phenotype with DA neuronal degeneration and an accumulation of Snca aggregates resistant to proteinase K digestion. Parkin deficiency does not affect MPTP toxicity in mice,[133] while it seems to increase susceptibility to lipopolysaccharide (LPS).[134]

7.2.4 LRRK2/PARK8

7.2.4.1 Human Model LRRK2/PARK8

Mutations of the *PARK8* gene (leucine-rich repeat kinase 2 or dardarin) were identified as a cause in familiar PD[135,136] and account for roughly 5% of cases of familial PD and 1–2% of cases of non-familiar PD.[135,136] In general the disease in these PD patients is characterized by a relative late age at onset, after 60 years of age (sometimes earlier), frequent presence of asymmetric tremor (dardarin derives from the onomatopoeic Basque word *dar-dar*, which means tremor) and an autosomal dominant genetic transmission with reduced penetrance. Several patients with different mutations of *PARK8* have shown neuropathological findings consistent with typical PD, such as atrophy of the Sn and other brain nuclei, loss of dopamine neurons and presence of Lewy bodies in the brainstem nuclei, olfactory bulb and cerebral cortex. To our knowledge the autopsy of one subject with the 'Basque' R1441G mutation has been reported.[137] The patient was a man with disease onset at 68 years of age, unilateral rest tremor and good response to medication for 15 years. He died at the age of 86, after 18 years of disease progression. His neuropathological examination disclosed mild neuronal loss in the Sn pars compacta without Snca, tau, Lrrk2 or ubiquitin cytoplasmic inclusions. Lewy bodies and Lewy neurites were also absent.

7.2.4.2 Animal Models of LRRK2/PARK8

Several KO models of *LRRK2* have been generated in *C. elegans*,[138] *Drosophila*[139] and mice,[140] and the results demonstrate that *LRRK2* is not essential for dopaminergic cell development and survival. However, it was recently shown that old *LRRK2* KO mice develop kidney atrophy with Snca aggregation and impairment of protein degradation by the autophagic pathway, indicating that probably the effects are less profound in the brain of these animals as a consequence of compensatory expression of Lrrk1 in brain.[141]

Transgenic *Drosophila* overexpressing the G2019S, Y1699C, or G2385R LRRK2 variants, but not the wild-type protein, show reduced lifespan and increased sensitivity to rotenone with loss of DA neurons in old flies.[142] The phenotype of G2019S transgenic flies can be rescued by overexpression of human Parkin.[142] Transgenic BAC mouse models of the *LRRK2* wild-type and R1441G mutant overexpressing 5–10 fold the basal kinase levels did not show any dopaminergic cell loss.[143] The R1441G transgenic mouse shows a hypokinetic phenotype not observed in the wild-type transgenic that can be suppressed by treatment with l-DOPA. By 10–12 months of age the R1441G mice had progressed to apparent immobility, reminiscent of akinesia in late PD patients. A deregulation of DA secretion, a possible pre-symptomatic event in PD patients, occurs in the R1441G transgenic showing in the striatum beaded and fragmented tyrosine hydroxylase positive axons.[143] Knock-in mice of the R1441C mutation also show no loss of dopaminergic neurons, while they have disturbances in D2 dopamine receptor transmission in response to amphetamine.[144]

None of the above models shows Snca aggregation or formation of inclusion bodies, while tau hyperphosphorylation has been reported in the R1441G transgenic mice but not in knock-in R1441C mice. A double inducible transgenic of LRRK2 and A53T Snca has been generated.[145] The A53T Snca transgenic shows massive neuronal loss in the frontal cortex and dorsal striatum, while no neuronal loss was observed in the wild-type or the G2019S LRRK2 transgenic mice. The double transgenic mouse showed accelerated striatum cell death with astrogliosis and microgliosis but the final number of neurons did not differ significantly. This acceleration of the 'toxic' effect of A53T Snca by LRRK2 wild-type and G2019S seem to be independent of the kinase activity of LRRK2, as similar effects were found with a kinase-death LRRK2 mutant.[145] Snca accumulation, aggregation, polyubiqutin accumulation, Golgi fragmentation and disturbances of microtubule dynamics were increased in the double transgenic mice, which is also not dependent on LRRK2 kinase activity.[145] As a consequence, other domains of LRRK2 apart from the kinase domain must be implicated. In this sense, the GTPase domain of LRRK2 is a good candidate. Two recent papers in yeast[146] and *C. elegans*[147] indicate that the GTPase domain of LRRK2 is responsible of its effect on Golgi distribution, vesicle trafficking and autophagy.

7.2.5 OMI/HTRA2/PARK13

7.2.5.1 Human Model

Four German patients in a extensive cohort of PD patients with typical clinical symptoms of PD and responders to treatment with L-DOPA were shown to carry a mutation (G399S) in the gene encoding for the mitochondrial located Omi/Htra2 protease and a variant polymorphism (A141S) was also heterozygously linked with typical features of PD.[148] Another study reported a missense mutation (R404W) in PD patients.[149] In contrast, two studies failed to demonstrate any correlation of HTRA2/OMI polymorphisms with PD.[150,151] Not surprisingly, OMI/HTRA2 immunoreactivity has also been found in Lewy bodies.[152]

7.2.5.2 Animal Model

In a two-hybrid screen in yeast with presinilin-1, it was found that presinilin-1 interacts with the product of the gene *PRSS25* that encodes a protease that was named Htra2/Omi because of its sequence homology with *E. coli HTRA* genes.

It contains a *C*-terminal PDZ domain and a central serine protease domain.[153,154] A spontaneous recessive mutation in mice, motor neuron degeneration-2 (mnd2), was identified to be caused by a missense mutation (S276C) in Omi/Htra2 that causes the loss of its proteolytic activity. These animals are smaller in size with atrophy of spleen and thymus, and develop muscle wasting and degeneration of striatal neurons.[155] Null *HTRA2/OMI* mice show a similar phenotype to nmd2 mice and the neurons derived from nmd2 and *OMI* null mice are more sensitive to mitochondrial and ER stress. Omi/HtrA2 clearly plays a role in apoptosis as it is released from the intermembrane mitochondrial space to the cytoplasm under those stress conditions.[156] Furthermore, Pink1 has been shown to interact with Omi that becomes phosphorylated through Pink1-dependent activation of p38 MAPK and leading to increased OMI activity.[157]

7.3   Environmental Factors in PD Pathology

Environmental factors are critical in the development of PD, as well as in other neurodegenerative diseases. The use of toxins to reproduce PD pathology is a classical approach.[9,10] Other environmental or stress factors may contribute to PD including inflammatory response and viruses. Below we review the human and animal models in relation to these pathogenic factors.

7.3.1   Toxins

7.3.1.1   *Human and Animal Models of PD Produced by Toxins*

The first patient who developed PD related to a toxin was due to a contamination with MPTP. This patient was a lab technician working with this compound at the beginning of the 1980s, at a time when MPTP was considered a powerful monoamine oxidase (MAO) inhibitor with potential usefulness in the treatment of PD.[158] Nobody paid attention to it until later when an epidemic of MPTP-induced PD was discovered by Langston in California. That epidemic was produced by a contamination by-product of the illegal synthesis of meperidine for recreational use. At that time, it was considered that up to 400 subjects had access to this toxic by-product.[159] Few of them died acutely, a few dozen were admitted in several hospitals in the area of San Francisco because of sudden onset of parkinsonism; others developed acute parkinsonism but did not request medical attention. The vast majority of the exposed population did not have acute symptoms but a few of them developed PD as late as 14 years after a single exposure.[160]

MPTP is certainly a good substrate of MaoB producing MPP+, the active toxic compound. MPP+ is actively taken up by the active Dat which concentrates this compound in DA neurons within vesicles by the vesicle monoamine transporter (Vmat2) and explains the exquisite sensitivity of the dopamine cells to this compound. MPP+ is a highly polar compound and therefore does not penetrate the blood–brain–barrier. Therefore MPP+ is mainly toxic for DA neurons in culture or in intact animals when injected into the brain. MPP+ binds to complex 1 of the mitochondrial respiratory chain, blocks it irreversibly, and produces cell death.[9,10]

MPTP produces different kinds of parkinsonism in monkeys depending on the species, the dose administered and the protocol of administration. *Maccaca fasciculata* (Cynomolgus) or *Mulata* (Rhesus) are much more sensitive than marmosets. Parenteral administration of single intermediate doses (below 1 mg/kg) produced persistent, but at least partially reversible, parkinsonism. Repetitive administration of small doses, 0.1–0.2 mg/kg weekly for several months, produces progressive parkinsonism. In addition, intracarotid injection of doses around 0.4–0.5 mg/kg produces a progressive, persistent, contralateral parkinsonism, and a mild and partially reversible ipsilateral deficit.[161]

Two criticisms have arisen against the quality of the MPTP in monkeys as a good model for PD. One was the lack of a progressive disease in monkeys, the reverse of what happen in humans; the second was the lack of Lewy bodies in MPTP-treated monkeys.[162] The first of these two criticisms is only partially justified. Progression of the disease could be expressed in two ways, as increasing disease severity or as spatial spreading of the pathology in brain and, therefore, subsequent appearance of new symptoms, related with the dysfunction of the new neuronal systems involved. Monkeys treated with single large doses of MPTP, such as after intracarotid injection or with multiple repetitive small doses, develop a progressive syndrome in terms of severity. The involvement of additional neuronal systems is spared by the requirement of a high affinity DA uptake system required to concentrate MPP+ into the cell. The issue of Lewy bodies is interesting; it appears that the formation of Lewy bodies in brain requires a very slow cooking since it has not been found in brains of patients with PD younger than 40 years of age, although it has being shown in transplanted foetal mid-brain implants of more than one decade of age.[31–34]

Nevertheless, the primate MPTP model reproduces many of the features of the human disease and is a good model to test symptomatic treatments, for dealing with complications and testing restorative therapies. MPTP is also used in other animal systems treated with different regimes; the sensitivity sometimes depends on the strain used. For example, C57BL/6 mice are sensitive while BALB/C mice are resistant.[163] The effect of MPTP on genetic models of PD has already been discussed above.

Paraquat is an analogue of MPP+ and widely used in agriculture. Paraquat was considered for a few years as a putative causative agent in patients with idiopathic PD, especially for those exposed by living in rural areas who were contaminated by inhalation. When injected into the brain, paraquat produces similar effects to those of MPP+.[164] However, in recent years, it has become clear that paraquat is unlikely to play a role in human PD, namely for two reasons. One, because it does not cross the blood–brain–barrier, and two, because most patients severely exposed die of pulmonary oedema before having the opportunity to develop parkinsonism.[165]

Guided by the effect of MPP+ on respiratory complex I activity, several groups have used rotenone, a classical complex I inhibitor, to study its effect on dopaminergic systems in rodents.[9,10] Rotenone's high toxicity and lack of reproducibility—probably related to its poor penetration across the blood–brain–barrier—make this surrogate model of little compliance to study PD pathology. There is no information about any 'primate rotenone model'. One of the authors (J.G.Y.), however, has treated a brilliant neuroscientist who developed typical PD at age 61, who was heavily exposed to rotenone for long periods of time more than 30 years earlier. Whether rotenone played a role in the pathogenesis of this case is presently unknown. Nevertheless, it would be of interest to evaluate whether the prevalence of PD in subjects exposed to rotenone is different than that of the age-matched population.

Exposure to hexanes has been considered causative for parkinsonism especially in subjects with abnormalities of the hexane-metabolizing enzymes[166] and in mice.[167]

Manganese is a known cause of parkinsonism in people exposed to high concentrations, specially Chilean miners[168] and induces in monkeys a levodopa-unresponsive parkinsonism due to damage of globus pallidus and the substantia nigra pars reticularis.[169] Manganese has recently attracted attention as PARK9/ATP13A2 (a vacuolar ATPase in yeast) seems to protect cells from manganese toxicity[170] and patients carrying *PARK9/ATP13A2* mutation have increase iron contents in their caudate and putamen.[171]

Accumulation of iron into the substantia nigra is known to be associated with PD and with parkinsonism.[172] Iron accumulation is found in Hallervorden–Spatz disease, an hereditary disorder related to mutations of the pantho-tenate kinase 2[173] that has been reproduced in *Drosophila*.[174] In mammals, iron could enter the brain during the first months of life but, once the mature blood–brain–barrier is operating, the only way of increasing iron concentrations in brain is by direct injection. In fact, direct injection of iron produces a model of lesion in the rat similar to that of 6-hydroxy-dopamine (6-OH-DA).[175]

6-OH-DA is a highly polar oxidized dopamine by-product which does not cross the blood–brain–barrier and which, when injected in brain, is stored in DA neurons, as well as in other monoaminergic neurons, producing cell death due to an excessive production of free radicals. In most cases, 6-OH-DA is given by stereotaxic administration in one substantia nigra or nigrostriatal pathway in animals, mostly rats, anesthetized and pre-treated with MAO inhibitors, which impair 6-OH-DA metabolism, and with nor-epinephrine reuptake inhibitors in order to avoid nor-epinephrine damage.[176] When the administration of 6-OH-DA is done properly, more than 95% of the dopamine neurons of the injected side disappear in an acute form that does not reproduce the slow progression o DA neurons death in PD patients. More progressive models with 6-OH-DA have been attempted.[177] Bilateral lesion of the nigrostriatal pathways with 6-OH-DA usually produce a non-viable model because the animals die of their inability to move and to eat. Unilateral lesions with 6-OH-DA produce almost normal animals which do not show baseline motor impairment. These animals, however, are a perfect model for denerva-tion supersensitivity of the DA receptors and can be used as a test for compounds with symptomatic effects on PD or neurorestorative therapies.

The potential therapeutical effects in this model are reducing the unbalanced rotation which is observed in these animals after treatment with dopamine receptor agonists, such as apomorphine, or with dopamine releasers, such as amphetamine.[178]

7.3.2 Inflammatory Response

Inflammatory mechanisms are a stress response that is commonly associated with chronic diseases and particularly with brain degeneration.[179] Neuro-inflammation has been increasingly associated with the characteristic selective and gradual death of DA neurons from the Sn and the loss of their projections to the striatum participating at the basis of the pathogenesis of PD.[180] Slight microenvironmental changes within the brain elicit pro-neuroinflammatory responses by activating glial cells (primarily, microglia and, to a lesser extent, astroglia). Inflammatory and neurotoxic mediators such as cytokines, lipids and free radicals, reactive oxygen (ROS) and nitrogen (NOS) species are produced, and their accumulation has a high impact on the survival of the extremely vulnerable nigral DA neurons, enhancing their progressive death.[180]

7.3.2.1 Human Model

The main evidence in favour of neuroinflammatory mechanisms in PD come from postmortem brain studies[181,182] and *in vivo* direct imaging of microglia reaction with [^{11}C]-(R) PK11195 by positron emission tomography (PET),[183-185] a ligand of mitochondrial peripheral benzodiazepine receptor. Chronic inflammatory disease of the gastrointestinal tract has also been correlated with idiopathic PD.[186] From genome analysis, SNPs associated with Crohn's disease like CARD15 are slightly more prevalent in PD patients;[187] a polymorphism close to *LRRK2* or *MUC19*[188] is also associated with Crohn's disease.[189] Finally, the role of microglia activation in the initiation (or extension) of PD pathology has also been suggested by the extensive microgliosis observed in postmortem studies of PD patients who have received treatment with grated embryonic neurons.[31-34]

7.3.2.2 Animal Model

Animal models to study the effect of neuroinflammation are mainly based on the use of the bacterial lipopolysaccharide (LPS). A single intranigral injection of LPS in rodents is a widely used approach to promote DA neurodegeneration. It has revealed the special vulnerability of DA neurons to LPS due to microglia activation, subsequent release of proinflammatory and neurotoxic factors from the activated cells.[190-194] Other regimes (small doses delivered by pumps) and alternative routes of administration (intraperitoneal) of LPS have also been used to promote a more progressive loss of DA neurons.[195-197] Even direct infusion of some of the cytokine mediators of LPS action like IL-1beta[198] and TNFalpha[199] produce DA neuronal death. LPS effect has also been studied in the context of SNCA gene dosage. SNCA null mice show less pronounced DA neurodegeneration that mice overexpressing Snca wild-type and mutant A53T; Snca aggregation, inclusion formation and DA degeneration can be prevented by reduction of nitric oxide and superoxide production by the microglia.[200]

7.3.3   Viruses

7.3.3.1   Human and Animal Models

During the first part of the 20th century the most frequent cause of parkinsonism was considered post-encephalitic parkinsonism, after the 1918 influenza pandemic, but was probably unrelated. The disease appears to be a complication of epidemic encephalitis of von Economo, a disease with an initial flu-like phase (pharyngitis) which in many patients was followed, frequently after a symptom-free interval, by sleep disorder, basal ganglia signs (particularly parkinsonism), dystonia, oculogyric crisis, and other neurological symptoms and neuropsychiatric sequelae. The pathogenesis mechanism of this disease is unknown, but the disease-free interval between the 'flu-like' episode and the development of the diseases suggests a delayed viral effect or an abnormal immune response.[201]

In addition to epidemic encephalitis of von Economo as a causative virus for parkinsonism, there are many reported secondary parkinsonism due to infection by a variety of virus. Recently, it has been found that highly pathogenic H5N1 influenza virus can enter the central nervous system (CNS) from the peripheral nervous system in mice and those affected regions have activation of microglia, Snca phosphorylation and aggregation with a significant loss of DA neurons in the Sn pars compacta 60 days after infection.[202]

7.4   Protein Aggregation (Proteostasis) in PD

A common neuropathological finding in PD and other neurodegenerative disorders is the presence of protein aggregates and formation of inclusion bodies named Lewy bodies in the case of PD and other synucleinopathies (DLB, MSA, etc.). The main protein component of Lewy bodies is Snca. Snca is a native unfolded protein prone to aggregation by a nucleation-dependent process. Cell proteostasis is mainly maintained by chaperons and protein degradation.

Chaperons have been found to be present in Lewy bodies; αB-crystallin[203,204] and Hsp90.[205] Furthermore, cell and *Drosophila* models that overexpressed Snca[61,206–208] can be rescued from the Snca toxic effects by concomitant expression of Hsp70, while Hsp70 overexpression has no beneficial effect in A53T transgenic mice.[64] The role of chaperons in protein folding and also in preventing protein misfolding is clear for 'normally' folded proteins (with secondary, tertiary and quaternary structures); in the case of Snca, a naturally unfolded protein, is conceptually more problematic. Mass-action, relative affinities and residence time of the Snca-chaperon complexes in the cell are critical to understand the cellular effects. Accordingly, the read-out on cell protection by chaperons may very well depend on cell context and is not too surprising to get contradictory results between different cells and species, *Drosophila* and mice. Snca has been described itself as chaperone-like protein with similarities to 14.3.3 proteins.[209,210] Snca's function as a chaperon-like protein is also suggested by experiments in mice defective in cysteine-string protein-alpha (CSPalpha) where Snca overexpression can rescue the rapidly progressive neurodegeneration phenotype of the CSPalpha KO mice, probably by restoring the SNARE [SNAP (soluble NSF attachment protein) REceptor] complex function in vesicular secretion through binding of Snca to phospholipids.[211]

Protein degradation is mainly achieved by two robust pathways in the cell—the ubiquitin-proteasome system and autophagy. Disruption of any of the two pathways leads to some sort of neurodegeneration. Depletion of the formation of 26S proteasome by conditional ablation of a 19S proteasomal subunit PSMC1 (Rpt2/S4) leads to intraneuronal Lewy-like inclusions and extensive neurodegeneration in the nigrostriatal pathway and forebrain regions; these inclusions contain mitochondria[212] imitating the so-called pale Lewy bodies. Loss of basal autophagy without affecting proteasome function also promotes neurodegeneration in mice.[213,214] Decreased autophagy is also observed in *DJ-1* KO.[215] According to an extended opinion that proteasome function is somehow impaired in neurodegenerative diseases, they have been several both positive[216,217] and negative[218–220] reports on promotion of DA cell degeneration in response to treatment of rat and mice with proteasome inhibitors. Even proteasome inhibitors have been reported to have protective effects on the death of dopaminergic cells in response to toxins.[221,222] The controversy is probably not settled. In any case, even if a reproducible animal model of DA cell death (or any other brain cells) is finally obtained by treatment with proteasome inhibitors, it can not be concluded that the DA cell death is due to the inhibition of the degradation of the accumulated 'toxic' proteins (Snca, polyQ proteins, *etc.*). Proteasome inhibitors have also effects at transcriptional and translational levels,[223] and promote a stress response that may even results in cytoprotection depending on cell context. These transcriptional and translational effects may prove to be more relevant in the generation of the 'neurodegenerative' phenotype or in its protection that the inhibition of the degradation of the proteins known to be accumulated and aggregated in neurodegenerative diseases.

7.5 Development and PD

A genetic programme is the basis of development that specifies a spatial and temporal frame for cell growth, division, differentiation and localization, and this programme is wired by regulation of gene expression at different levels. As mentioned in section 7.1, it is conceivable that a similar programme may exist for aging (another phase of development) and there are indications that the same transcription factors involved in early development can play a role in the aging process, as demonstrated in *C. elegans*.[224] Accordingly, transcription and growth factors implicated in mesencephalic DA (mesDA) neuronal development and maintenance may play a role in PD pathogenesis. During mice embryogenesis, mesDA neurons are derived from the ventral midline of the mesencephalon through the intersection of Shh expressed along the ventral neural tube, and Fgf8, locally released at the mid/hindbrain boundary and in the rostral forebrain,[225] and by Wnt1[226] signalling. MesDA neurons originate in the floor plate of the mesencephalon; Otx2 which controls anteroposterior patterning induces Lmx1a expression that upregulates Msx-1 which, in turn, regulates Ngn2 to generate mesDA neurons. The pathway Otx2-Lmx1a-Msh1 is also involved in specification of mesDA identity.[227] Foxa1 and 2 mediate the effect of Shh and also participate in the above process both upstream and downstream of Lmx1a.[228] Afterwards, mesDA neurons acquire the expression of transcription factors selective for mesDA neurons including Nurr1 (Nr4a2/Not1/Rnr-1/Hzf-3/Tinur), Lmx1b, Pitx3, En1/2 and Foxa1/2,[227] activating the expression of specific DA neuronal markers (TH, DAT, Vmar2) and promoting the terminal differentiation and maintenance of mesDA neurons.[227] A similar transcriptional programme seems to operate during embryonic human development of mesDA neurons.[229] Apart from the growth factors mentioned above (Shh, Fgf8 and Wnt1), many growth factors also participate in the process of growth control (TGF a/b), and in the survival and maintenance of mesDA neurons like GDNF and MANF/CDBF families.[227,230]

7.5.1 Human Model

Several studies have been performed to determine whether genetic variations in the genes of transcription factors involved in the development and maintenance of mesDA neurons play a role in sporadic and early onset PD. Two recent studies have identified the existence of genetic variations with low significance in *LMX1A* and *LMX1B* genes in patients with PD[231] and an association of a SNP in the *EN2* promoter with young-onset of PD.[232] In contrast, polymorphisms of the *PITX3* gene have been associated with sporadic and early onset PD. An exhaustive genetic association study using 54 SNPs covering *LMX1B*, *EN1/2* and *PITX3* genes only identified a strong association of the *PITX3* promoter SNP rs3758549, *i.e.* C > T with PD.[233]

Since the *PITX3* C-allele appears to be a recessive risk allele with a high estimated population frequency (83%), the authors suggest that an allele-dependent dysregulation of PITX3 expression might contribute to the susceptibility to PD. However, the above association could not be replicated in another study.[234] Bergman *et al.* found that two other alleles, the A-allele of the rs4919621 SNP of the PITX3 gene and the C-allele of the SNP rs2281983, are significantly more common in PD patients with an early age of onset (≤50 years) than in controls or in those patients with a late age of PD onset (>50 years). Clearly, further studies are needed to demonstrate a significant association of genetic variations in *LMX1A*, *LMX1B*, *EN1/2* and *PITX3* with PD.

The important role that Nurr1 plays in DA neurons has been underscored by the identification of several changes in this gene that are associated with PD. Two monoallelic mutations in the 5′ region of Nurr1 gene (c.-291delinsT and c.-245T >G) have been shown to be associated with PD; these mutations reduced the expression of Nurr1.[235] A homozygous polymorphism was found in intron 6 of the Nurr1 gene in association with PD.[236] A missense mutation (S125C) in Nurr1 has been described in a PD patient[237] and a novel single base substitution in the 5′-UTR (c.-309C > T) correlating with a decrease in Nurr1 mRNA expression has also been described in PD patients.[238] Furthermore, Nurr1 expression is reduced in neurons with pathological signs in brains of PD patients[239] and a decrease in Nurr1 activity is observed in peripheral blood lymphocytes of PD patients.[240]

The most prominent group of neurotrophic factors involved in DA neuron survival is the glial cell line-derived neurotrophic factor (GDNF) family.[241,242] Gdnf family members, Gdnf, neurturin (Nrtn), artemin (Artn) and persephin (Pspn), have been shown to mediate their actions by binding to the coreceptor GFRa1 to activate RET protein tyrosine kinase receptor acting through signalling pathways poorly studied in DA neurons.[243,244] Mutations in the RET receptor gene (and much less frequently in the genes for its ligands) are the cause of aganglionic megacolon or Hirschsprung disease[245] due to agenesis of the enteric ganglia along a variable length of the intestine. Cerebral dopamine neurotrophic factor (CDNF) and mesencephalic astrocyte-derived neurotrophic factor (MANF) is a novel evolutionary conserved family of neurotrophic factors, with two members in vertebrates, MANF[246] and CDNF,[247] while there is only one homologue in invertebrates.[248]

7.5.2 Animal Model

Transcription factors needed for mesDA neuronal specification, differentiation and/or maintenance were shown to be crucial after analysing mice where these genes have been genetically inactivated.[227,230] Mice with single null-alleles of *OTX2*, *LMX1a*, *MSX1/2*, *NGN2* and *FOXA1/2*, that are implicated in the proliferation of progenitor mesDA neurons, show strong neurogenic deficits especially in mesDA lineage.[227] *EN1* and *EN2* control essential mechanisms for the survival of mesDA neurons both during development and in the adult.[249,250] Double (EN1/EN2) KO mice show the generation of a small set of DA neurons in the ventral midbrain area early in development which differentiate but die by apoptosis at later stages.[249,250] This apoptotic death has been recently shown to be linked to alterations in the prosurvival neurotrophin signalling pathways.[251] Heterozygous mice, *EN1* +/− ;*EN2*+/− and *EN1*+/− ; *EN2* −/− , exhibit slow and progressive loss of nigral DA neurons which leads to decreases in DA levels in the striatum.[252,253] Importantly, *EN1* +/− ; *EN2* −/− adult mice show motor deficits such as akinesia and bradykinesia, a phenotype that resembles key pathological features of PD.[252] Pitx3 has highly restricted and constitutive expression in mesencephalic/midbrain DA neurons[254] and its gene is disrupted in the aphakia mouse.[255–257] Aphakia/Pitx3-null newborn mice show a dramatic selective loss of the DA neurons at the SN with a decrease in the production of TH.[258–260]

Nurr1 is an orphan receptor that belongs to the nuclear receptor superfamily of transcription factors[261] and regulates the expression of genes essential for DA synthesis and storage.[262–265] MesDA neurons express high levels of Nurr1 during development and throughout adulthood, and it is required for their establishment, as well as for the maintenance and survival of maturing and adult DA neurons.[266–271] Mice lacking Nurr1 (Nurr1-/-) die within 24 h after birth. In these animals DA neurons are totally absent in the Sn. Mesencephalic precursors fail to undergo terminal differentiation and adopt a mature DA phenotype, but the striatal projection of Sn neurons is preserved.[272–275] Heterozygous Nurr1-deficient (*NURR1*+/−) mice, which survive to adulthood,[268,273,276] show increased vulnerability to MPTP[276] and lactacystin, an irreversible proteasome inhibitor.[277] As they grow older, *NURR1*+/− mice show a lower number of DA neurons in the nigra, a reduction in nigrostriatal DA levels and a decrease in DAT expression in the nigra.[278] Concomitantly, old *NURR1*+/− mice display an age-dependent DA dysfunction associated with motor impairment that is analogous to parkinsonian, showing decreases in rotarod performance and locomotor activities.[278] More recently, a mouse strain with conditional targeting of the Nurr1 gene has been generated.[271] In these mice when Nurr1 is ablated at late stages of mesDA neuron development, there is a rapid loss of striatal DA, loss of DA neuronal markers and neuron degeneration. The ablation of *NURR1* in the adult brain results in a slow and progressive loss of DA neurons, similar to PD patients, DA neurons at the Sn being more vulnerable in this model.[271]

Mutant mice deficient for GDNF, GFRα1 and RET die at birth.[279] However, mice where RET is absent in DA neurons from mid-embryonic development only start to present a low but significant loss of DA neurons after one year of age.[280] Heterozygous GDNF (+/−) deficient mice show a slight age-dependent loss of DA neurons.[281–283] In contrast, tamoxifen inducible KO of *GDNF* in adult (two-month-old) mice promotes a dramatic cell death of cathecolaminergic neurons affecting the locus coeruleus, the Sn and the VTA after seven months of treatment, resulting in an hypokinetic phenotype.[284]

The GDNF conditional model suggests an absolute requirement of GDNF for the maintenance of the adult cathecolaminergic neurons.[284] Although further studies should be done with Ret and GDNF-family members mouse mutants, the mouse model of conditional GDNF deficiency in the adulthood developed by Pascual *et al.*[284] constitutes so far one of the best models available to study the higher vulnerability of DA neurons that may appears with aging. The CDNF/MANF family plays a neuroprotective role of DA neurons in the 6-OH-DA experimental model in rats.[248,285,286] Since the mechanisms underlying the actions of neurotrophic factors and their signalling cascades in DA neurons are mostly unknown, the development of conditional-deletion animals, in the line of Pascual *et al.*,[284] will help to elucidate their role in DA neuron survival, their connection with other intrinsic or extrinsic factors that increase DA neuron vulnerability in aging, and their specific contribution to PD pathogenesis.

7.6 Mitochondrial Dysfunction and PD

A mitochondrial dysfunction is probably operating in many, if not all, neurodegenerative diseases. Here we concentrate on autonomous mitochondria cell degeneration mechanism which has to be interpreted through previous comments on the role *PINK1* and *PARKIN* in mitochondrial function. The mitochondrial dysfunctions that have been associated with PD are reduced functioning of the mitochondrial electron transport in the respiratory chain affecting oxidative phosphorylation, mitochondrial DNA damage, impaired calcium buffering, and anomalies in mitochondrial morphology and dynamics.[287]

7.6.1 Human Model

The toxic effects of complex I inhibitors in human and animals have been discussed above and it is clear that MPTP is able to selectively kill DA neurons. Decrease activity of the mitochondrial respiratory complex have been described in postmortem brain from PD patients due to reduced activity of complex I in Sn and also in the frontal cortex; similar complex I deficits have been described in skeletal muscle and platelets from PD patients.[288] In contrast, none of the hereditary dysfunctions of complex I have clinical features of PD, probably because there are more severe and manifested very early, allowing some compensatory mechanisms.[289]

Mitochondrial DNA deletions have been shown to be accumulated in mesDA neurons of Sn, and not in other brain regions and cell types, from both PD patients and normal age-matched controls.[290,291] Those somatic mutations have been attributed, but not demonstrated, to be a consequence of ROS damage, while it is more likely due to mitochondrial DNApol γ mutational rate.[292] Rare DNApol γ mutations have been demonstrated in early onset PD without ophtalmoplegia.[293]

Impairment in calcium homeostasis is strongly suggested as possibly responsible for the death of DA neurons; Sn neurons have L-type Ca(v)1.3 Ca2+ channels necessary for their pacemaking activity.[294] The buffering of cytoplasmic Ca levels by ER and mitochondria is more critical for normal function of mesDA neurons in Sn than in other neurons.[295] Accordingly, specific calcium blockers are predicted as neuroprotectors for PD. Epidemiological evidence in cohorts and case-control studies of patients that are treated with calcium blockers for hypertension have shown no correlation with PD, while a recent case-control study of the use of specific L-type dihydropyridine calcium blockers in Denmark shows a reduced risk of PD (see ref. 296 and references therein). Mitochondrial ultrastructural changes and dynamics are likely to be present in human PD patients both in familiar (PINK1, PARKIN) and idiopathic forms.

7.6.2 Animal Model

The mtDNA Mutator and the MitoPark mouse are the two main models developed to study the role of autonomous mitochondrial deficits in general pathology and PD. The mtDNA Mutator mice are a knock-in mouse harbouring the D257A mutation in DNA POLG leading to a defective proof-reading activity of the mitochondrial polymerase.[297] Homozygous knock-in mice are normal until 25 weeks of age when they develop a premature aging phenotype greying of hair and alopecia, weight loss, hearing loss, heart enlargement, osteoporosis, *etc.*; a similar phenotype has been obtained in a second independent strain of Mutator mice.[298]

Apart from an increase in the number of point mutations accumulated with increasing age, the mitochondrial DNA of these animals also contains extended deletions both in linear and circular forms of mitDNA.[297] The mitochondrial respiratory deficit in these mice occurs with a normal rate of translation of mitochondrial-encoded respiratory proteins. Accordingly, a defect in the assembly of the respiratory complex due to point mutations is more likely responsible for the phenotype.[299] Other authors claimed that the phenotype is better correlated with mitDNA deletions[300] in spite of the fact the mice with big deletions of mitDNA do not show a progeroid syndrome.[300,301]

The MitoPark mouse was produced by inactivation in DA cells of the nuclear-encoded mitochondrial transcription factor A (TFAM); these mice show some clinical features of PD with adult-onset and progressive degeneration of nigrostriatal DA circuitry, motor deficit sensitive to L-Dopa and altered response to treatment as the disease progresses.[302] Notably, protein inclusion also developed in this DAT-driven Tfam mice that are immunoreactive with h116–131 anti-Snca antibodies but not with other anti-Snca antibodies. Those protein inclusions were demonstrated not to contain Snca, as they also appear (and react with h116–131 antibody) in double transgenic Snca/TFAM null mice.[302] Furthermore, TFAM null mice have extensive cell death by apoptosis, but without ROS generation.[303]

7.7 Learning from Models to Master a Complex Disease: Channels and Challenges

A conceptual framework for complex chronic diseases is essential to master their comprehension and treatment, not only for symptomatic treatment but also to interfere significantly with their pathogenetic mechanisms. Common channels or pathways are operating in many, if not all, chronic complex diseases. A schematic representation of the pathogenetic mechanisms involved in brain degeneration is illustrated in Figure 7.1, which is based on the scheme

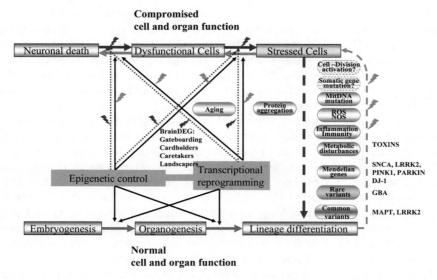

Figure 7.1 Conceptual framework of pathogenetic mechanisms of PD. Normal cell and organ function in the adult proceed by a transcriptional reprogramming in different steps from embryogenesis to late differentiated state of the cells in the adult which is also controlled by epigenetic mechanisms (bottom part of the figure). A similar programme is postulated to occur, but in reverse, during neurodegeneration (upper part of the figure). These events are promoted by aging, postulated to be a developmental programme to which brain degeneration (BDEG) genes contribute decisively. Four classes of BDEG genes are postulated: gateboarding, boarding cardholders, caretakers and landscapers (see text for a full description). The main pathogenetic channels leading to degeneration are indicated by red boxes. Two are postulated, but unknown if they contribute to neurodegeneration; activation of cell division and somatic nuclear gene mutations (dotted boxes). The other channels going from normal to the disease state are indicated with the genes products implicated in PD on the right-hand side of the figure. Protein aggregation is indicated as a channel that may be 'unique' for brain degenerative diseases including PD. Dark blue lines represent reversibility by compensatory mechanisms or by action of therapeutical intervention. For a detailed description see main text. Based on the figure used by Garraway and Sellars[304] to illustrate cancer development.

used by Garraway and Sellars[304] to illustrate cancer development and adapted for PD pathology.

In brain degeneration, like in cancer, the general mechanisms of growth, differentiation and cell-cycle control are also likely to be implicated. Affected neurons and/or non-neuronal cells in brain degeneration are probably being re-programmed through transcriptional regulation to an active programme of growth (possibly by activation of the mTor pathway) and de-differentiation (a reverse of the normal development) along the aging process, similar to what happens in cancer cells. This aging re-programming may lead to activation of cell division in brain cells (not demonstrated), but the consequence in brain is cell death. Note that neurons are post-mitotic cells, but they have high activity of the mitotic E3 ligases, APC/C-Cdh1 and –Cdc20 complexes that participate in axon and dendrite morphogenesis, remodelling and synapse differentiation.[305] The possible de-differentiation programme is illustrated by the phenotypes of *EN-1/2* and *NURR-1* heterozygous and conditional mice, and the conditional GDNF mouse (see section 7.5).

Another channel listed in Figure 7.1 is nuclear somatic mutations, but at present also undemonstrated in brain degenerative diseases. If those nuclear somatic mutations are found in brain degenerative diseases, there would be a change of paradigm in the study of brain degenerative disease, as happened with cancer a long time ago. After an initial methodological discussion and proof-of-concept project, it could be interesting to launch a massive whole genome sequencing of samples from different regions of brain from normal and diseased individuals with different brain degenerative pathologies co-ordinated with study of epigenetic modifications of the somatic genome and analysis of samples from males and females at 30, 50, 65 and 75 years of age.

The main feature that seems to distinguish brain degeneration from other chronic diseases is protein aggregation. The other channels listed in Figure 7.1 (genetics and epigenetics, aging, metabolic disturbances, inflammation and immunity, generation of ROS and NOS, mitochondrial DNA mutation) are probably common to many complex chromic diseases.

The blue lines in Figure 7.1 indicate compensatory mechanisms or therapeutic interventions that may lead to the recovery of cell and organ function. The good thing about the therapy of chronic complex diseases is that those common channels that induce the different types of stress to the cells of an organ can be easily tackled and reduced, producing beneficial outcomes (even if the effects are small) for many chronic diseases. Just an example, metformin, a complex I inhibitor, is used worldwide for the treatment of type 2 diabetes, and epidemiological studies suggest it may also protect against different types of cancer.[306] Metformin's mechanism of action is still not completely understood, because its anti-gluconeogenic liver effects are not explained by its activation of the LKB1/AMPK pathway but by a 'simple' decrease in hepatic energy state.[307] A simple question: does metformin treatment have any epidemiological relationship with PD or other chronic brain degenerative disease? With no report in PubMed on metformin and PD, it may be interesting to carry out this epidemiological study if it has not been done.

The most precise delineation of those common channels not unexpectedly comes from simple organisms that are 'less sophisticated' and with their own limitations show basic chronic mechanisms. Expression of Snca in yeast cells at low levels does not produce any significant effect on yeast viability but its overexpression kills yeast cells. The elegant work of Lindquist's group has shown that several pathways can modify Snca toxicity both positively and negatively, even with specificity for Snca.[170,308–311] Neutral lipids accumulate in lipid droplets in yeast cells overexpressing Snca,[308] the inhibition with statins of the mevalonate–ergosterol biosynthesis pathway as well as inhibition with rapamycin of the mTOR pathway, increases Snca toxicity in yeast cells.[170] In contrast to yeast results, mice treated with rapamycin[312] or simvastatin are protected from MPTP-induced nigral degeneration[313] and treatment with lovastatin reduces Snca accumulation and oxidation in Snca transgenic mice.[314] In the human model, the macrolide rapamicycin seems not to cross the blood–brain–barrier well and we could not find any epidemiological study on PD and rapamycin. There are conflicting results on the brain accessibility of the different statins used to control cholesterol metabolism for prevention of atherosclerosis; clearly more epidemiological data are needed to clarify the possible relationship between therapies with different statins and PD.[315]

Even if the results can be contradictory between different models, it is worth pursuing this approach because it discloses targets that may be already in use for the treatment of other human chronic disease, and well-performed cohort and case-control studies in humans can give fast answers for possible beneficial or detrimental effects with respect to PD (and other neurodegenerative diseases).

This cross-talk can also be useful in the bed-to-bench transfer. For example, proteasome inhibitors have been in use as treatment for multiple myeloma and mantle cell lymphoma for more than decade and are now indicated for other types of cancer.[316] An epidemiological study on those treated patients in relation to PD and other brain degenerative diseases may give us some clues to understand the conflicting results of the bench experiments with proteasome inhibitors.

Continuing with the channels depicted in Figure 7.1, and going upwards on the right-hand side of the figure, the 'less influential channel' for the risk to developing PD is the common SNPs. Common SNPs, or copy number variants, can be shared between different types of neurological diseases and with other chronic diseases. Just two examples, *MAPT* and *LRRK2* SNPs, are also found associated with Alzheimer disease[20,21] and Crohn's disease,[316] respectively. Those common variants with low odds ratio (OR), together with common disease multiple rare alleles with higher OR, like the glucocerebrosidase mutant alleles for PD,[317,318] are telling us about how individual genomes respond to the continuous challenge of the environment.

Continuing with the genome, epigenetic control of genome expression is still in its infancy, specially in brain degenerative disorders,[319] but it could be predicted to have also a great impact in understanding the plasticity in genome response to environmental conditions and aging, and as a consequence in chronic diseases, as clearly demonstrated for cancer.[320]

The Mendelian mutated genes are the culprits of the genetic determinism of a complex disease. These quantitatively rare Mendelian mutations as a cause of brain degeneration are highly informative about critical pathways that play a role in the disease, and they offer us a natural human model where to assay protective therapies, for example, LRRK2 patients in PD. The basic problem is to know the pathways where they participate and what their hierarchy is in those pathways. Combined efforts with different animal systems are providing us with certain clues for PD, but again as expected those genes participate in other chronic diseases like cancer. Some examples:

- DJ-1 (initially isolated as an oncogene cooperating with Ras) is connected to PTEN[321] and HIFα;[322]
- PINK1 just by its name with PTEN inducible kinase-1 and Parkin whose mutations in cancer cells promote a defect in the ubiquitylation and degradation of cyclin E.[323]

The conceptual framework presented in Figure 7.1 is based on cancer as a model of chronic human disease, so let us borrow for brain degeneration some concepts used to understand the complexity of cancer. By genetic evidence there are brain degeneration suppressors (recessive genes) like *DJ-1*, *PINK1* and *PARKIN* and activators (dominant genes) like *SNCA* and *LRRK2*. Following Vogelstein classification,[324] we may also have in brain degenerative diseases— gatekeepers, caretakers and landscapers genes.

Brain degeneration (BDEG) gatekeepers in PD probably have not yet been described, unless in brain degeneration we have to include dominant genes under gatekeeper functions like *SNCA* and *LRRK2*. A possible different scenario may be operative in neurodegeneration. The travel to neurodegeneration may require the inactivation (partial or total) of gateboarding genes that are critical to start the 'travel' to neurodegeneration and the activation of boarding cardholder genes (dominant genes) that promotes the 'travel' to degeneration. *SNCA* and *LRRK2* will belong in this classification to boarding cardholder genes, and Parkin Q311X and R275W mutant forms may also belong to this class. The BDEG gateboarding (gatekeepers) genes will be part of the network that keeps continuous check and balance of mesDA integrated function (or any type of neuronal cell) under normal conditions. In any case, the number of BDEG gateboarding genes must be small; some of them may be cell-type specific and downstream nodes or effectors that keep terminal differentiation, growth and cell-cycle control under check, preventing a catastrophic death, are likely. Those gene products will also inform us of the spatial location and temporal interplay, where and when the pathological changes begin to occur (synapses, axon, dendrites and soma).

BDEG caretakers include those genes whose function is 'protective' as a part of the robust systems that brain cells have to cope with in their normal activity and sustainability. Those caretaker genes include (see discussion below):

- genes implicated in the DNA repair systems, whose mutation also cause brain degeneration[325] and whose efficiency decrease with aging;[326]
- the metabolic network that cope with generated ROS and NOS orchestrated in part by Nrf1[327] and Nrf2[328] transcription factors;
- the inflammatory stress response;[329]
- the specific transcriptional machinery and growth factors required for phenotype maintenance;
- cell proteostasis mechanisms.

Compensatory metabolic adaptations to ROS stress are clearly illustrated in animal models of PD by the increase of glutathione levels observed in the PARKIN null-mice.[330]

The role of all these BDEG caretakers is important both in a cell-autonomous and non-autonomous context, especially in the metabolic context as neurons are highly dependent on glia for metabolic adaptations and in neurotrophin production for survival. Most of the genes identified in familiar forms of PD are BDEG caretakers—*DJ-1, PINK-1, PARKIN* and their mutation increase the risk of suffering PD with an early-onset.

BDEG landscapers would be genes that acting in a non-autonomous cell context, glial cells and vasculature, can promote or prevent neuronal degeneration. Pure BDEG landscaper genes are probably rare in brain. A beautiful illustration of BDEG landscaper genes is the case of amyotrophic lateral sclerosis where motor neuron pathology can be reproduced by overexpression of mutant Cu-Zn SOD in glial cells.[331] These non-cell autonomous effects probably play a role in other brain degenerative diseases including PD and in the generation of other synucleinopathies like MSA.

Finally in the pathogenic mechanism, mitochondria as central organelle in energy production, calcium intracellular homeostasis, generation of ROS and NOS, activation of cell death and by mito-autonomous mechanisms (accumulated mutations due to aging) play an important role in PD neurodegeneration (see above under *PINK1/PARKIN* and mitochondria).

All the different channels of pathogenesis discussed above and some not discussed like endoplasmic reticulum (ER) stress (also common in many chronic human diseases) can be considered as typical of complex chronic diseases, while in this case with some specificity for promoting brain dysfunction in PD. The 'most salient feature' that differentiates chronic brain degenerative diseases from other chronic human diseases is the presence of aggregated proteins in most, if not all, neurodegenerative diseases. One possibility is that the main protein that aggregates and produces the inclusion bodies seen in PD is a non-pathogenic consequence of the pathophysiological mechanism of brain degeneration and has nothing to do with the pathogenesis. Definitive evidence in the case of PD is missing, except for the demonstration in a conditional Snca transgenic mouse that the pathological changes do not progress if Snca expression is shut-down.[332] Nevertheless, there is a clear demonstration in the case of a transgenic conditional mouse model of Huntington's disease[333] and in prion diseases[334] that expression of the protein that becomes aggregated is essential for developing the neuropathology. Accordingly, the proteinopathic hypothesis of neurodegeneration can be reasonably sustained and considered a distinct 'quality' of the neurodegenerative diseases. As a consequence, the following question is very pertinent in PD, is neuronal DA degeneration inevitably linked to Snca aggregation? The answer seems to be 'no'; other proteins can do it, as suggested by the γ-synuclein transgenic[68] and the MitoPark mice.[302] Probably many proteins from the 'amylome' can do it;[335] exploring the amylome by the generation of different transgenic mice will clarify the issue of protein specific *versus* unspecific effects of protein aggregation in DA neurodegeneration and in other brain degenerative diseases—a critical issue.

Animal modelling of PD by Snca overexpression in mice is not satisfactory. Certainly, mice are resistant to developing chronic diseases. In contrast, flies developed a 'typical' model of PD by Snca overexpression probably because, at their evolutionary stage, the functions that the naturally unfolded Snca play in vertebrates and the network that keeps natively unfolded proteins under tight control along the animal's lifespan have not yet appeared or are not robust enough. No animal has been reported to suffer from a syndrome similar to PD, while Snca expression is present in all vertebrates.

Provisionally, we can conclude that PD only affects humans. There are three possibilities, at least, to explain this paradox of man's uniqueness in showing Snca protein aggregation and PD:

1. Humans have lost (or reduced) the protective mechanisms that are operating in other vertebrates during evolution and/or in the specific circumstances of mesDA differentiation and maintenance in humans.
2. Those protective mechanisms are present in humans but deteriorate with aging.
3. There is something else–the prionoid hypothesis.

We know part of those BDEG caretakers of the protecting network regulating proteostasis: chaperons and protein degradation, the ubiquitin–proteasome system and autophagy and new ones like the MOAG-4/SERF.[336] PARKIN and LRRK2 are likely to be participating in this protective network working against aggregation.

The first possibility that human evolution has made brain cells more vulnerable is unlikely. The second part of the first hypothesis can be tested experimentally by the use of human inducible pluripotent stem (iPS), taking mice or monkey iPS cells for comparison. The networks found in those experiments could then be validated in human brain samples.

The second hypothesis is more likely and we have good tools to explore it, but we probably need to get further knowledge of the aging process itself. The system biology approach would help to fish, especially if there is a genetic programme of aging. A test of this hypothesis could be to make inducible Snca (or other PARK genes) transgenic mice engineered into mice with accelerated aging produced by different mechanisms.

The third hypothesis is to postulate that the 'uniqueness' is due to the transmission of altered protein conformation to intracellular proteins producing the disease in a similar fashion to prions—the prionoid hypothesis;[337] the main difference being that PrP (prionic) diseases are not 'unique' to humans as many vertebrates suffer similar diseases. Some preliminary results indicate that such a possibility of a 'contagious conformation' can not be discarded even if the environmental trigger is elusive. Aggregated Snca has been found in gastrointestinal ganglia of PD patients[338] and the recent BAC Snca transgenic mice developed inclusions also in the colonic ganglia.[55]

Taking the prionoid hypothesis further will mean trying to find Snca aggregates in the lymphoreticular system including spleen and demonstrating the possible ascending route of Snca aggregation transmission within the central nervous system. The strong spinal cord degeneration of Thy1-driven Snca (and other mice models) transgenic mice could be an indication of an early manifestation of the disease progression that in mice becomes the staging ground, because of its limited lifespan, for aggregated Snca in its journey to upper brain structures. Other routes for the trigger of misfolding are also worth to be explored as some of the transgenic Snca animals seem to present olfactory impairments like PD patients.[339] Another indirect support of the role of extracellular misfolded Snca in cell neuronal damage comes from the amelioration of pathological signs of Snca transgenic mice by immunization with Snca.[340] Finally, the prionoid hypothesis could be part of the mechanism involved in the development of Lewy body pathology in the grafts of embryonic cells used to treat PD patients. All evidence is suggestive, but the prionoid hypothesis required more experimental evidence.

Further work is certainly needed, but the challenges and channels to explore are exciting and the final outcome will give us a better understanding of PD and other brain degenerative diseases.

7.8 Conclusions

While no animal model reproduces all the cardinal features of PD pathology, we have learned a lot from available models, and as much as we learn, more questions open up to be investigated. Conceptual frameworks are also important; they could be wrong and may not accommodate all the available data, but they are useful if they put forward tests to be falsified. Learning from other chronic human diseases is helpful and we presented such an approach relating cancer and brain degeneration, but those common channels extend to other chronic human disease like type 2 diabetes and atherosclerosis.

Our knowledge and the possibility of a more rational therapeutic intervention for the treatment of PD and other chronic brain degenerative diseases will be accelerated by:

- the study using conditional mice of the different pathways implicated in neurodegeneration, also engineered to premature aging;
- analysis of somatic mutations in brains from PD patients;
- test of the prionoid 'infective' hypothesis;
- better characterization of the BDEG gateboarding, cardholders, caretakers, landscapers gene networks;
- a close bidirectional bench–bed interaction.

Acknowledgments

This work was supported by grants from MICINN SAF2008-00766, CM SAL-0202/2006 and CIBERNED to J.G.C.; from MICINN SAF2008-01951,

CM SAL-0202/2006, and CIBERNED to T.I.; and from the Spanish Ministry of Health, FIS 2007/PI07037, CM SAL-0202/2006, Lain Entralgo NDG07/4 and CIBERNED to J.G.Y. We apologize to authors whose research was unable to be cited or who may feel their work has not been adequately considered. This review reflects our view and the proposal by one of the authors (J.G.C.) of a conceptual framework that may be useful to foster future directions of research in brain degenerative disorders and specifically in Parkinson's disease.

References

1. C. M. Muller, R. A. De Vos, C. A. Maurage, D. R. Thal, M. Tolnay and H. Braak, *J. Neuropathol. Exp. Neurol.*, 2005, **64**(7), 623.
2. J. E. Ahlskog, *Parkinsonism Relat. Disord.*, 2000, **7**(1), 63.
3. M. V. Alvarez, V. G. Evidente and E. D. Driver-Dunckley, *Semin. Neurol.*, 2007, **27**(4), 356.
4. S. Lesage and A. Brice, *Hum. Mol. Genet.*, 2009, **18**(R1), R4.
5. A. J. Lees, J. Hardy and T. Revesz, *Lancet*, 2009, **373**(9680), 2055.
6. L. Ishihara and C. Brayne, *Acta Neurol. Scand.*, 2006, **113**(4), 211.
7. H. Braak, J. R. Bohl, C. M. Muller, U. Rub, R. A. De Vos and K. Del Tredici, *Mov. Disord.*, 2006, **21**(12), 2042.
8. K. A. Jellinger, *Mov. Disord.*, 2003, **18**(Suppl 6), S2.
9. A. Schober, *Cell Tissue Res.*, 2004, **318**(1), 215.
10. J. Bove, D. Prou, C. Perier and S. Przedborski, *NeuroRx*, 2005, **2**(3), 484.
11. M. F. Chesselet, S. Fleming, F. Mortazavi and B. Meurers, *Parkinsonism Relat. Disord.*, 2008, **14**(Suppl 2), S84–S87.
12. K. L. Lim and C. H. Ng, *Biochim. Biophys. Acta*, 2009, **1792**(7), 604.
13. T. M. Dawson, H. S. Ko and V. L. Dawson, *Neuron*, 2010, **66**(5), 646.
14. D. J. Moore, A. B. West, V. L. Dawson and T. M. Dawson, *Annu. Rev. Neurosci.*, 2005, **28**, 57.
15. M. R. Cookson, *Annu. Rev. Biochem.*, 2005, **74**, 29.
16. H. A. Hirsch, D. Iliopoulos, A. Joshi, Y. Zhang, S. A. Jaeger, M. Bulyk, P. N. Tsichlis, L. X. Shirley and K. Struhl, *Cancer Cell*, 2010, **17**(4), 348.
17. B. A. Kudlow, B. K. Kennedy and R. J. Monnat, *Nat. Rev. Mol. Cell Biol.*, 2007, **8**(5), 394.
18. M. V. Blagosklonny and M. N. Hall, *Aging*, 2009, **1**(4), 357.
19. W. Cookson, L. Liang, G. Abecasis, M. Moffatt and M. Lathrop, *Nat. Rev. Genet.*, 2009, **10**(3), 184.
20. W. Satake, Y. Nakabayashi, I. Mizuta, Y. Hirota, C. Ito, M. Kubo, T. Kawaguchi, T. Tsunoda, M. Watanabe, A. Takeda, H. Tomiyama, K. Nakashima, K. Hasegawa, F. Obata, T. Yoshikawa, H. Kawakami, S. Sakoda, M. Yamamoto, N. Hattori, M. Murata, Y. Nakamura and T. Toda, *Nat. Genet.*, 2009, **41**(12), 1303.

21. J. Simon-Sanchez, C. Schulte, J. M. Bras, M. Sharma, J. R. Gibbs, D. Berg, C. Paisan-Ruiz, P. Lichtner, S. W. Scholz, D. G. Hernandez, R. Kruger, M. Federoff, C. Klein, A. Goate, J. Perlmutter, M. Bonin, M. A. Nalls, T. Illig, C. Gieger, H. Houlden, M. Steffens, M. S. Okun, B. A. Racette, M. R. Cookson, K. D. Foote, H. H. Fernandez, B. J. Traynor, S. Schreiber, S. Arepalli, R. Zonozi, K. Gwinn, B. M. van der, G. Lopez, S. J. Chanock, A. Schatzkin, Y. Park, A. Hollenbeck, J. Gao, X. Huang, N. W. Wood, D. Lorenz, G. Deuschl, H. Chen, O. Riess, J. A. Hardy, A. B. Singleton and T. Gasser, *Nat. Genet.*, 2009, **41**(12), 1308.

22. K. Ueda, H. Fukushima, E. Masliah, Y. Xia, A. Iwai, M. Yoshimoto, D. A. Otero, J. Kondo, Y. Ihara and T. Saitoh, *Proc. Natl. Acad. Sci. U. S. A.*, 1993, **90**(23), 11282.

23. M. H. Polymeropoulos, C. Lavedan, E. Leroy, S. E. Ide, A. Dehejia, A. Dutra, B. Pike, H. Root, J. Rubenstein, R. Boyer, E. S. Stenroos, S. Chandrasekharappa, A. Athanassiadou, T. Papapetropoulos, W. G. Johnson, A. M. Lazzarini, R. C. Duvoisin, G. Di Iorio, L. I. Golbe and R. L. Nussbaum, *Science*, 1997, **276**(5321), 2045.

24. R. Kruger, W. Kuhn, T. Muller, D. Woitalla, M. Graeber, S. Kosel, H. Przuntek, J. T. Epplen, L. Schols and O. Riess, *Nat. Genet.*, 1998, **18**(2), 106.

25. K. Seidel, L. Schols, S. Nuber, E. Petrasch-Parwez, K. Gierga, Z. Wszolek, D. Dickson, W. P. Gai, A. Bornemann, O. Riess, A. Rami, W. F. Den Dunnen, T. Deller, U. Rub and R. Kruger, *Ann. Neurol.*, 2010, **67**(5), 684.

26. J. J. Zarranz, J. Alegre, J. C. Gomez-Esteban, E. Lezcano, R. Ros, I. Ampuero, L. Vidal, J. Hoenicka, O. Rodriguez, B. Atares, V. Llorens, T. E. Gomez, T. Del Ser, D. G. Munoz and J. G. de Yebenes, *Ann. Neurol.*, 2004, **55**(2), 164.

27. M. C. Chartier-Harlin, J. Kachergus, C. Roumier, V. Mouroux, X. Douay, S. Lincoln, C. Levecque, L. Larvor, J. Andrieux, M. Hulihan, N. Waucquier, L. Defebvre, P. Amouyel, M. Farrer and A. Destee, *Lancet*, 2004, **364**(9440), 1167.

28. P. Ibanez, A. M. Bonnet, B. Debarges, E. Lohmann, F. Tison, P. Pollak, Y. Agid, A. Durr and A. Brice, *Lancet*, 2004, **364**(9440), 1169.

29. A. B. Singleton, M. Farrer, J. Johnson, A. Singleton, S. Hague, J. Kachergus, M. Hulihan, T. Peuralinna, A. Dutra, R. Nussbaum, S. Lincoln, A. Crawley, M. Hanson, D. Maraganore, C. Adler, M. R. Cookson, M. Muenter, M. Baptista, D. Miller, J. Blancato, J. Hardy and K. Gwinn-Hardy, *Science*, 2003, **302**(5646), 841.

30. A. Jowaed, I. Schmitt, O. Kaut and U. Wullner, *J. Neurosci.*, 2010, **30**(18), 6355.

31. J. H. Kordower, Y. Chu, R. A. Hauser, C. W. Olanow and T. B. Freeman, *Mov. Disord.*, 2008, **23**(16), 2303.

32. J. H. Kordower, Y. Chu, R. A. Hauser, T. B. Freeman and C. W. Olanow, *Nat. Med.*, 2008, **14**(5), 504.

33. J. Y. Li, E. Englund, J. L. Holton, D. Soulet, P. Hagell, A. J. Lees, T. Lashley, N. P. Quinn, S. Rehncrona, A. Bjorklund, H. Widner, T. Revesz, O. Lindvall and P. Brundin, *Nat. Med.*, 2008, **14**(5), 501.

34. I. Mendez, A. Vinuela, A. Astradsson, K. Mukhida, P. Hallett, H. Robertson, T. Tierney, R. Holness, A. Dagher, J. Q. Trojanowski and O. Isacson, *Nat. Med.*, 2008, **14**(5), 507.

35. M. B. Feany and W. W. Bender, *Nature*, 2000, **404**(6776), 394.

36. Y. Pesah, H. Burgess, B. Middlebrooks, K. Ronningen, J. Prosser, V. Tirunagaru, J. Zysk and G. Mardon, *Genesis*, 2005, **41**(4), 154.

37. M. Lakso, S. Vartiainen, A. M. Moilanen, J. Sirvio, J. H. Thomas, R. Nass, R. D. Blakely and G. Wong, *J. Neurochem.*, 2003, **86**(1), 165.

38. T. Kuwahara, A. Koyama, K. Gengyo-Ando, M. Masuda, H. Kowa, M. Tsunoda, S. Mitani and T. Iwatsubo, *J. Biol. Chem.*, 2006, **281**(1), 334.

39. P. Cao, Y. Yuan, E. A. Pehek, A. R. Moise, Y. Huang, K. Palczewski and Z. Feng, *PLoS. One*, 2010, **5**(2), e9312.

40. S. Vartiainen, V. Aarnio, M. Lakso and G. Wong, *Exp. Gerontol.*, 2006, **41**(9), 871.

41. R. Ved, S. Saha, B. Westlund, C. Perier, L. Burnam, A. Sluder, M. Hoener, C. M. Rodrigues, A. Alfonso, C. Steer, L. Liu, S. Przedborski and B. Wolozin, *J. Biol. Chem.*, 2005, **280**(52), 42655.

42. A. Abeliovich, Y. Schmitz, I. Farinas, D. Choi-Lundberg, W. H. Ho, P. E. Castillo, N. Shinsky, J. M. Verdugo, M. Armanini, A. Ryan, M. Hynes, H. Phillips, D. Sulzer and A. Rosenthal, *Neuron*, 2000, **25**(1), 239.

43. C. G. Specht and R. Schoepfer, *BMC Neurosci.*, 2001, **2**, 11.

44. P. H. van der, K. H. Wiederhold, A. Probst, S. Barbieri, C. Mistl, S. Danner, S. Kauffmann, K. Hofele, W. P. Spooren, M. A. Ruegg, S. Lin, P. Caroni, B. Sommer, M. Tolnay and G. Bilbe, *J. Neurosci.*, 2000, **20**(16), 6021.

45. M. Neumann, P. J. Kahle, B. I. Giasson, L. Ozmen, E. Borroni, W. Spooren, V. Muller, S. Odoy, H. Fujiwara, M. Hasegawa, T. Iwatsubo, J. Q. Trojanowski, H. A. Kretzschmar and C. Haass, *J. Clin. Invest.*, 2002, **110**(10), 1429.

46. B. I. Giasson, J. E. Duda, S. M. Quinn, B. Zhang, J. Q. Trojanowski and V. M. Lee, *Neuron*, 2002, **34**(4), 521.

47. M. K. Lee, W. Stirling, Y. Xu, X. Xu, D. Qui, A. S. Mandir, T. M. Dawson, N. G. Copeland, N. A. Jenkins and D. L. Price, *Proc. Natl. Acad. Sci. U. S. A.*, 2002, **99**(13), 8968.

48. S. Gispert, D. Del Turco, L. Garrett, A. Chen, D. J. Bernard, J. Hamm-Clement, H. W. Korf, T. Deller, H. Braak, G. Auburger and R. L. Nussbaum, *Mol. Cell. Neurosci.*, 2003, **24**(2), 419.

49. E. Sotiriou, D. K. Vassilatis, M. Vila and L. Stefanis, *Neurobiol. Aging*, 2010, **31**(10), 2013.

50. T. Gomez-Isla, M. C. Irizarry, A. Mariash, B. Cheung, O. Soto, S. Schrump, J. Sondel, L. Kotilinek, J. Day, M. A. Schwarzschild,

J. H. Cha, K. Newell, D. W. Miller, K. Ueda, A. B. Young, B. T. Hyman and K. H. Ashe, *Neurobiol. Aging*, 2003, **24**(2), 245.

51. E. Masliah, E. Rockenstein, I. Veinbergs, M. Mallory, M. Hashimoto, A. Takeda, Y. Sagara, A. Sisk and L. Mucke, *Science*, 2000, **287**(5456), 1265.

52. E. Rockenstein, M. Mallory, M. Hashimoto, D. Song, C. W. Shults, I. Lang and E. Masliah, *J. Neurosci. Res.*, 2002, **68**(5), 568.

53. Y. Matsuoka, M. Vila, S. Lincoln, A. McCormack, M. Picciano, J. Lafrancois, X. Yu, D. Dickson, W. J. Langston, E. McGowan, M. Farrer, J. Hardy, K. Duff, S. Przedborski and D. A. Di Monte, *Neurobiol. Dis.*, 2001, **8**(3), 535.

54. D. E. Cabin, S. Gispert-Sanchez, D. Murphy, G. Auburger, R. R. Myers and R. L. Nussbaum, *Neurobiol. Aging*, 2005, **26**(1), 25.

55. Y. M. Kuo, Z. Li, Y. Jiao, N. Gaborit, A. K. Pani, B. M. Orrison, B. G. Bruneau, B. I. Giasson, R. J. Smeyne, M. D. Gershon and R. L. Nussbaum, *Hum. Mol. Genet.*, 2010, **19**(9), 1633.

56. Y. Lim, V. Kehm, C. Li, J. Trojanowski and V. M. Lee, *Exp. Neurol.*, 2010, **221**(1), 86.

57. B. Schneider, R. Zufferey and P. Aebischer, *Parkinsonism Relat. Disord.*, 2008, **14**(Suppl 2), S169–S171.

58. B. C. Lo, J. L. Ridet, B. L. Schneider, N. Deglon and P. Aebischer, *Proc. Natl. Acad. Sci. U. S. A.*, 2002, **99**(16), 10813.

59. D. Kirik, L. E. Annett, C. Burger, N. Muzyczka, R. J. Mandel and A. Bjorklund, *Proc. Natl. Acad. Sci. U. S. A.*, 2003, **100**(5), 2884.

60. A. Eslamboli, M. Romero-Ramos, C. Burger, T. Bjorklund, N. Muzyczka, R. J. Mandel, H. Baker, R. M. Ridley and D. Kirik, *Brain*, 2007, **130**(3), 799.

61. P. K. Auluck, H. Y. Chan, J. Q. Trojanowski, V. M. Lee and N. M. Bonini, *Science*, 2002, **295**(5556), 865.

62. G. Du, X. Liu, X. Chen, M. Song, Y. Yan, R. Jiao and C. C. Wang, *Mol. Biol. Cell*, 2010, **21**(13), 2128.

63. T. F. Outeiro, E. Kontopoulos, S. M. Altmann, I. Kufareva, K. E. Strathearn, A. M. Amore, C. B. Volk, M. M. Maxwell, J. C. Rochet, P. J. McLean, A. B. Young, R. Abagyan, M. B. Feany, B. T. Hyman and A. G. Kazantsev, *Science*, 2007, **317**(5837), 516.

64. D. R. Shimshek, M. Mueller, C. Wiessner, T. Schweizer and P. H. van der Putten, *PLoS One*, 2010, **5**(4), e10014.

65. J. M. George, *Genome Biol.*, 2002, **3**(1), 3002.

66. M. Hashimoto, E. Rockenstein, M. Mante, M. Mallory and E. Masliah, *Neuron*, 2001, **32**(2), 213.

67. Y. Fan, P. Limprasert, I. V. Murray, A. C. Smith, V. M. Lee, J. Q. Trojanowski, B. L. Sopher and A. R. La Spada, *Hum. Mol. Genet.*, 2006, **15**(20), 3002.

68. N. Ninkina, O. Peters, S. Millership, H. Salem, P. H. van der and V. L. Buchman, *Hum. Mol. Genet.*, 2009, **18**(10), 1779.

69. W. Dauer, N. Kholodilov, M. Vila, A. C. Trillat, R. Goodchild, K. E. Larsen, R. Staal, K. Tieu, Y. Schmitz, C. A. Yuan, M. Rocha, V. Jackson-Lewis, S. Hersch, D. Sulzer, S. Przedborski, R. Burke and R. Hen, *Proc. Natl. Acad. Sci. U. S. A.*, 2002, **99**(22), 14524.

70. F. Fornai, O. M. Schluter, P. Lenzi, M. Gesi, R. Ruffoli, M. Ferrucci, G. Lazzeri, C. L. Busceti, F. Pontarelli, G. Battaglia, A. Pellegrini, F. Nicoletti, S. Ruggieri, A. Paparelli and T. C. Sudhof, *Proc. Natl. Acad. Sci. U. S. A.*, 2005, **102**(9), 3413.

71. M. Cano-Jaimez, F. Perez-Sanchez, M. Milan, P. Buendia, S. Ambrosio and I. Farinas, *Neurobiol. Dis.*, 2010, **38**(1), 92.

72. E. Junn, K. W. Lee, B. S. Jeong, T. W. Chan, J. Y. Im and M. M. Mouradian, *Proc. Natl. Acad. Sci. U. S. A.*, 2009, **106**(31), 13052.

73. S. Rathke-Hartlieb, P. J. Kahle, M. Neumann, L. Ozmen, S. Haid, M. Okochi, C. Haass and J. B. Schulz, *J. Neurochem.*, 2001, **77**(4), 1181.

74. G. K. Tofaris, R. P. Garcia, T. Humby, S. L. Lambourne, M. O'Connell, B. Ghetti, H. Gossage, P. C. Emson, L. S. Wilkinson, M. Goedert and M. G. Spillantini, *J. Neurosci.*, 2006, **26**(15), 3942.

75. M. Wakamatsu, A. Ishii, S. Iwata, J. Sakagami, Y. Ukai, M. Ono, D. Kanbe, S. Muramatsu, K. Kobayashi, T. Iwatsubo and M. Yoshimoto, *Neurobiol. Aging*, 2008, **29**(4), 574.

76. J. P. Daher, M. Ying, R. Banerjee, R. S. McDonald, M. D. Hahn, L. Yang, B. M. Flint, B. Thomas, V. L. Dawson, T. M. Dawson and D. J. Moore, *Mol. Neurodegener.*, 2009, **4**, 34.

77. P. J. Kahle, M. Neumann, L. Ozmen, V. Muller, H. Jacobsen, W. Spooren, B. Fuss, B. Mallon, W. B. Macklin, H. Fujiwara, M. Hasegawa, T. Iwatsubo, H. A. Kretzschmar and C. Haass, *EMBO Rep.*, 2002, **3**(6), 583.

78. C. W. Shults, E. Rockenstein, L. Crews, A. Adame, M. Mante, G. Larrea, M. Hashimoto, D. Song, T. Iwatsubo, K. Tsuboi and E. Masliah, *J. Neurosci.*, 2005, **25**(46), 10689.

79. K. Ubhi, P. H. Lee, A. Adame, C. Inglis, M. Mante, E. Rockenstein, N. Stefanova, G. K. Wenning and E. Masliah, *J. Neurosci. Res.*, 2009, **87**(12), 2728.

80. K. Ubhi, E. Rockenstein, M. Mante, C. Inglis, A. Adame, C. Patrick, K. Whitney and E. Masliah, *J. Neurosci.*, 2010, **30**(18), 6236.

81. I. Yazawa, B. I. Giasson, R. Sasaki, B. Zhang, S. Joyce, K. Uryu, J. Q. Trojanowski and V. M. Lee, *Neuron*, 2005, **45**(6), 847.

82. V. Bonifati, P. Rizzu, M. J. van Baren, O. Schaap, G. J. Breedveld, E. Krieger, M. C. Dekker, F. Squitieri, P. Ibanez, M. Joosse, J. W. van Dongen, N. Vanacore, J. C. van Swieten, A. Brice, G. Meco, C. M. van Duijn, B. A. Oostra and P. Heutink, *Science*, 2003, **299**(5604), 256.

83. P. M. Abou-Sleiman, D. G. Healy and N. W. Wood, *Cell Tissue Res.*, 2004, **318**(1), 185.

84. A. Skoneczna, A. Micialkiewicz and M. Skoneczny, *Free Radic. Biol. Med.*, 2007, **42**(9), 1409.

85. R. Ved, S. Saha, B. Westlund, C. Perier, L. Burnam, A. Sluder, M. Hoener, C. M. Rodrigues, A. Alfonso, C. Steer, L. Liu, S. Przedborski and B. Wolozin, *J. Biol. Chem.*, 2005, **280**(52), 42655.

86. F. M. Menzies, S. C. Yenisetti and K. T. Min, *Curr. Biol.*, 2005, **15**(17), 1578.

87. M. Meulener, A. J. Whitworth, C. E. Armstrong-Gold, P. Rizzu, P. Heutink, P. D. Wes, L. J. Pallanck and N. M. Bonini, *Curr. Biol.*, 2005, **15**(17), 1572.

88. Y. Yang, S. Gehrke, M. E. Haque, Y. Imai, J. Kosek, L. Yang, M. F. Beal, I. Nishimura, K. Wakamatsu, S. Ito, R. Takahashi and B. Lu, *Proc. Natl. Acad. Sci. U. S. A.*, 2005, **102**(38), 13670.

89. J. Park, S. Y. Kim, G. H. Cha, S. B. Lee, S. Kim and J. Chung, *Gene*, 2005, **361**, 133.

90. M. C. Meulener, K. Xu, L. Thomson, H. Ischiropoulos and N. M. Bonini, *Proc. Natl. Acad. Sci. U. S. A.*, 2006, **103**(33), 12517.

91. E. Lavara-Culebras and N. Paricio, *Gene*, 2007, **400**, 158.

92. L. Chen, B. Cagniard, T. Mathews, S. Jones, H. C. Koh, Y. Ding, P. M. Carvey, Z. Ling, U. J. Kang and X. Zhuang, *J. Biol. Chem.*, 2005, **280**(22), 21418.

93. R. H. Kim, P. D. Smith, H. Aleyasin, S. Hayley, M. P. Mount, S. Pownall, A. Wakeham, T. You, S. K. Kalia, P. Horne, D. Westaway, A. M. Lozano, H. Anisman, D. S. Park and T. W. Mak, *Proc. Natl. Acad. Sci. U. S. A.*, 2005, **102**(14), 5215.

94. M. S. Goldberg, A. Pisani, M. Haburcak, T. A. Vortherms, T. Kitada, C. Costa, Y. Tong, G. Martella, A. Tscherter, A. Martins, G. Bernardi, B. L. Roth, E. N. Pothos, P. Calabresi and J. Shen, *Neuron*, 2005, **45**(4), 489.

95. A. B. Manning-Bog, W. M. Caudle, X. A. Perez, S. H. Reaney, R. Paletzki, M. Z. Isla, V. P. Chou, A. L. McCormack, G. W. Miller, J. W. Langston, C. R. Gerfen and D. A. Dimonte, *Neurobiol. Dis.*, 2007, **27**(2), 141.

96. J. S. Chandran, X. Lin, A. Zapata, A. Hoke, M. Shimoji, S. O. Moore, M. P. Galloway, F. M. Laird, P. C. Wong, D. L. Price, K. R. Bailey, J. N. Crawley, T. Shippenberg and H. Cai, *Neurobiol. Dis.*, 2008, **29**(3), 505.

97. L. Y. Hao, B. I. Giasson and N. M. Bonini, *Proc. Natl. Acad. Sci. U. S. A.*, 2010, **107**(21), 9747.

98. T. T. Pham, F. Giesert, A. Rothig, T. Floss, M. Kallnik, K. Weindl, S. M. Holter, U. Ahting, H. Prokisch, L. Becker, T. Klopstock, K. Beyer, K. Gorner, P. J. Kahle, D. M. Vogt Weisenhorn and W. Wurst, *Genes Brain Behav.*, 2010, **9**(3), 305.

99. C. P. Ramsey, E. Tsika, H. Ischiropoulos and B. I. Giasson, *Hum. Mol. Genet.*, 2010, **19**(8), 1425.

100. T. Kitada, Y. Tong, C. A. Gautier and J. Shen, *J. Neurochem.*, 2009, **111**(3), 696.

101. L. Aron, P. Klein, T. T. Pham, E. R. Kramer, W. Wurst and R. Klein, *PLoS Biol.*, 2010, **8**(4), e1000349.

102. A. Brice, A. Durr and C. Lucking, GeneReviews, 1993; http://www.ncbi.nlm.nih.gov/bookshelf/br.fcgi?book = gene&part = jpd [last updated 1 October 2007].

103. T. Kitada, S. Asakawa, N. Hattori, H. Matsumine, Y. Yamamura, S. Minoshima, M. Yokochi, Y. Mizuno and N. Shimizu, Nature, 1998, 392(6676), 605.

104. J. F. Guo, B. Xiao, B. Liao, X. W. Zhang, L. L. Nie, Y. H. Zhang, L. Shen, H. Jiang, K. Xia, Q. Pan, X. X. Yan and B. S. Tang, Mov. Disord., 2008, 23(14), 2074.

105. T. M. Dawson and V. L. Dawson, Mov. Disord., 2010, 25(Suppl 1), S32–S39.

106. L. Silvestri, V. Caputo, E. Bellacchio, L. Atorino, B. Dallapiccola, E. M. Valente and G. Casari, Hum. Mol. Genet., 2005, 14(22), 3477.

107. C. Zhou, Y. Huang, Y. Shao, J. May, D. Prou, C. Perier, W. Dauer, E. A. Schon and S. Przedborski, Proc. Natl. Acad. Sci. U. S. A., 2008, 105(33), 12022.

108. A. Beilina, B. M. van der, R. Ahmad, S. Kesavapany, D. W. Miller, G. A. Petsko and M. R. Cookson, Proc. Natl. Acad. Sci. U. S. A., 2005, 102(16), 5703.

109. M. E. Haque, K. J. Thomas, C. D'Souza, S. Callaghan, T. Kitada, R. S. Slack, P. Fraser, M. R. Cookson, A. Tandon and D. S. Park, Proc. Natl. Acad. Sci. U. S. A., 2008, 105(5), 1716.

110. S. Takatori, G. Ito and T. Iwatsubo, Neurosci. Lett., 2008, 430(1), 13.

111. A. Weihofen, B. Ostaszewski, Y. Minami and D. J. Selkoe, Hum. Mol. Genet., 2008, 17(4), 602.

112. J. C. Greene, A. J. Whitworth, I. Kuo, L. A. Andrews, M. B. Feany and L. J. Pallanck, Proc. Natl. Acad. Sci. U. S. A., 2003, 100(7), 4078.

113. A. J. Whitworth, D. A. Theodore, J. C. Greene, H. Benes, P. D. Wes and L. J. Pallanck, Proc. Natl. Acad. Sci. U. S. A., 2005, 102(22), 8024.

114. I. E. Clark, M. W. Dodson, C. Jiang, J. H. Cao, J. R. Huh, J. H. Seol, S. J. Yoo, B. A. Hay and M. Guo, Nature, 2006, 441(7097), 1162.

115. J. Park, S. B. Lee, S. Lee, Y. Kim, S. Song, S. Kim, E. Bae, J. Kim, M. Shong, J. M. Kim and J. Chung, Nature, 2006, 441(7097), 1157.

116. Y. Yang, I. Nishimura, Y. Imai, R. Takahashi and B. Lu, Neuron, 2003, 37(6), 911.

117. Y. Yang, Y. Ouyang, L. Yang, M. F. Beal, A. McQuibban, H. Vogel and B. Lu, Proc. Natl. Acad. Sci. U. S. A., 2008, 105(19), 7070.

118. H. Deng, M. W. Dodson, H. Huang and M. Guo, Proc. Natl. Acad. Sci. U. S. A., 2008, 105(38), 14503.

119. D. P. Narendra, S. M. Jin, A. Tanaka, D. F. Suen, C. A. Gautier, J. Shen, M. R. Cookson and R. J. Youle, PLoS. Biol., 2010, 8(1), e1000298.

120. S. Geisler, K. M. Holmstrom, D. Skujat, F. C. Fiesel, O. C. Rothfuss, P. J. Kahle and W. Springer, Nat. Cell Biol., 2010, 12(2), 119.

121. C. T. Chu, Hum. Mol. Genet., 2010, 19(R1), R28–R37.

122. N. Matsuda, S. Sato, K. Shiba, K. Okatsu, K. Saisho, C. A. Gautier, Y. S. Sou, S. Saiki, S. Kawajiri, F. Sato, M. Kimura, M. Komatsu, N. Hattori and K. Tanaka, *J. Cell Biol.*, 2010, **189**(2), 211.

123. J. M. Itier, P. Ibanez, M. A. Mena, N. Abbas, C. Cohen-Salmon, G. A. Bohme, M. Laville, J. Pratt, O. Corti, L. Pradier, G. Ret, C. Joubert, M. Periquet, F. Araujo, J. Negroni, M. J. Casarejos, S. Canals, R. Solano, A. Serrano, E. Gallego, M. Sanchez, P. Denefle, J. Benavides, G. Tremp, T. A. Rooney, A. Brice and D. Y. Garcia, *Hum. Mol. Genet.*, 2003, **12**(18), 2277.

124. M. S. Goldberg, S. M. Fleming, J. J. Palacino, C. Cepeda, H. A. Lam, A. Bhatnagar, E. G. Meloni, N. Wu, L. C. Ackerson, G. J. Klapstein, M. Gajendiran, B. L. Roth, M. F. Chesselet, N. T. Maidment, M. S. Levine and J. Shen, *J. Biol. Chem.*, 2003, **278**(44), 43628.

125. R. von Coelln, B. Thomas, J. M. Savitt, K. L. Lim, M. Sasaki, E. J. Hess, V. L. Dawson and T. M. Dawson, *Proc. Natl. Acad. Sci. U. S. A.*, 2004, **101**(29), 10744.

126. F. A. Perez and R. D. Palmiter, *Proc. Natl. Acad. Sci. U. S. A.*, 2005, **102**(6), 2174.

127. T. Kitada, A. Pisani, D. R. Porter, H. Yamaguchi, A. Tscherter, G. Martella, P. Bonsi, C. Zhang, E. N. Pothos and J. Shen, *Proc. Natl. Acad. Sci. U. S. A.*, 2007, **104**(27), 11441.

128. C. A. Gautier, T. Kitada and J. Shen, *Proc. Natl. Acad. Sci. U. S. A.*, 2008, **105**(32), 11364.

129. S. Gispert, F. Ricciardi, A. Kurz, M. Azizov, H. H. Hoepken, D. Becker, W. Voos, K. Leuner, W. E. Muller, A. P. Kudin, W. S. Kunz, A. Zimmermann, J. Roeper, D. Wenzel, M. Jendrach, M. Garcia-Arencibia, J. Fernandez-Ruiz, L. Huber, H. Rohrer, M. Barrera, A. S. Reichert, U. Rub, A. Chen, R. L. Nussbaum and G. Auburger, *PLoS One*, 2009, **4**(6), e5777.

130. T. K. Sang, H. Y. Chang, G. M. Lawless, A. Ratnaparkhi, L. Mee, L. C. Ackerson, N. T. Maidment, D. E. Krantz and G. R. Jackson, *J. Neurosci.*, 2007, **27**(5), 981.

131. C. Wang, R. Lu, X. Ouyang, M. W. Ho, W. Chia, F. Yu and K. L. Lim, *J. Neurosci.*, 2007, **27**(32), 8563.

132. C. L. Kragh, L. B. Lund, F. Febbraro, H. D. Hansen, W. P. Gai, C. Richter-Landsberg and P. H. Jensen, *J. Biol. Chem.*, 2009, **284**(15), 10211.

133. X. R. Zhu, L. Maskri, C. Herold, V. Bader, C. C. Stichel, O. Gunturkun and H. Lubbert, *Eur. J. Neurosci.*, 2007, **26**(7), 1902.

134. T. C. Frank-Cannon, T. Tran, K. A. Ruhn, T. N. Martinez, J. Hong, M. Marvin, M. Hartley, I. Trevino, D. E. O'Brien, B. Casey, M. S. Goldberg and M. G. Tansey, *J. Neurosci.*, 2008, **28**(43), 10825.

135. C. Paisan-Ruiz, S. Jain, E. W. Evans, W. P. Gilks, J. Simon, B. M. van der, D. M. Lopez, S. Aparicio, A. M. Gil, N. Khan, J. Johnson, J. R. Martinez, D. Nicholl, I. M. Carrera, A. S. Pena, R. de Silva, A. Lees,

J. F. Marti-Masso, J. Perez-Tur, N. W. Wood and A. B. Singleton, *Neuron*, 2004, **44**(4), 595.

136. A. Zimprich, S. Biskup, P. Leitner, P. Lichtner, M. Farrer, S. Lincoln, J. Kachergus, M. Hulihan, R. J. Uitti, D. B. Calne, A. J. Stoessl, R. F. Pfeiffer, N. Patenge, I. C. Carbajal, P. Vieregge, F. Asmus, B. Muller-Myhsok, D. W. Dickson, T. Meitinger, T. M. Strom, Z. K. Wszolek and T. Gasser, *Neuron*, 2004, **44**(4), 601.

137. J. F. Marti-Masso, J. Ruiz-Martinez, M. J. Bolano, I. Ruiz, A. Gorostidi, F. Moreno, I. Ferrer and A. L. de Munain, *Mov. Disord.*, 2009, **24**(13), 1998.

138. A. Sakaguchi-Nakashima, J. Y. Meir, Y. Jin, K. Matsumoto and N. Hisamoto, *Curr. Biol.*, 2007, **17**(7), 592.

139. D. Wang, B. Tang, G. Zhao, Q. Pan, K. Xia, R. Bodmer and Z. Zhang, *Mol. Neurodegener.*, 2008, **3**, 3.

140. L. Wang, C. Xie, E. Greggio, L. Parisiadou, H. Shim, L. Sun, J. Chandran, X. Lin, C. Lai, W. J. Yang, D. J. Moore, T. M. Dawson, V. L. Dawson, G. Chiosis, M. R. Cookson and H. Cai, *J. Neurosci.*, 2008, **28**(13), 3384.

141. Y. Tong, H. Yamaguchi, E. Giaime, S. Boyle, R. Kopan, R. J. Kelleher III and J. Shen, *Proc. Natl. Acad. Sci. U. S. A.*, 2010, **107**(21), 9879.

142. C. H. Ng, S. Z. Mok, C. Koh, X. Ouyang, M. L. Fivaz, E. K. Tan, V. L. Dawson, T. M. Dawson, F. Yu and K. L. Lim, *J. Neurosci.*, 2009, **29**(36), 11257.

143. Y. Li, W. Liu, T. F. Oo, L. Wang, Y. Tang, V. Jackson-Lewis, C. Zhou, K. Geghman, M. Bogdanov, S. Przedborski, M. F. Beal, R. E. Burke and C. Li, *Nat. Neurosci.*, 2009, **12**(7), 826.

144. Y. Tong, A. Pisani, G. Martella, M. Karouani, H. Yamaguchi, E. N. Pothos and J. Shen, *Proc. Natl. Acad. Sci. U. S. A.*, 2009, **106**(34), 14622.

145. X. Lin, L. Parisiadou, X. L. Gu, L. Wang, H. Shim, L. Sun, C. Xie, C. X. Long, W. J. Yang, J. Ding, Z. Z. Chen, P. E. Gallant, J. H. Tao-Cheng, G. Rudow, J. C. Troncoso, Z. Liu, Z. Li and H. Cai, *Neuron*, 2009, **64**(6), 807.

146. Y. Xiong, C. E. Coombes, A. Kilaru, X. Li, A. D. Gitler, W. J. Bowers, V. L. Dawson, T. M. Dawson and D. J. Moore, *PLoS Genet.*, 2010, **6**(4), e1000902.

147. C. Yao, R. El Khoury, W. Wang, T. A. Byrd, E. A. Pehek, C. Thacker, X. Zhu, M. A. Smith, A. L. Wilson-Delfosse and S. G. Chen, *Neurobiol. Dis.*, 2010, **40**(1), 73.

148. K. M. Strauss, L. M. Martins, H. Plun-Favreau, F. P. Marx, S. Kautzmann, D. Berg, T. Gasser, Z. Wszolek, T. Muller, A. Bornemann, H. Wolburg, J. Downward, O. Riess, J. B. Schulz and R. Kruger, *Hum. Mol. Genet.*, 2005, **14**(15), 2099.

149. V. Bogaerts, K. Nuytemans, J. Reumers, P. Pals, S. Engelborghs, B. Pickut, E. Corsmit, K. Peeters, J. Schymkowitz, P. P. De Deyn,

P. Cras, F. Rousseau, J. Theuns and C. Van Broeckhoven, *Hum. Mutat.*, 2008, **29**(6), 832.

150. O. A. Ross, A. I. Soto, C. Vilarino-Guell, M. G. Heckman, N. N. Diehl, M. M. Hulihan, J. O. Aasly, S. Sando, J. M. Gibson, T. Lynch, A. Krygowska-Wajs, G. Opala, M. Barcikowska, K. Czyzewski, R. J. Uitti, Z. K. Wszolek and M. J. Farrer, *Parkinsonism Relat. Disord.*, 2008, **14**(7), 539.

151. J. Simon-Sanchez and A. B. Singleton, *Hum. Mol. Genet.*, 2008, **17**(13), 1988.

152. Y. Kawamoto, Y. Kobayashi, Y. Suzuki, H. Inoue, H. Tomimoto, I. Akiguchi, H. Budka, L. M. Martins, J. Downward and R. Takahashi, *J. Neuropathol. Exp. Neurol.*, 2008, **67**(10), 984.

153. C. W. Gray, R. V. Ward, E. Karran, S. Turconi, A. Rowles, D. Viglienghi, C. Southan, A. Barton, K. G. Fantom, A. West, J. Savopoulos, N. J. Hassan, H. Clinkenbeard, C. Hanning, B. Amegadzie, J. B. Davis, C. Dingwall, G. P. Livi and C. L. Creasy, *Eur. J. Biochem.*, 2000, **267**(18), 5699.

154. L. Faccio, C. Fusco, A. Chen, S. Martinotti, J. V. Bonventre and A. S. Zervos, *J. Biol. Chem.*, 2000, **275**(4), 2581.

155. J. M. Jones, P. Datta, S. M. Srinivasula, W. Ji, S. Gupta, Z. Zhang, E. Davies, G. Hajnoczky, T. L. Saunders, M. L. Van Keuren, T. Fernandes-Alnemri, M. H. Meisler and E. S. Alnemri, *Nature*, 2003, **425**(6959), 721.

156. L. M. Martins, A. Morrison, K. Klupsch, V. Fedele, N. Moisoi, P. Teismann, A. Abuin, E. Grau, M. Geppert, G. P. Livi, C. L. Creasy, A. Martin, I. Hargreaves, S. J. Heales, H. Okada, S. Brandner, J. B. Schulz, T. Mak and J. Downward, *Mol. Cell Biol.*, 2004, **24**(22), 9848.

157. H. Plun-Favreau, K. Klupsch, N. Moisoi, S. Gandhi, S. Kjaer, D. Frith, K. Harvey, E. Deas, R. J. Harvey, N. McDonald, N. W. Wood, L. M. Martins and J. Downward, *Nat. Cell Biol.*, 2007, **9**(11), 1243.

158. J. W. Langston and P. A. Ballard, Jr., *N. Engl. J. Med.*, 1983, **309**(5), 310.

159. J. W. Langston, *Neurology*, 1996, **47**(6 Suppl 3), S153–S160.

160. F. J. Vingerhoets, B. J. Snow, J. W. Tetrud, J. W. Langston, M. Schulzer and D. B. Calne, *Ann. Neurol.*, 1994, **36**(5), 765.

161. M. E. Emborg, *ILAR J.*, 2007, **48**(4), 339.

162. G. Halliday, M. T. Herrero, K. Murphy, H. McCann, F. Ros-Bernal, C. Barcia, H. Mori, F. J. Blesa and J. A. Obeso, *Mov. Disord.*, 2009, **24**(10), 1519.

163. N. M. Filipov, A. B. Norwood and S. C. Sistrunk, *Neuroreport*, 2009, **20**(7), 713.

164. W. S. Choi, G. Abel, H. Klintworth, R. A. Flavell and Z. Xia, *J. Neuropathol. Exp. Neurol.*, 2010, **69**(5), 511.

165. R. J. Dinis-Oliveira, J. A. Duarte, A. Sanchez-Navarro, F. Remiao, M. L. Bastos and F. Carvalho, *Crit. Rev. Toxicol.*, 2008, **38**(1), 13.

166. G. Pezzoli, S. Barbieri, C. Ferrante, A. Zecchinelli and V. Foa, *Lancet*, 1989, **2**(8667), 874.

167. G. Pezzoli, S. Ricciardi, C. Masotto, C. B. Mariani and A. Carenzi, *Brain Res.*, 1990, **531**(1–2), 355.

168. I. Mena, *Ann. Clin. Lab Sci.*, 1974, **4**(6), 487.

169. C. W. Olanow, P. F. Good, H. Shinotoh, K. A. Hewitt, F. Vingerhoets, B. J. Snow, M. F. Beal, D. B. Calne and D. P. Perl, *Neurology*, 1996, **46**(2), 492.

170. A. D. Gitler, A. Chesi, M. L. Geddie, K. E. Strathearn, S. Hamamichi, K. J. Hill, K. A. Caldwell, G. A. Caldwell, A. A. Cooper, J. C. Rochet and S. Lindquist, *Nat. Genet.*, 2009, **41**, 308.

171. S. A. Schneider, C. Paisan-Ruiz, N. P. Quinn, A. J. Lees, H. Houlden, J. Hardy and K. P. Bhatia, *Mov. Disord.*, 2010, **25**(8), 979.

172. W. R. Martin, M. Wieler and M. Gee, *Neurology*, 2008, **70**(16 Pt 2), 1411.

173. A. Gregory, B. J. Polster and S. J. Hayflick, *J. Med. Genet.*, 2009, **46**(2), 73.

174. Z. Wu, C. Li, S. Lv and B. Zhou, *Hum. Mol. Genet.*, 2009, **18**(19), 3659.

175. G. J. Sengstock, C. W. Olanow, R. A. Menzies, A. J. Dunn and G. W. Arendash, *J. Neurosci. Res.*, 1993, **35**(1), 67.

176. K. Fuxe and U. Ungerstedt, *Pharmacol. Ther. B*, 1976, **2**(1), 41.

177. A. K. Wright, M. Garcia-Munoz and G. W. Arbuthnott, *Rev. Neurosci.*, 2009, **20**(2), 85.

178. R. J. Carey, *Psychopharmacology (Berl)*, 1986, **89**(3), 269.

179. C. K. Glass, K. Saijo, B. Winner, M. C. Marchetto and F. H. Gage, *Cell*, 2010, **140**(6), 918.

180. M. G. Tansey and M. S. Goldberg, *Neurobiol. Dis.*, 2010, **37**(3), 510.

181. P. L. McGeer, S. Itagaki, B. E. Boyes and E. G. McGeer, *Neurology*, 1988, **38**(8), 1285.

182. E. Croisier, L. B. Moran, D. T. Dexter, R. K. Pearce and M. B. Graeber, *J. Neuroinflammation*, 2005, **2**, 14.

183. Y. Ouchi, E. Yoshikawa, Y. Sekine, M. Futatsubashi, T. Kanno, T. Ogusu and T. Torizuka, *Ann. Neurol.*, 2005, **57**(2), 168.

184. A. Gerhard, N. Pavese, G. Hotton, F. Turkheimer, M. Es, A. Hammers, K. Eggert, W. Oertel, R. B. Banati and D. J. Brooks, *Neurobiol. Dis.*, 2006, **21**(2), 404.

185. Y. Ouchi, S. Yagi, M. Yokokura and M. Sakamoto, *Parkinsonism. Relat. Disord.*, 2009, **15**(Supp 3), S200–S204.

186. C. Weller, N. Oxlade, S. M. Dobbs, R. J. Dobbs, A. Charlett and I. T. Bjarnason, *FEMS Immunol. Med. Microbiol.*, 2005, **44**(2), 129.

187. M. Bialecka, M. Kurzawski, G. Klodowska-Duda, G. Opala, S. Juzwiak, G. Kurzawski, E. K. Tan and M. Drozdzik, *Neurosci. Res.*, 2007, **57**(3), 473.

188. J. C. Barrett, S. Hansoul, D. L. Nicolae, J. H. Cho, R. H. Duerr, J. D. Rioux, S. R. Brant, M. S. Silverberg, K. D. Taylor, M. M. Barmada, A. Bitton, T. Dassopoulos, L. W. Datta, T. Green, A. M. Griffiths,

E. O. Kistner, M. T. Murtha, M. D. Regueiro, J. I. Rotter, L. P. Schumm, A. H. Steinhart, S. R. Targan, R. J. Xavier, C. Libioulle, C. Sandor, M. Lathrop, J. Belaiche, O. Dewit, I. Gut, S. Heath, D. Laukens, M. Mni, P. Rutgeerts, A. Van Gossum, D. Zelenika, D. Franchimont, J. P. Hugot, M. de Vos, S. Vermeire, E. Louis, L. R. Cardon, C. A. Anderson, H. Drummond, E. Nimmo, T. Ahmad, N. J. Prescott, C. M. Onnie, S. A. Fisher, J. Marchini, J. Ghori, S. Bumpstead, R. Gwilliam, M. Tremelling, P. Deloukas, J. Mansfield, D. Jewell, J. Satsangi, C. G. Mathew, M. Parkes, M. Georges and M. J. Daly, *Nat. Genet.*, 2008, **40**(8), 955.
189. J. Van Limbergen, D. C. Wilson and J. Satsangi, *Annu. Rev. Genomics Hum. Genet.*, 2009, **10**, 89.
190. A. Castano, A. J. Herrera, J. Cano and A. Machado, *J. Neurochem.*, 1998, **70**(4), 1584.
191. A. J. Herrera, A. Castano, J. L. Venero, J. Cano and A. Machado, *Neurobiol. Dis.*, 2000, **7**(4), 429.
192. H. Arai, T. Furuya, T. Yasuda, M. Miura, Y. Mizuno and H. Mochizuki, *J. Biol. Chem.*, 2004, **279**(49), 51647.
193. R. M. de Pablos, A. J. Herrera, R. F. Villaran, J. Cano and A. Machado, *FASEB J.*, 2005, **19**(3), 407.
194. M. M. Iravani, C. C. Leung, M. Sadeghian, C. O. Haddon, S. Rose and P. Jenner, *Eur. J. Neurosci.*, 2005, **22**(2), 317.
195. H. M. Gao, J. Jiang, B. Wilson, W. Zhang, J. S. Hong and B. Liu, *J. Neurochem.*, 2002, **81**(6), 1285.
196. C. Cunningham, D. C. Wilcockson, S. Campion, K. Lunnon and V. H. Perry, *J. Neurosci.*, 2005, **25**(40), 9275.
197. L. Qin, X. Wu, M. L. Block, Y. Liu, G. R. Breese, J. S. Hong, D. J. Knapp and F. T. Crews, *Glia*, 2007, **55**(5), 453.
198. C. C. Ferrari, M. C. Pott Godoy, R. Tarelli, M. Chertoff, A. M. Depino and F. J. Pitossi, *Neurobiol. Dis.*, 2006, **24**(1), 183.
199. A. L. Lella Ezcurra, M. Chertoff, C. Ferrari, M. Graciarena and F. Pitossi, *Neurobiol. Dis.*, 2010, **37**(3), 630.
200. H. M. Gao, P. T. Kotzbauer, K. Uryu, S. Leight, J. Q. Trojanowski and V. M. Lee, *J. Neurosci.*, 2008, **28**(30), 7687.
201. R. C. Dale, A. J. Church, R. A. Surtees, A. J. Lees, J. E. Adcock, B. Harding, B. G. Neville and G. Giovannoni, *Brain*, 2004, **127**(Pt 1), 21.
202. H. Jang, D. Boltz, K. Sturm-Ramirez, K. R. Shepherd, Y. Jiao, R. Webster and R. J. Smeyne, *Proc. Natl. Acad. Sci. U. S. A.*, 2009, **106**(33), 14063.
203. S. Murayama, K. Arima, Y. Nakazato, J. Satoh, M. Oda and T. Inose, *Acta Neuropathol.*, 1992, **84**(1), 32.
204. D. L. Pountney, T. M. Treweek, T. Chataway, Y. Huang, F. Chegini, P. C. Blumbergs, M. J. Raftery and W. P. Gai, *Neurotox. Res.*, 2005, **7**(1–2), 77.
205. K. Uryu, C. Richter-Landsberg, W. Welch, E. Sun, O. Goldbaum, E. H. Norris, C. T. Pham, I. Yazawa, K. Hilburger, M. Micsenyi,

B. I. Giasson, N. M. Bonini, V. M. Lee and J. Q. Trojanowski, *Am. J. Pathol.*, 2006, **168**(3), 947.

206. P. J. McLean, J. Klucken, Y. Shin and B. T. Hyman, *Biochem. Biophys. Res. Commun.*, 2004, **321**(3), 665.

207. J. Klucken, Y. Shin, B. T. Hyman and P. J. McLean, *Biochem. Biophys. Res. Commun.*, 2004, **325**(1), 367.

208. M. M. Dedmon, J. Christodoulou, M. R. Wilson and C. M. Dobson, *J. Biol. Chem.*, 2005, **280**(15), 14733.

209. N. Ostrerova, L. Petrucelli, M. Farrer, N. Mehta, P. Choi, J. Hardy and B. Wolozin, *J. Neurosci.*, 1999, **19**(14), 5782.

210. J. M. Souza, B. I. Giasson, V. M. Lee and H. Ischiropoulos, *FEBS Lett.*, 2000, **474**(1), 116.

211. S. Chandra, G. Gallardo, R. Fernandez-Chacon, O. M. Schluter and T. C. Sudhof, *Cell*, 2005, **123**(3), 383.

212. L. Bedford, D. Hay, A. Devoy, S. Paine, D. G. Powe, R. Seth, T. Gray, I. Topham, K. Fone, N. Rezvani, M. Mee, T. Soane, R. Layfield, P. W. Sheppard, T. Ebendal, D. Usoskin, J. Lowe and R. J. Mayer, *J. Neurosci.*, 2008, **28**(33), 8189.

213. M. Komatsu, S. Waguri, T. Chiba, S. Murata, J. Iwata, I. Tanida, T. Ueno, M. Koike, Y. Uchiyama, E. Kominami and K. Tanaka, *Nature*, 2006, **441**(7095), 880.

214. T. Hara, K. Nakamura, M. Matsui, A. Yamamoto, Y. Nakahara, R. Suzuki-Migishima, M. Yokoyama, K. Mishima, I. Saito, H. Okano and N. Mizushima, *Nature*, 2006, **441**(7095), 885.

215. G. Krebiehl, S. Ruckerbauer, L. F. Burbulla, N. Kieper, B. Maurer, J. Waak, H. Wolburg, Z. Gizatullina, F. N. Gellerich, D. Woitalla, O. Riess, P. J. Kahle, T. Proikas-Cezanne and R. Kruger, *PLoS One*, 2010, **5**(2), e9367.

216. K. S. McNaught, D. P. Perl, A. L. Brownell and C. W. Olanow, *Ann. Neurol.*, 2004, **56**(1), 149.

217. H. Miwa, T. Kubo, A. Suzuki, K. Nishi and T. Kondo, *Neurosci. Lett.*, 2005, **380**(1–2), 93.

218. J. H. Kordower, N. M. Kanaan, Y. Chu, B. R. Suresh, J. Stansell III, B. T. Terpstra, C. E. Sortwell, K. Steece-Collier and T. J. Collier, *Ann. Neurol.*, 2006, **60**(2), 264.

219. J. Bove, C. Zhou, V. Jackson-Lewis, J. Taylor, Y. Chu, H. J. Rideout, D. C. Wu, J. H. Kordower, L. Petrucelli and S. Przedborski, *Ann. Neurol.*, 2006, **60**(2), 260.

220. A. M. Landau, E. Kouassi, R. Siegrist-Johnstone and J. Desbarats, *Mov. Disord.*, 2007, **22**(3), 403.

221. H. Sawada, R. Kohno, T. Kihara, Y. Izumi, N. Sakka, M. Ibi, M. Nakanishi, T. Nakamizo, K. Yamakawa, H. Shibasaki, N. Yamamoto, A. Akaike, M. Inden, Y. Kitamura, T. Taniguchi and S. Shimohama, *J. Biol. Chem.*, 2004, **279**(11), 10710.

222. M. Inden, J. Kondo, Y. Kitamura, K. Takata, K. Nishimura, T. Taniguchi, H. Sawada and S. Shimohama, *J. Pharmacol. Sci.*, 2005, **97**(2), 203.

223. B. Alvarez-Castelao, I. Martin-Guerrero, A. Garcia-Orad and J. G. Castano, *J. Biol. Chem.*, 2009, **284**(41), 28253.
224. Y. V. Budovskaya, K. Wu, L. K. Southworth, M. Jiang, P. Tedesco, T. E. Johnson and S. K. Kim, *Cell*, 2008, **134**(2), 291.
225. W. Ye, K. Shimamura, J. L. Rubenstein, M. A. Hynes and A. Rosenthal, *Cell*, 1998, **93**(5), 755.
226. N. Prakash, C. Brodski, T. Naserke, E. Puelles, R. Gogoi, A. Hall, M. Panhuysen, D. Echevarria, L. Sussel, D. M. Weisenhorn, S. Martinez, E. Arenas, A. Simeone and W. Wurst, *Development*, 2006, **133**(1), 89.
227. M. P. Smidt and J. P. Burbach, *Nat. Rev. Neurosci.*, 2007, **8**(1), 21.
228. W. Lin, E. Metzakopian, Y. E. Mavromatakis, N. Gao, N. Balaskas, H. Sasaki, J. Briscoe, J. A. Whitsett, M. Goulding, K. H. Kaestner and S. L. Ang, *Dev. Biol.*, 2009, **333**(2), 386.
229. J. Nelander, J. B. Hebsgaard and M. Parmar, *Gene Expr. Patterns*, 2009, **9**(8), 555.
230. J. O. Andressoo and M. Saarma, *Curr. Opin. Neurobiol.*, 2008, **18**(3), 297.
231. O. Bergman, A. Hakansson, L. Westberg, A. C. Belin, O. Sydow, L. Olson, B. Holmberg, L. Fratiglioni, L. Backman, E. Eriksson and H. Nissbrandt, *J. Neural Transm.*, 2009, **116**(3), 333.
232. I. Rissling, K. Strauch, C. Hoft, W. H. Oertel and J. C. Moller, *Neurodegener. Dis.*, 2009, **6**(3), 102.
233. J. Fuchs, J. C. Mueller, P. Lichtner, C. Schulte, M. Munz, D. Berg, U. Wullner, T. Illig, M. Sharma and T. Gasser, *Neurobiol. Aging*, 2009, **30**(5), 731.
234. O. Bergman, A. Hakansson, L. Westberg, K. Nordenstrom, B. A. Carmine, O. Sydow, L. Olson, B. Holmberg, E. Eriksson and H. Nissbrandt, *Neurobiol. Aging*, 2010, **31**(1), 114.
235. W. D. Le, P. Xu, J. Jankovic, H. Jiang, S. H. Appel, R. G. Smith and D. K. Vassilatis, *Nat. Genet.*, 2003, **33**(1), 85.
236. P. Y. Xu, R. Liang, J. Jankovic, C. Hunter, Y. X. Zeng, T. Ashizawa, D. Lai and W. D. Le, *Neurology*, 2002, **58**(6), 881.
237. D. A. Grimes, F. Han, M. Panisset, L. Racacho, F. Xiao, R. Zou, K. Westaff and D. E. Bulman, *Mov. Disord.*, 2006, **21**(7), 906.
238. P. M. Sleiman, D. G. Healy, M. M. Muqit, Y. X. Yang, B. M. van der, J. L. Holton, T. Revesz, N. P. Quinn, K. Bhatia, J. K. Diss, A. J. Lees, M. R. Cookson, D. S. Latchman and N. W. Wood, *Neurosci. Lett.*, 2009, **457**(2), 75.
239. Y. Chu, W. Le, K. Kompoliti, J. Jankovic, E. J. Mufson and J. H. Kordower, *J. Comp. Neurol.*, 2006, **494**(3), 495.
240. W. Le, T. Pan, M. Huang, P. Xu, W. Xie, W. Zhu, X. Zhang, H. Deng and J. Jankovic, *J. Neurol. Sci.*, 2008, **273**(1–2), 29.
241. L. F. Lin, D. H. Doherty, J. D. Lile, S. Bektesh and F. Collins, *Science*, 1993, **260**(5111), 1130.
242. T. F. Oo, N. Kholodilov and R. E. Burke, *J. Neurosci.*, 2003, **23**(12), 5141.
243. M. M. Bespalov and M. Saarma, *Trends Pharmacol. Sci.*, 2007, **28**(2), 68.

244. Y. A. Sidorova, K. Matlik, M. Paveliev, M. Lindahl, E. Piranen, J. Milbrandt, U. Arumae, M. Saarma and M. M. Bespalov, *Mol. Cell Neurosci.*, 2010, **44**(3), 223.

245. J. Amiel, E. Sproat-Emison, M. Garcia-Barcelo, F. Lantieri, G. Burzynski, S. Borrego, A. Pelet, S. Arnold, X. Miao, P. Griseri, A. S. Brooks, G. Antinolo, L. de Pontual, M. Clement-Ziza, A. Munnich, C. Kashuk, K. West, K. K. Wong, S. Lyonnet, A. Chakravarti, P. K. Tam, I. Ceccherini, R. M. Hofstra and R. Fernandez, *J. Med. Genet.*, 2008, **45**(1), 1.

246. P. Petrova, A. Raibekas, J. Pevsner, N. Vigo, M. Anafi, M. K. Moore, A. E. Peaire, V. Shridhar, D. I. Smith, J. Kelly, Y. Durocher and J. W. Commissiong, *J. Mol. Neurosci.*, 2003, **20**(2), 173.

247. P. Lindholm, M. H. Voutilainen, J. Lauren, J. Peranen, V. M. Leppanen, J. O. Andressoo, M. Lindahl, S. Janhunen, N. Kalkkinen, T. Timmusk, R. K. Tuominen and M. Saarma, *Nature*, 2007, **448**(7149), 73.

248. M. Palgi, R. Lindstrom, J. Peranen, T. P. Piepponen, M. Saarma and T. I. Heino, *Proc. Natl Acad. Sci. U. S. A.*, 2009, **106**(7), 2429.

249. H. H. Simon, H. Saueressig, W. Wurst, M. D. Goulding and D. D. O'Leary, *J. Neurosci.*, 2001, **21**(9), 3126.

250. L. Alberi, P. Sgado and H. H. Simon, *Development*, 2004, **131**(13), 3229.

251. K. N. Alavian, P. Sgado, L. Alberi, S. Subramaniam and H. H. Simon, *Neural Dev.*, 2009, **4**, 11.

252. P. Sgado, L. Alberi, D. Gherbassi, S. L. Galasso, G. M. Ramakers, K. N. Alavian, M. P. Smidt, R. H. Dyck and H. H. Simon, *Proc. Natl. Acad. Sci. U. S. A.*, 2006, **103**(41), 15242.

253. P. Sgado, C. Viaggi, C. Fantacci and G. U. Corsini, *Parkinsonism. Relat. Disord.*, 2008, **14**(Suppl 2), S103–S106.

254. M. P. Smidt, H. S. van Schaick, C. Lanctot, J. J. Tremblay, J. J. Cox, A. A. van der Kleij, G. Wolterink, J. Drouin and J. P. Burbach, *Proc. Natl. Acad. Sci. U. S. A.*, 1997, **94**(24), 13305.

255. D. S. Varnum and L. C. Stevens, *J. Hered.*, 1968, **59**(2), 147.

256. E. V. Semina, R. S. Reiter and J. C. Murray, *Hum. Mol. Genet.*, 1997, **6**(12), 2109.

257. E. V. Semina, J. C. Murray, R. Reiter, R. F. Hrstka and J. Graw, *Hum. Mol. Genet.*, 2000, **9**(11), 1575.

258. I. Nunes, L. T. Tovmasian, R. M. Silva, R. E. Burke and S. P. Goff, *Proc. Natl. Acad. Sci. U. S. A.*, 2003, **100**(7), 4245.

259. M. P. Smidt, S. M. Smits, H. Bouwmeester, F. P. Hamers, A. J. van der Linden, A. J. Hellemons, J. Graw and J. P. Burbach, *Development*, 2004, **131**(5), 1145.

260. S. L. Maxwell, H. Y. Ho, E. Kuehner, S. Zhao and M. Li, *Dev. Biol.*, 2005, **282**(2), 467.

261. S. W. Law, O. M. Conneely, F. J. DeMayo and B. W. O'Malley, *Mol. Endocrinol.*, 1992, **6**(12), 2129.

262. T. Iwawaki, K. Kohno and K. Kobayashi, *Biochem. Biophys. Res. Commun.*, 2000, **274**(3), 590.

263. P. Sacchetti, T. R. Mitchell, J. G. Granneman and M. J. Bannon, *J. Neurochem.*, 2001, **76**(5), 1565.
264. E. Hermanson, B. Joseph, D. Castro, E. Lindqvist, P. Aarnisalo, A. Wallen, G. Benoit, B. Hengerer, L. Olson and T. Perlmann, *Exp. Cell Res.*, 2003, **288**(2), 324.
265. S. M. Smits, T. Ponnio, O. M. Conneely, J. P. Burbach and M. P. Smidt, *Eur. J. Neurosci.*, 2003, **18**(7), 1731.
266. R. H. Zetterstrom, R. Williams, T. Perlmann and L. Olson, *Brain Res. Mol. Brain Res.*, 1996, **41**(1–2), 111.
267. O. Saucedo-Cardenas and O. M. Conneely, *J. Mol. Neurosci.*, 1996, **7**(1), 51.
268. R. H. Zetterstrom, L. Solomin, L. Jansson, B. J. Hoffer, L. Olson and T. Perlmann, *Science*, 1997, **276**(5310), 248.
269. A. Wallen, R. H. Zetterstrom, L. Solomin, M. Arvidsson, L. Olson and T. Perlmann, *Exp. Cell Res.*, 1999, **253**(2), 737.
270. J. B. Eells, B. K. Lipska, S. K. Yeung, J. A. Misler and V. M. Nikodem, *Behav. Brain Res.*, 2002, **136**(1), 267.
271. B. Kadkhodaei, T. Ito, E. Joodmardi, B. Mattsson, C. Rouillard, M. Carta, S. Muramatsu, C. Sumi-Ichinose, T. Nomura, D. Metzger, P. Chambon, E. Lindqvist, N. G. Larsson, L. Olson, A. Bjorklund, H. Ichinose and T. Perlmann, *J. Neurosci.*, 2009, **29**(50), 15923.
272. S. O. Castillo, J. S. Baffi, M. Palkovits, D. S. Goldstein, I. J. Kopin, J. Witta, M. A. Magnuson and V. M. Nikodem, *Mol. Cell Neurosci.*, 1998, **11**(1–2), 36.
273. O. Saucedo-Cardenas, J. D. Quintana-Hau, W. D. Le, M. P. Smidt, J. J. Cox, F. De Mayo, J. P. Burbach and O. M. Conneely, *Proc. Natl. Acad. Sci. U. S. A.*, 1998, **95**(7), 4013.
274. J. Witta, J. S. Baffi, M. Palkovits, E. Mezey, S. O. Castillo and V. M. Nikodem, *Brain Res. Mol. Brain Res.*, 2000, **84**(1–2), 67.
275. J. B. Eells, B. K. Lipska, S. K. Yeung, J. A. Misler and V. M. Nikodem, *Behav. Brain Res.*, 2002, **136**(1), 267.
276. W. Le, O. M. Conneely, Y. He, J. Jankovic and S. H. Appel, *J. Neurochem.*, 1999, **73**(5), 2218.
277. T. Pan, W. Zhu, H. Zhao, H. Deng, W. Xie, J. Jankovic and W. Le, *Brain Res.*, 2008, **1222**, 222.
278. C. Jiang, X. Wan, Y. He, T. Pan, J. Jankovic and W. Le, *Exp. Neurol.*, 2005, **191**(1), 154.
279. M. S. Airaksinen and M. Saarma, *Nat. Rev. Neurosci.*, 2002, **3**(5), 383.
280. E. R. Kramer, L. Aron, G. M. Ramakers, S. Seitz, X. Zhuang, K. Beyer, M. P. Smidt and R. Klein, *PLoS Biol.*, 2007, **5**(3), e39.
281. W. C. Griffin III, H. A. Boger, A. C. Granholm and L. D. Middaugh, *Brain Res.*, 2006, **1068**(1), 257.
282. H. A. Boger, L. D. Middaugh, P. Huang, V. Zaman, A. C. Smith, B. J. Hoffer, A. C. Tomac and A. C. Granholm, *Exp. Neurol.*, 2006, **202**(2), 336.
283. H. A. Boger, L. D. Middaugh, V. Zaman, B. Hoffer and A. C. Granholm, *Brain Res.*, 2008, **1241**, 18.

284. A. Pascual, M. Hidalgo-Figueroa, J. I. Piruat, C. O. Pintado, R. Gomez-Diaz and J. Lopez-Barneo, *Nat. Neurosci.*, 2008, **11**(7), 755.

285. P. Lindholm, M. H. Voutilainen, J. Lauren, J. Peranen, V. M. Leppanen, J. O. Andressoo, M. Lindahl, S. Janhunen, N. Kalkkinen, T. Timmusk, R. K. Tuominen and M. Saarma, *Nature*, 2007, **448**(7149), 73.

286. M. H. Voutilainen, S. Back, E. Porsti, L. Toppinen, L. Lindgren, P. Lindholm, J. Peranen, M. Saarma and R. K. Tuominen, *J. Neurosci.*, 2009, **29**(30), 9651.

287. R. Banerjee, A. A. Starkov, M. F. Beal and B. Thomas, *Biochim. Biophys. Acta*, 2009, **1792**(7), 651.

288. A. H. Schapira, *Lancet Neurol.*, 2008, **7**(1), 97.

289. S. Papa, V. Petruzzella, S. Scacco, A. M. Sardanelli, A. Iuso, D. Panelli, R. Vitale, R. Trentadue, D. De Rasmo, N. Capitanio, C. Piccoli, F. Papa, M. Scivetti, E. Bertini, T. Rizza and G. De Michele, *Biochim. Biophys. Acta*, 2009, **1787**(5), 502.

290. A. Bender, K. J. Krishnan, C. M. Morris, G. A. Taylor, A. K. Reeve, R. H. Perry, E. Jaros, J. S. Hersheson, J. Betts, T. Klopstock, R. W. Taylor and D. M. Turnbull, *Nat. Genet.*, 2006, **38**(5), 515.

291. Y. Kraytsberg, E. Kudryavtseva, A. C. McKee, C. Geula, N. W. Kowall and K. Khrapko, *Nat. Genet.*, 2006, **38**(5), 518.

292. N. G. Larsson, *Annu. Rev. Biochem.*, 2010, **79**, 683.

293. G. Davidzon, P. Greene, M. Mancuso, K. J. Klos, J. E. Ahlskog, M. Hirano and S. DiMauro, *Ann. Neurol.*, 2006, **59**(5), 859.

294. C. S. Chan, J. N. Guzman, E. Ilijic, J. N. Mercer, C. Rick, T. Tkatch, G. E. Meredith and D. J. Surmeier, *Nature*, 2007, **447**(7148), 1081.

295. C. S. Chan, T. S. Gertler and D. J. Surmeier, *Trends Neurosci.*, 2009, **32**(5), 249.

296. B. Ritz, S. L. Rhodes, L. Qian, E. Schernhammer, J. H. Olsen and S. Friis, *Ann. Neurol.*, 2010, **67**(5), 600.

297. A. Trifunovic, A. Wredenberg, M. Falkenberg, J. N. Spelbrink, A. T. Rovio, C. E. Bruder, Y. Bohlooly, S. Gidlof, A. Oldfors, R. Wibom, J. Tornell, H. T. Jacobs and N. G. Larsson, *Nature*, 2004, **429**(6990), 417.

298. G. C. Kujoth, A. Hiona, T. D. Pugh, S. Someya, K. Panzer, S. E. Wohlgemuth, T. Hofer, A. Y. Seo, R. Sullivan, W. A. Jobling, J. D. Morrow, H. Van Remmen, J. M. Sedivy, T. Yamasoba, M. Tanokura, R. Weindruch, C. Leeuwenburgh and T. A. Prolla, *Science*, 2005, **309**(5733), 481.

299. A. Trifunovic, A. Hansson, A. Wredenberg, A. T. Rovio, E. Dufour, I. Khvorostov, J. N. Spelbrink, R. Wibom, H. T. Jacobs and N. G. Larsson, *Proc. Natl. Acad. Sci. U. S. A.*, 2005, **102**(50), 17993.

300. M. Vermulst, J. Wanagat, G. C. Kujoth, J. H. Bielas, P. S. Rabinovitch, T. A. Prolla and L. A. Loeb, *Nat. Genet.*, 2008, **40**(4), 392.

301. K. Inoue, K. Nakada, A. Ogura, K. Isobe, Y. Goto, I. Nonaka and J. I. Hayashi, *Nat. Genet.*, 2000, **26**(2), 176.

302. M. I. Ekstrand, M. Terzioglu, D. Galter, S. Zhu, C. Hofstetter, E. Lindqvist, S. Thams, A. Bergstrand, F. S. Hansson, A. Trifunovic,

B. Hoffer, S. Cullheim, A. H. Mohammed, L. Olson and N. G. Larsson, *Proc. Natl. Acad. Sci. U. S. A.*, 2007, **104**(4), 1325.
303. J. Wang, J. P. Silva, C. M. Gustafsson, P. Rustin and N. G. Larsson, *Proc. Natl. Acad. Sci. U. S. A.*, 2001, **98**(7), 4038.
304. L. A. Garraway and W. R. Sellers, *Nat. Rev. Cancer*, 2006, **6**(8), 593.
305. Y. Yang, A. H. Kim and A. Bonni, *Curr. Opin. Neurobiol.*, 2010, **20**(1), 92.
306. N. Papanas, E. Maltezos and D. P. Mikhailidis, *Expert Opin. Investig. Drugs*, 2010, **19**(8), 913.
307. M. Foretz, S. Hebrard, J. Leclerc, E. Zarrinpashneh, M. Soty, G. Mithieux, K. Sakamoto, F. Andreelli and B. Viollet, *J. Clin. Invest.*, 2010, **120**(7), 2355.
308. T. F. Outeiro and S. Lindquist, *Science*, 2003, **302**(5651), 1772.
309. A. A. Cooper, A. D. Gitler, A. Cashikar, C. M. Haynes, K. J. Hill, B. Bhullar, K. Liu, K. Xu, K. E. Strathearn, F. Liu, S. Cao, K. A. Caldwell, G. A. Caldwell, G. Marsischky, R. D. Kolodner, J. Labaer, J. C. Rochet, N. M. Bonini and S. Lindquist, *Science*, 2006, **313**(5785), 324.
310. A. D. Gitler, B. J. Bevis, J. Shorter, K. E. Strathearn, S. Hamamichi, L. J. Su, K. A. Caldwell, G. A. Caldwell, J. C. Rochet, J. M. McCaffery, C. Barlowe and S. Lindquist, *Proc. Natl. Acad. Sci. U. S. A.*, 2008, **105**(1), 145.
311. E. Yeger-Lotem, L. Riva, L. J. Su, A. D. Gitler, A. G. Cashikar, O. D. King, P. K. Auluck, M. L. Geddie, J. S. Valastyan, D. R. Karger, S. Lindquist and E. Fraenkel, *Nat. Genet.*, 2009, **41**, 316.
312. C. Malagelada, Z. H. Jin, V. Jackson-Lewis, S. Przedborski and L. A. Greene, *J. Neurosci.*, 2010, **30**(3), 1166.
313. A. Ghosh, A. Roy, J. Matras, S. Brahmachari, H. E. Gendelman and K. Pahan, *J. Neurosci.*, 2009, **29**(43), 13543.
314. A. O. Koob, K. Ubhi, J. F. Paulsson, J. Kelly, E. Rockenstein, M. Mante, A. Adame and E. Masliah, *Exp. Neurol.*, 2010, **221**(2), 267.
315. C. Becker and C. R. Meier, *Expert Opin. Drug Saf.*, 2009, **8**(3), 261.
316. R. Z. Orlowski and D. J. Kuhn, *Clin. Cancer Res.*, 2008, **14**(6), 1649.
317. J. Mitsui, I. Mizuta, A. Toyoda, R. Ashida, Y. Takahashi, J. Goto, Y. Fukuda, H. Date, A. Iwata, M. Yamamoto, N. Hattori, M. Murata, T. Toda and S. Tsuji, *Arch. Neurol.*, 2009, **66**(5), 571.
318. E. Sidransky, M. A. Nalls, J. O. Aasly, J. Aharon-Peretz, G. Annesi, E. R. Barbosa, A. Bar-Shira, D. Berg, J. Bras, A. Brice, C. M. Chen, L. N. Clark, C. Condroyer, E. V. De Marco, A. Durr, M. J. Eblan, S. Fahn, M. J. Farrer, H. C. Fung, Z. Gan-Or, T. Gasser, R. Gershoni-Baruch, N. Giladi, A. Griffith, T. Gurevich, C. Januario, P. Kropp, A. E. Lang, G. J. Lee-Chen, S. Lesage, K. Marder, I. F. Mata, A. Mirelman, J. Mitsui, I. Mizuta, G. Nicoletti, C. Oliveira, R. Ottman, A. Orr-Urtreger, L. V. Pereira, A. Quattrone, E. Rogaeva, A. Rolfs, H. Rosenbaum, R. Rozenberg, A. Samii, T. Samaddar, C. Schulte, M. Sharma,

A. Singleton, M. Spitz, E. K. Tan, N. Tayebi, T. Toda, A. R. Troiano, S. Tsuji, M. Wittstock, T. G. Wolfsberg, Y. R. Wu, C. P. Zabetian, Y. Zhao and S. G. Ziegler, *N. Engl. J. Med.*, 2009, **361**(17), 1651.

319. R. G. Urdinguio, J. V. Sanchez-Mut and M. Esteller, *Lancet Neurol.*, 2009, **8**(11), 1056.

320. V. Davalos and M. Esteller, *Curr. Opin. Oncol.*, 2010, **22**(1), 35.

321. R. H. Kim, M. Peters, Y. Jang, W. Shi, M. Pintilie, G. C. Fletcher, C. DeLuca, J. Liepa, L. Zhou, B. Snow, R. C. Binari, A. S. Manoukian, M. R. Bray, F. F. Liu, M. S. Tsao and T. W. Mak, *Cancer Cell*, 2005, **7**(3), 263.

322. S. Vasseur, S. Afzal, J. Tardivel-Lacombe, D. S. Park, J. L. Iovanna and T. W. Mak, *Proc. Natl. Acad. Sci. U. S. A.*, 2009, **106**(4), 1111.

323. S. Veeriah, B. S. Taylor, S. Meng, F. Fang, E. Yilmaz, I. Vivanco, M. Janakiraman, N. Schultz, A. J. Hanrahan, W. Pao, M. Ladanyi, C. Sander, A. Heguy, E. C. Holland, P. B. Paty, P. S. Mischel, L. Liau, T. F. Cloughesy, I. K. Mellinghoff, D. B. Solit and T. A. Chan, *Nat. Genet.*, 2010, **42**(1), 77.

324. K. W. Kinzler and B. Vogelstein, *Cell*, 1996, **87**(2), 159.

325. K. W. Caldecott, *Nat. Rev. Genet.*, 2008, **9**(8), 619.

326. G. A. Garinis, G. T. van der Horst, J. Vijg and J. H. Hoeijmakers, *Nat. Cell Biol.*, 2008, **10**(11), 1241.

327. M. Biswas and J. Y. Chan, *Toxicol. Appl. Pharmacol.*, 2010, **244**(1), 16.

328. T. W. Kensler, N. Wakabayashi and S. Biswal, *Annu. Rev. Pharmacol. Toxicol.*, 2007, **47**, 89.

329. C. K. Glass, K. Saijo, B. Winner, M. C. Marchetto and F. H. Gage, *Cell*, 2010, **140**(6), 918.

330. J. M. Itier, P. Ibanez, M. A. Mena, N. Abbas, C. Cohen-Salmon, G. A. Bohme, M. Laville, J. Pratt, O. Corti, L. Pradier, G. Ret, C. Joubert, M. Periquet, F. Araujo, J. Negroni, M. J. Casarejos, S. Canals, R. Solano, A. Serrano, E. Gallego, M. Sanchez, P. Denefle, J. Benavides, G. Tremp, T. A. Rooney, A. Brice and D. Y. Garcia, *Hum. Mol. Genet.*, 2003, **12**(18), 2277.

331. H. Ilieva, M. Polymenidou and D. W. Cleveland, *J. Cell Biol.*, 2009, **187**(6), 761.

332. S. Nuber, E. Petrasch-Parwez, B. Winner, J. Winkler, S. von Horsten, T. Schmidt, J. Boy, M. Kuhn, H. P. Nguyen, P. Teismann, J. B. Schulz, M. Neumann, B. J. Pichler, G. Reischl, C. Holzmann, I. Schmitt, A. Bornemann, W. Kuhn, F. Zimmermann, A. Servadio and O. Riess, *J. Neurosci.*, 2008, **28**(10), 2471.

333. A. Yamamoto, J. J. Lucas and R. Hen, *Cell*, 2000, **101**(1), 57.

334. A. Sailer, H. Bueler, M. Fischer, A. Aguzzi and C. Weissmann, *Cell*, 1994, **77**(7), 967.

335. L. Goldschmidt, P. K. Teng, R. Riek and D. Eisenberg, *Proc. Natl. Acad. Sci. U. S. A.*, 2010, **107**(8), 3487.

336. T. J. van Ham, M. A. Holmberg, A. T. van der Goot, E. Teuling, M. Garcia-Arencibia, H. E. Kim, D. Du, K. L. Thijssen, M. Wiersma,

R. Burggraaff, P. van Bergeijk, J. van Rheenen, V. Jerre, V. R. M. Hofstra, D. C. Rubinsztein and E. A. Nollen, *Cell*, 2010, **142**(4), 601.
337. B. Frost and M. I. Diamond, *Nat. Rev. Neurosci.*, 2010, **11**(3), 155.
338. H. Braak, R. A. De Vos, J. Bohl and K. Del Tredici, *Neurosci. Lett.*, 2006, **396**(1), 67.
339. F. Marxreiter, S. Nuber, M. Kandasamy, J. Klucken, R. Aigner, R. Burgmayer, S. Couillard-Despres, O. Riess, J. Winkler and B. Winner, *Eur. J. Neurosci.*, 2009, **29**(5), 879.
340. E. Masliah, E. Rockenstein, A. Adame, M. Alford, L. Crews, M. Hashimoto, P. Seubert, M. Lee, J. Goldstein, T. Chilcote, D. Games and D. Schenk, *Neuron*, 2005, **46**(6), 857.

CHAPTER 8

Neuroprotection in Parkinson's Disease

ALBERTO PASCUAL, JAVIER VILLADIEGO, MARÍA HIDALGO-FIGUEROA, SIMÓN MÉNDEZ-FERRER, RAQUEL GÓMEZ-DÍAZ, JUAN JOSÉ TOLEDO-ARAL AND JOSÉ LOPEZ-BARNEO

Instituto de Biomedicina de Sevilla, Hospital Universitario Virgen del Rocio/CSIC/Universidad de Sevilla, Sevilla, Spain; Centro de Investigación Biomédica en Red sobre Enfermedades Neurodegenerativas (CIBERNED)

8.1 Neurotrophism and Neurotrophic Factors

Neurotrophic factors (NTFs) are small natural proteins necessary for the development and survival of nerve cells as well as for the maintenance of their morphological and functional phenotype. The 'neurotrophic hypothesis' enunciates that the establishment and maintenance of neuronal networks require the release at the target structures of NTFs, which are taken up by the nerve terminals and retrogradely transported to the soma of the projecting neurons. On reaching the nucleus, NTFs induce a gene programme that promotes neuronal survival and maintenance of phenotype.

Although the existence of 'chemotactic' influences between the growth cones of axons and their targets had already been postulated by Cajal's group,[1] the modern concept of neurotrophism is based on the work of Hamburguer and Levi-Montalcini[2] who reported that the phenomenon of naturally occurring

RSC Drug Discovery Series No. 6
Animal Models for Neurodegenerative Disease
Edited by Jesús Avila, Jose J. Lucas and Félix Hernández
© Royal Society of Chemistry 2011
Published by the Royal Society of Chemistry, www.rsc.org

cell death observed during development was dependent on the target where these dying neurons were projecting. Removing a prospective target early in development could dramatically increase neuronal loss. In the absence of target, the number of initially existing neurons did not change, indicating that the target influences survival and not the number of neurons generated. These seminal observations suggested that cells acting as a target of developing neurons produce limited amount of specific molecules which are required for their survival. The first molecule identified with these specific characteristics was the nerve growth factor (NGF).[3,4] As indicated above, the current concept of NTF applies not only to molecules regulating neuronal number during development but also to agents necessary for the maintenance of neuronal populations in adulthood.[5,6]

During the last decades, several proteins have been classified as NTFs because their effect on neuronal survival, differentiation (including synapto-genesis and neurite branching *in vitro*), maturation of electrophysiological properties, and plasticity. However, the role of these molecules as 'canonical' NTFs is not always well demonstrated. This is true even in the case of NGF (the 'prototypical NTF'), which is essential for the survival of sympathetic, sensory and central cholinergic neurons. Nevertheless, the dependence of these neuronal populations on NGF is not always conserved by the end of the developmental critical period. Indeed, some postnatal neurons (as it is the case of a subset of DRG neurons) can switch dependence from one (NGF) to another (glial cell line-derived neurotrophic factor) trophic factor.[7,8] Moreover, although deprivation of NGF in the adult compromises sympathetic neurons viability, it only affects gene expression in sensory neurons without decreasing their cell number.[9] However, it has not been definitely established whether, in the adult nervous system, neurons depend on one or several trophic factors with overlapping effects.

Because of their potent role in neuronal survival, NTFs have aroused clinical interest as potential neuroprotective agents that could prevent or retard the progression of neurodegenerative diseases. In this chapter we summarize current knowledge on the role of NTFs in the pathogenesis of Parkinson's disease (PD), emphasizing the data obtained from animal models of NTF deficiency. We also discuss the possible clinical applicability of NTFs as neuroprotective agents in PD.

8.2 Neuroprotection of Mesencephalic Dopaminergic Neurons: Role of GDNF

Parkinson's disease (PD), a neurodegenerative disorder that affects over one million Europeans,[10,11] is characterized by motor symptoms (tremor, brady-kinesia, rigidity and alteration of gait).[12,13] The aetiology and pathogenesis of PD are essentially unknown, although several causative mechanisms have been proposed including alterations of protein folding/degradation, mitochondrial dysfunction and oxidative stress, neuroinflammation, and Ca^{2+} excitotoxicity.[14,15]

PD is caused by the progressive loss of specific sets of neurons both in the brainstem and in the peripheral nervous system. From a clinical standpoint, the most critical neuronal population affected corresponds to the mesencephalic dopaminergic (DA) neurons in the substantia nigra (SN) pars compacta projecting to the striatum (nigrostriatal neurons), thus leading to dysfunction of the neuronal circuits in the basal ganglia and alteration of motor control. DA neurons in the neighbouring ventral tegmental area (VTA) are less affected than SN neurons. Although the loss of nigrostriatal DA neurons is the most apparent pathological hallmark of PD, other cell types are affected even before SN cell death. Among those are noradrenergic neurons in the locus coeruleus (LC) and cells in the dorsal nucleus of the vagus or in the sympathetic ganglia. Peripheral sympathetic denervation (loss of cardiac or celiac fibres) has been proposed to be an early marker for PD.[16]

For the last decades, intrastriatal transplantation of dopamine-producing cells (most frequently foetal mesencephalic DA neurons) has been considered as an experimental therapeutic approach to advanced PD once pharmaceutical drugs have ceased to provide clinical benefit.[17–19] However, allogenic cell replacement therapies have recently been almost completely abandoned (for discussion, see refs. 20,21) due to the scarcity of tissue available for transplantation and because they do not always produce the expected beneficial effects (and in some cases can even induce the appearance of abnormal movements called dyskinesias[22,23]). Recently, intrastriatal delivery of NTFs, which could 'protect' nigrostriatal neurons and thus halt or retard PD progression, is being considered as an alternative therapeutic strategy to dopamine cell replacement.[24] The prototypical 'neuroprotective' agent used in most preclinical and clinical studies is glial cell line-derived neurotrophic factor (GDNF), which after isolation demonstrated a remarkable trophic effect on mesencephalic DA neurons *in vitro*.[25]

8.2.1   Biology of GDNF and Other Dopaminotrophic Factors

GDNF belongs to a family of ligands, which include artemin (ARTN), neurturin (NRTN) and persephin (PSPN), all being distantly related members of the transforming growth factor-beta superfamily. GDNF has attracted special attention due to its potent effect on dopaminergic and noradrenergic neuron survival[26,27] (see ref. 28 for a comprehensive review). Collectively, the members of the GDNF family of trophic factors are often grouped as 'dopaminotrophic' factors. GDNF is expressed in several regions of the adult brain (particularly in the striatum, anteroventral thalamus and septum)[29,30] and signals through an extracellular GPI-anchored receptor (GFRα1) that activates a tyrosine kinase transmembrane protein (c-ret).[26,31] A non-canonical form of GDNF signalling also exists in the adult rodent brain, which is independent of Ret and is mediated by neural cell adhesion molecule (NCAM).[32] Like other NTFs, GDNF is taken up by axon terminals of projecting neurons and transported to cell soma (for a review on retrograde transport, see ref. 33). Injection of

^{125}I-GDNF in striatum results in labelled cells in the ipsilateral SN and VTA, thus suggesting a trophic role of GNDF in adult nigrostriatal neurons.[34]

GDNF has been shown to activate pathways associated with the promotion of antioxidant defence[35] and neuronal survival.[36] Among these pathways, the PI3K/AKT cascade[37] can protect neurons through several mechanisms including inactivation of apoptotic proteins.[38,39] GDNF overexpression (using lentiviral infection or engineered GDNF-producing cells) protects catecholaminergic neurons from toxic damage and induces fibre outgrowth *in vivo*.[26,27,40,41] However, the molecular mechanisms underlying these functional effects are still unknown. Recently, it has been reported that lentiviral GDNF delivered to the rat striatum induces gene expression in the SN, notably tyrosine hydroxylase (TH), GTP cyclohydrolase-I (which catalyses the synthesis of a cofactor of TH, tetrahydrobiopterin), GDNF receptors, and Dlk-1 (a factor involved in cell proliferation and differentiation). Since GDNF receptors are located in terminals of SN neurons, it is expected that these genes are upregulated in DA SN neurons in response to GDNF activation.[42] In any event, GDNF-dependent signalling pathways are as yet poorly studied and the role of GDNF in the maintenance of adult DA neurons remains essentially unknown.

8.2.2 Dopaminotrophic Factor-Mediated Neuronal Protection

As indicated above, GDNF has a well-recognized potent neurotrophic effect on DA neurons *in vitro*. Addition of GDNF to primary cultures of midbrain neurons favours the survival of the culture, and increases dopamine uptake, cell size and neurite length.[25] It has been demonstrated that GDNF exerts a protective role on DA neurons exposed to neurotoxic agents.[14] GDNF increases the survival of mesencephalic DA neurons treated with either 1-methyl-4-phenylpyridinium (MPP$^+$) or 6-hydroxydopamine (6-OHDA) and promotes the regrowth of damaged dopaminergic fibres.[43] A neuroprotective effect of GDNF against MPP$^+$ toxicity has been also shown in organotypic cultures of ventral mesencephalon, a preparation in which the integrity of DA neurons is better preserved than in enzymatically dispersed preparations.[44]

As it occurs with GDNF, the other trophic factors members of the GDNF family, NRTN, ARTN and PSPN, also increase the survival of DA neurons *in vitro* and show a neuroprotective effect on cells treated with 6-OHDA, although they seem to be less efficient than GDNF. The neuroprotective action of NRNT[45] and ARTN[46] is mediated by cross-reactive stimulation of the canonical GDNF receptor GFRα1-Ret that is expressed on DA neurons. Curiously, PSPN was believed to be unable to stimulate the GFRα1-Ret receptor[47] so the mechanism by which it exerted the protective action on dopaminergic neurons was uncertain. However, it has have been recently shown that PSPN can activate the GFRα1-Ret receptor as well,[48] explaining the protective effect that PSPN exerts on DA neurons.

Among the different NTFs tested *in vivo*, GDNF has shown the most potent and robust effect on animal (rodent and primate) models of parkinsonism.[28]

In the initial *in vivo* studies, the direct infusion of recombinant protein was used to deliver GDNF to the brain parenchyma. These experiments demonstrated that GDNF protects nigral DA neurons by activating their metabolism and dopamine turnover. GDNF also induced axonal sprouting and striatal reinnervation.[49]

As direct intrastriatal infusion of GDNF has potential complications derived from the chronically implanted infusion device or the diffusion of the trophic factor, an alternative approach for brain GDNF delivery is the administration of replication-deficient viral vectors. Three viral vector systems (adenovirus, adeno-associated virus and lentivirus) engineered to produce continuous expression of GDNF have been used with good experimental results.[40,50,51] The transfection based on adeno-associated virus and lentivirus is particularly interesting because it produces a long-term expression of GDNF without detectable cellular pathology or immune reaction (for a review, see ref. 28). However, the use of viral vectors in patients raises safety concerns because of their potential immunogenicity and the risk of mutagenesis by insertion into the genome of host cells. Another strategy to produce continuous intrastriatal GDNF delivery is the use of genetically modified cells that synthesize and release GDNF. In this regard, several groups have reported protective effects of GDNF-secreting genetically modified cells in animal models of PD.[52–55] The long-term survival of the engineered cells and the prevention of the immune reaction that they can trigger are two of the major hurdles that this technology must overcome before it can be considered for clinical application.[56]

Intrastriatal grafting of carotid body (CB) cells is a methodology developed in our laboratory that also seems to produce GDNF-mediated neuroprotection in PD.[57,58] The CB, a bilateral organ located in the bifurcation of the carotid artery, is a major arterial chemoreceptor organ responsible for the detection of changes in O_2 concentration in the blood. In conditions presenting hypoxemia, CB sensory cells release transmitters that activate afferent sensory fibres terminating in the respiratory centre to induce a compensatory hyperventilatory response.[59] CB sensory cells (called glomus cells) are of neural-crest origin and contain high levels of dopamine as well as GDNF (Figure 8.1)[60,61] and are thus well-suited to be used as donor tissue in transplantation studies in PD. Indeed, intrastriatal CB grafts can produce a significant histological and functional recovery in rodent and primate models of parkinsonism[57,58,60] (Figure 8.2). It has been shown that the benefit induced by rat CB grafts is mainly due to a trophic effect on the nigrostriatal neurons rather than to the release of dopamine from the transplanted cells.[60] Once in the brain, grafted CB cells remain metabolically active during the entire animal lifespan and maintain their ability to produce GDNF (Figures 8.2A and 8.2B).[61] The reason for the long-lasting cell survival observed in CB transplants might be related to the fact that glomus cells are activated by hypoxia, an environmental condition presumed to be present inside the graft that is possibly deleterious for other cells types. Another advantage for the clinical application of CB cell therapy is that it allows autografts to be performed in PD patients, since unilateral CB surgical resection has no significant side effects in humans.[62]

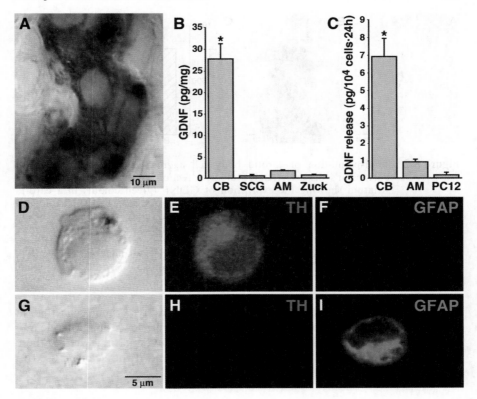

Figure 8.1 Synthesis and release of GDNF in dopaminergic carotid body glomus cells. (**A**) Histological section of mouse carotid body tissue showing the expression of GDNF (green colour aggregasomes, X-gal staining) in tyrosine hydroxylase (TH) positive glomus cells (brown colour). (**B**) and (**C**) Measurement of GDNF content (**B**) and release to the medium (**C**) in cultures of rat carotid body (CB), superior cervical ganglion (SCG), adrenal medulla (AM), organ of Zuckerkandl (Zuck) and PC12 cells. (**D–I**) Immunohistochemical analysis showing that the GDNF promoter is active in mouse type I carotid body cells (TH positive and glial fibrillary acidic protein (GFAP) negative cells). Modified from ref. 61. Animals GDNF/ LacZ were used to study activity of the GFAP promoter (see refs. 60,71).

Two phase I/II clinical trials have shown that CB autotransplantation induces clinical benefits in PD patients, particularly in those who are not in an advanced stage of the disease.[63,64] The CB is small organ and therefore a limitation of CB-based cell therapy is the scarcity of tissue available for transplantation. To overcome this obstacle we have started a programme to expand CB cells *in vitro*. It is well-known that, in conditions of chronic hypoxemia (*i.e.* in high altitude residents or in patients with chronic obstructive pulmonary diseases), the CB undergoes a compensatory hypertrophy.[65–67] This led us to hypothesize that adult mammalian CB could contain latent neural progenitors activated by low O_2 tension. In accord with this proposal, we have

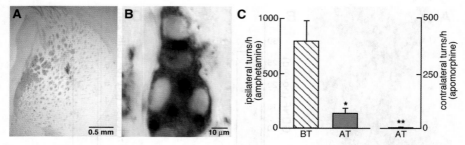

Figure 8.2 Transplantation of carotid body cell aggregates in rodent models of
Parkinson's disease. (**A**) Histological section at the level of the mouse
striatum showing the expression of GDNF (green colour) in a carotid
body transplant from a GDNF/LacZ mouse (see ref. 71). (**B**) Histological
section of a striatum containing a carotid body transplant. Cells in
the graft show the expression of GDNF (green colour aggregasomes,
X-gal staining) in tyrosine hydroxylase (TH) positive glomus cells (brown
colour). (**C**) Rotational behavior of a hemiparkinsonian rat before (BT)
and after (AT) intrastriatal transplantation of carotid body tissue.
Modified from refs. 60,61.

identified in the CB a population of neural crest-derived stem cells that in
response to hypoxia can proliferate and differentiate in new dopaminergic and
GDNF-producing glomus cells.[68] We are currently investigating whether these
newly identified progenitors are active in humans and if they can be used to
expand CB tissue before transplantation.

8.3 Genetic Models of NTF Depletion: Conditional GDNF Knock-out Mice

Most of the knowledge available on the physiological function of GDNF has
come from the analysis of genetically modified mice models. It is more than 15
years since three groups independently showed that ablation of the GDNF gene
($Gdnf^{-/-}$ mice) results in animal death after birth due to renal agenesis and the
absence of then enteric plexus.[69–71] Unexpectedly, the $Gdnf^{-/-}$ mice had an
apparently normal number and organization of mesencephalic DA neurons.
These observations suggested that the trophic dependence of nigrostriatal
neurons on GDNF, supported by the pharmacological experiments (exogenous
administration of the trophic factor), must be acquired during postnatal
maturation. Heterozygous $Gdnf$ mice are fertile and develop normally,
although they manifest an accelerated decline in spontaneous motor activity
and coordination with age.[72] Nevertheless, this embryonic GDNF deficit seems
to have little impact on the adult, since at 20 months of age, the mice show only
a 15% decrease of TH-positive SN neurons and no difference in striatal TH^{+}
fibre density with respect to controls.[72] $Gdnf^{+/-}$ mice, however, seemed to show
a higher susceptibility to neurotoxin-induced long-term degeneration of
monoaminergic neurons than wild-type littermates.[73]

Region-specific genetic deletion (driven by the promoter of the dopamine transporter gene) of Ret (the canonical GDNF receptor) in DA neurons has provided conflicting results regarding the role of this pathway in the maintenance of adult neurons. No differences in adult DA nigrostriatal neurons have been shown in Ret-null mice *versus* controls, as determined by comparative morphometric and biochemical analysis.[31] However, another group has reported that embryonic deletion of Ret in catecholaminergic neurons results in a significant decrease of TH$^+$ SN neurons and striatal nerve terminals of aged mice although, unexpectedly, neurons in the VTA and LC remain unaffected.[74] The variable results obtained with the regional Ret-null mouse might be related to the fact that, besides the Grfα-1/Ret pathway, GDNF can signal through 'non canonical' N-CAM receptors,[32] that may compensate for the absence of Ret.

We have recently reported the generation of a conditional GDNF-null mouse in which GDNF expression was markedly reduced in adulthood.[29] This model avoids the establishment of developmental compensatory modifications, which could mask the true physiologic action of GDNF in the adult nervous system. The conditional GDNF deficient mice show selective and extensive catecholaminergic neuronal death, most notably in the LC, SN and VTA (Figure 8.3). GABAergic and cholinergic pathways appear to be unaffected. The neurochemical and histological alterations in GDNF-deprived mice induce the appearance of behavioural motor disturbances characterized by a progressive akinetic syndrome (Figure 8.4). These data have demonstrated that endogenous GDNF is absolutely required for trophic maintenance of mesencephalic

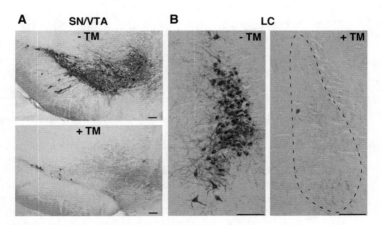

Figure 8.3   Mesencephalic catecholaminergic neuronal death in the adult conditional GDNF knout out mouse. (**A**) Mesencephalic dopaminergic neurons in substantia nigra (SN) and ventral tegmental area (VTA) of a conditional GDNF-null animal before (−TM) and seven months after (+TM) deletion of the GDNF gene. (**B**) Same experiment showing the disappearance of cells in the locus coeruleus (LC) after GDNF depletion. Modified from ref. 29.

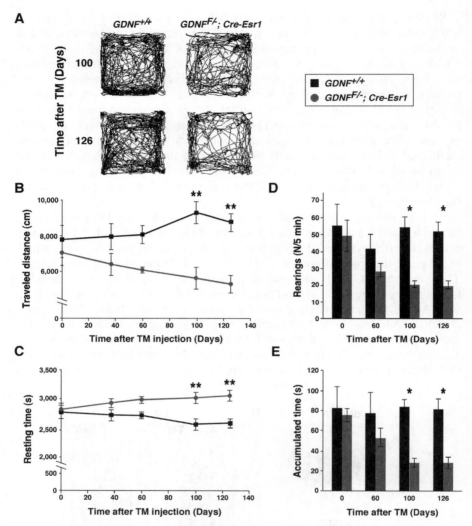

Figure 8.4   Motor abnormalities in the adult conditional GDNF knock-out mouse parkinsonian model. Spontaneous activity (animals were recorded during 60 min
in an open field chamber) measured in wild-type (black bars: $GDNF^{+/+}$; +/+
mice treated with TM; $n = 13$) and GDNF depleted animals (grey bars:
$GDNF^{F/-}$; *Cre-Esr1* mice treated with TM; $n = 7$) three days before and 37,
60, 100, and 126 days after TM injection. (**A**) Four traces from wild-type and
GDNF-depleted animals are shown as representative of the time points 100
and 126 days after TM injection. Activity trace during minutes 16–30 is presented. (**B**) Travelled distance (cm) was calculated by following the centre of
gravity of the subject. (**C**) Resting time (s) was the time spent in resting state
(with reference to the default velocity threshold of 2.57 cm/s). (**D,E**) Vertical
movements quantified in three periods of five minutes from each animal and
time point. Averaged readings (**D**) and the accumulated time spent with both
forepaws without contacting the floor (**E**) are plotted. Each individual point
represents 3–8 animals. One-way ANOVA followed by the Tukey test. Means
± S.E.M.; *, $p < 0.05$; **, $p < 0.01$. Modified from ref. 29.

dopaminergic and noradrenergic neurons. The GDNF-deficient mouse is a well-defined model in which to study neuroprotection in experimental PD. It remains to be investigated in the future which essential GDNF-controlled targets are required for mammalian SN, VTA and LC neuronal survival.

8.4 Clinical Effects of GDNF

The good results obtained in the preclinical studies on the neuroprotective role of GDNF have stimulated the development of clinical studies designed to test the therapeutic effects of GDNF in advanced PD patients. Besides the clinical studies with CB tissue referred to above, several clinical trials have been performed using direct intracerebral infusion of GDNF.

In a controlled clinical trial, monthly intraventricular GDNF injection failed to provide clinical benefit in advanced PD patients and instead resulted in frequent adverse events.[75] A postmortem examination in one patient suggested that GDNF did not reach the target cells *via* this route. However, encouraging clinical and neurochemical results were observed with continuous intraputaminal GDNF infusion on PD patients in two independent open-label clinical trials. One of the trials performed on five PD patients reported beneficial clinical outcomes at one year, while [^{18}F]-dopa PET studies showed an increase in putaminal uptake around the tip of each catheter.[76] The second study on 10 patients using a different delivery protocol also reported positive results at six months.[77] However, a randomized placebo-controlled trial involving 34 PD patients showed no significant clinical differences between groups at six months despite increased [^{18}F]-dopa uptake in the recombinant GDNF-treated group.[78] The open-label extension of this study was halted due to safety concerns: three patients developed neutralizing antibodies, which could potentially cross-react with endogenous GDNF, while in a parallel toxicology study some monkeys developed cerebellar damage. Besides GDNF, other members of the same protein family (particularly neurturin) are also being assayed in pilot clinical trials with yet inconclusive results.

8.5 Conclusions and Perspectives

NTFs exert a potent effect on the survival and maintenance of phenotype in adult neurons. Therefore intracerebral administration of these factors is a promising therapeutic strategy in neurodegenerative disorders, such as PD, presenting progressive neuronal death. There is a vast scientific literature supporting the neuroprotective role of exogenous GDNF on the nigrostriatal pathway. However, most of the clinical trials performed to test the efficacy of NTF-based therapies in advanced PD patients have been quite discouraging. The generation of the conditional GDNF-null mouse model has recently allowed us to show the absolute requirement of GDNF for the survival of dopaminergic and noradrenergic mesencephalic neurons in adult brain. These data unequivocally demonstrate a major physiological neuroprotective effect of GDNF and thus it should revive interest in GDNF-based therapies.

Clinical application of NTFs is confronted with several technological and scientific challenges that should be addressed in future preclinical and clinical research. Before new clinical trials are performed, a safe and efficacious route of GDNF delivery (produced in cells, purified, or encoded in viral vectors) must be clearly established. In this regard, diffusion of GDNF in the brain parenchyma and the appropriate concentration of GDNF delivered to cells are critical issues that might determine the clinical outcome. It must be also investigated whether the administration of appropriate cocktails of several trophic factors offer advantages over the use of GDNF alone. Besides these technologically oriented studies, much research should be done to unravel the actual physiological role of NTFs and their molecular mechanism of action on adult neurons. This work might eventually lead to the identification of new signalling pathways that will provide targets accessible to small molecules amenable for their use as pharmaceutical drugs. The molecular physiology and pharmacology of neuroprotection are still in their infancy. Therefore, it can be presumed that the development of these fields will surely offer new opportunities for a more effective fight against PD and other neurodegenerative diseases.

Acknowledgements

Research in the authors' laboratories is supported by grants from the Marcelino Botin Foundation, the Ministry of Science and Innovation, and the Andalusian Government. Authors are members of TERCEL (ISCIII and FEDER).

References

1. S. R. Cajal, *Degeneration and regeneration of the nervous system*, Oxford University Press: London, 1928.
2. V. Hamburger and R. Levi-Montalcini, *J. Exp. Zool.*, 1949, **111**, 457.
3. S. Cohen, *Proc. Natl. Acad. Sci. USA*, 1960, **46**, 302.
4. R. Levi-Montalcini, *Science*, 1987, **237**, 1154.
5. E. J. Huang and L. F. Reichardt, *Annu. Rev. Neurosci.*, 2001, **24**, 677.
6. B. Lu, P. T. Pang and N. H. Woo, *Nat. Rev. Neurosci.*, 2005, **6**, 603.
7. W. Luo, S. R. Wickramasinghe, J. M. Savitt, J. W. Griffin, T. M. Dawson and D. D. Ginty, *Neuron*, 2007, **54**, 739.
8. D. C. Molliver, D. E. Wright, M. L. Leitner, A. S. Parsadanian, K. Doster, D. Wen, Q. Yan and W. D. Snider, *Neuron*, 1997, **19**, 849.
9. P. D. Gorin and E. M. Johnson Jr., *Brain Res.*, 1980, **198**, 27.
10. L. M. de Lau and M. M. Breteler, *Lancet Neurol.*, 2006, **5**, 525.
11. M. C. de Rijk, C. Tzourio, M. M. Breteler, J. F. Dartigues, L. Amaducci, S. Lopez-Pousa, J. M. Manubens-Bertran, A. Alpérovitch and W. A. Rocca, *J. Neurol. Neurosurg. Psychiatry.*, 1997, **62**, 10.

12. S. Fahn, *Ann. N. Y. Acad. Sci.*, 2003, **991**, 1.
13. A. E. Lang and A. M. Lozano, *N. Engl. J. Med.*, 1998, **339**, 1130.
14. W. Dauer and S. Przedborski, *Neuron*, 2003, **39**, 889.
15. M. J. Farrer, *Nat. Rev. Genet.*, 2006, **7**, 306.
16. H. Braak, K. Del Tredici, U. Rüb, R. A. de Vos, E. N. Jansen Steur and E. Braak, *Neurobiol. Aging*, 2003, **24**, 197.
17. C. R. Freed, R. E. Breeze, N. L. Rosenberg, S. A. Schneck, E. Kriek, J.-X. Qi, T. Lone, Y.-B. Zhang, J. A. Snyder, T. H. Wells, L. Olson Ramig, L. Thompson, J. C. Mazziotta, S. C. Huang, S. T. Grafton, D. Brooks, G. Sawle, G. Schroter and A. A. Ansari, *N. Engl. J. Med.*, 1992, **327**, 1549.
18. O. Lindvall, P. Brundin, H. Widner, S. Rehncrona, B. Gustavii, R. Frackowiak, K. L. Leenders, G. Sawle, J. C. Rothwell, C. D. Marsden and A. Björklund, *Science*, 1990, **247**, 574.
19. P. Piccini, D. J. Brooks, A. Björklund, R. N. Gunn, P. M. Grasby, O. Rimoldi, P. Brundin, P. Hagell, S. Rehncrona, H. Widner and O. Lindvall, *Nat. Neurosci.*, 1999, **2**, 1137.
20. C. W. Olanow, J. H. Kordower, A. E. Lang and J. A. Obeso, *Ann. Neurol.*, 2009, **66**, 591.
21. B. J. Snyder and C. W. Olanow, *Curr. Opin. Neurol.*, 2005, **18**, 376.
22. C. R. Freed, P. E. Greene, R. E. Breeze, W. Y. Tsai, W. DuMouchel, R. Kao, S. Dillon, H. Winfield, S. Culver, J. Q. Trojanowski, D. Eidelberg and S. Fahn, *N. Engl. J. Med.*, 2001, **344**, 710.
23. C. W. Olanow, C. G. Goet, J. H. Kordower, A. J. Stoessl, V. Sossi, M. F. Brin, K. M. Shannon, G. M. Nauert, D. P. Perl, J. Godbold and T. B. Freeman, *Ann. Neurol.*, 2003, **54**, 403.
24. T. Deierborg, D. Soulet, L. Roybon, V. Hall and P. Brundin, *Prog. Neurobiol.*, 2008, **85**, 407.
25. L. F. Lin, D. H. Doherty, J. D. Lile, S. Bektesh and F. Collins, *Science*, 1993, **260**, 1130.
26. E. Arenas, M. Trupp, P. Akerud and C. F. Ibanez, *Neuron*, 1995, **15**, 1465.
27. D. M. Gash, Z. Zhang, A. Ovadia, W. A. Cass, A. Yi, L. Simmerman, D. Russell, D. Martin, P. A. Lapchak, F. Collins, B. J. Hoffer and G. A. Gerhardt, *Nature*, 1996, **380**, 252.
28. D. Kirik, B. Georgievska and A. Bjorklund, *Nat. Neurosci.*, 2004, **7**, 105.
29. A. Pascual, M. Hidalgo-Figueroa, J. I. Piruat, C. O. Pintado, R. Gómez-Díaz and J. López-Barneo, *Nat. Neurosci.*, 2008, **11**, 755.
30. M. Trupp, N. Belluardo, H. Funakoshi and C. F. Ibanez, *J. Neurosci.*, 1997, **17**, 3554.
31. S. Jain, J. P. Golden, D. Wozniak, E. Pehek, E. M. Johnson Jr. and J. Milbrandt, *et al.*, *J. Neurosci.*, 2006, **26**, 11230.
32. G. Paratcha, F. Ledda and C. F. Ibanez, *Cell*, 2003, **113**, 867.
33. C. F. Ibanez, *Trends Cell Biol.*, 2007, **17**, 519.
34. A. Tomac, J. Widenfalk, L. F. Lin, T. Kohno, T. Ebendal, B. J. Hoffer and L. Olson, *Proc. Natl. Acad. Sci. USA*, 1995, **92**, 8274.
35. C. C. Chao and E. H. Lee, *Neuropharmacology*, 1999, **38**, 913.

36. T. Pawson and T. M. Saxton, *Cell*, 1999, **97**, 675.
37. F. Neff, C. Noelker, K. Eggert and J. Schlegel, *Ann. N. Y. Acad. Sci.*, 2002, **973**, 70.
38. H. Dudek, S. R. Datta, T. F. Franke, M. J. Birnbaum, R. Yao, G. M. Cooper, R. A. Segal, D. R. Kaplan and M. E. Greenberg, *Science*, 1997, **275**, 661.
39. R. M. Soler, X. Dolcet, M. Encinas, J. Egea, J. R. Bayascas and J. X. Comella, *J. Neurosci.*, 1999, **19**, 9160.
40. D. L. Choi-Lundberg, Q. Lin, Y. N. Chang, Y. L. Chiang, C. M. Hay, H. Mohajeri, B. L. Davidson and M. C. Bohn, *Science*, 1997, **275**, 838.
41. A. Tomac, E. Lindqvist, L. F. Lin, S. O. Ogren, D. Young, B. J. Hoffer and L. Olson, *Nature*, 1995, **373**, 335.
42. N. S. Christophersen, M. Grønborg, T. N. Petersen, L. Fjord-Larsen, J. R. Jørgensen, B. Juliusson, N. Blom, C. Rosenblad and P. Brundin, *Exp. Neurol.*, 2007, **204**, 791.
43. J. G. Hou, L. F. Lin and C. Mytilineou, *J. Neurochem.*, 1996, **66**, 74.
44. B. Jakobsen, J. B. Gramsbergen, A. Moller Dall, C. Rosenblad and J. Zimmer, *Eur. J. Neurosci.*, 2005, **21**, 2939.
45. B. A. Horger, M. C. Nishimura, M. P. Armanini, L. C. Wang, K. T. Poulsen, C. Rosenblad, D. Kirik, B. Moffat, L. Simmons, E. Johnson, Jr, J. Milbrandt, A. Rosenthal, A. Bjorklund, R. A. Vandlen, M. A. Hynes and H. S. Phillips, *J. Neurosci.*, 1998, **18**, 4929.
46. R. H. Baloh, M. G. Tansey, P. A. Lampe, T. J. Fahrner, H. Enomoto, K. S. Simburger, M. L. Leitner, T. Araki, E. M. Johnson, Jr and J. Milbrandt, *Neuron*, 1998, **21**, 1291.
47. J. Milbrandt, F. J. de Sauvage, T. J. Fahrner, R. H. Baloh, M. L. Leitner, M. G. Tansey, P. A. Lampe, R. O. Heuckeroth, P. T. Kotzbauer, K. S. Simburger, J. P. Golden, J. A. Davies, R. Vejsada, A. C. Kato, M. Hynes, D. Sherman, M. Nishimura, L. C. Wang, R. Vandlen, B. Moffat, R. D. Klein, K. Poulsen, C. Gray, A. Garces, C. E. Henderson, H. S. Phillips and E. M. Johnson, Jr, *Neuron*, 1998, **20**, 245.
48. Y. A. Sidorova, K. Mätlik, M. Paveliev, M. Lindahl, E. Piranen, J. Milbrandt, U. Arumäe, M. Saarma and M. M. Bespalov, *Mol. Cell. Neurosci.*, 2010, **44**, 223.
49. A. Bjorklund, C. Rosenblad, C. Winkler and D. Kirik, *Neurobiol. Dis.*, 1997, **4**, 186.
50. J. H. Kordower, M. E. Emborg, J. Bloch, S. Y. Ma, Y. Chu, L. Leventhal, J. McBride, E. Y. Chen, S. Palfi, B. Z. Roitberg, W. D. Brown, J. E. Holden, R. Pyzalski, M. D. Taylor, P. Carvey, Z. Ling, D. Trono, P. Hantraye, N. Déglon and P. Aebischer, *Science*, 2000, **290**, 767.
51. R. J. Mandel, S. K. Spratt, R. O. Snyder and S. E. Leff, *Proc. Natl. Acad. Sci. USA*, 1997, **94**, 14083.
52. P. Akerud, J. M. Canals, E. Y. Snyder and E. Arenas, *J. Neurosci.*, 2001, **21**, 8108.
53. N. Nakao, H. Yokote, K. Nakai and T. Itakura, *J. Neurosurg.*, 2000, **92**, 659.

54. A. Sajadi, J. C. Bensadoun, B. L. Schneider, C. Lo Bianco and P. Aebischer, *Neurobiol. Dis.*, 2006, **22**, 119.
55. T. Yasuhara, T. Shingo, A. Takeuchi, T. Yasuhara, K. Kobayashi, K. Takahashi, K. Muraoka, T. Matsui, Y. Miyoshi, H. Hamada and I. Date, *J. Neurosurg.*, 2005, **102**, 80.
56. P. Aebischer and J. Ridet, *Trends Neurosci.*, 2001, **24**, 533.
57. E. F. Espejo, R. J. Montoro, J. A. Armengol and J. Lopez-Barneo, *Neuron*, 1998, **20**, 197.
58. M. R. Luquin, R. J. Montoro, J. Guillén, L. Saldise, R. Insausti, J. Del Río and J. López-Barneo, *Neuron*, 1999, **22**, 743.
59. J. Lopez-Barneo, P. Ortega-Saenz, R. Pardal, A. Pascual and J. I. Piruat, *Eur. Respir. J.*, 2008, **32**, 1386.
60. J. J. Toledo-Aral, S. Mendez-Ferrer, R. Pardal, M. Echevarria and J. Lopez-Barneo, *J. Neurosci.*, 2003, **23**, 141.
61. J. Villadiego, S. Méndez-Ferrer, T. Valdés-Sánchez, I. Silos-Santiago, I. Fariñas, J. López-Barneo and J. J. Toledo-Aral, *J. Neurosci.*, 2005, **25**, 4091.
62. Y. Honda, *J. Appl. Physiol.*, 1992, **73**, 1.
63. V. Arjona, A. Mínguez-Castellanos, R. J. Montoro, A. Ortega, F. Escamilla, J. J. Toledo-Aral, R. Pardal, S. Méndez-Ferrer, J. M. Martín, M. Pérez, M. J. Katati, E. Valencia, T. García and J. López-Barneo, *Neurosurgery*, 2003, **53**, 321.
64. A. Minguez-Castellanos, F. Escamilla-Sevilla, G. R. Hotton, J. J. Toledo-Aral, A. Ortega-Moreno, S. Méndez-Ferrer, J. M. Martín-Linares, M. J. Katati, P. Mir, J. Villadiego, M. Meersmans, M. Pérez-García, D. J. Brooks, V. Arjona and J. López-Barneo, *J. Neurol. Neurosurg. Psychiatry*, 2007, **78**, 825.
65. J. Arias-Stella and J. Valcarcel, *Hum. Pathol.*, 1976, **7**, 361.
66. E. E. Lack, *Am. J. Pathol.*, 1978, **91**, 497.
67. K. H. McGregor, J. Gil and S. Lahiri, *J. Appl. Physiol.*, 1984, **57**, 1430.
68. R. Pardal, P. Ortega-Saenz, R. Duran and J. Lopez-Barneo, *Cell*, 2007, **131**, 364.
69. M. W. Moore, R. D. Klein, I. Fariñas, H. Sauer, M. Armanini, H. Phillips, L. F. Reichardt, A. M. Ryan, K. Carver-Moore and A. Rosenthal, *Nature*, 1996, **382**, 76.
70. J. G. Pichel, L. Shen, H. Z. Sheng, A. C. Granholm, J. Drago, A. Grinberg, E. J. Lee, S. P. Huang, M. Saarma, B. J. Hoffer, H. Sariola and H. Westphal, *Nature*, 1996, **382**, 73.
71. M. P. Sanchez, I. Silos-Santiago, J. Frisén, B. He, S. A. Lira and M. Barbacid, *Nature*, 1996, **382**, 70.
72. H. A. Boger, L. D. Middaugh, P. Huang, V. Zaman, A. C. Smith, B. J. Hoffer, A. C. Tomac and A. C. Granholm, *Exp. Neurol.*, 2006, **202**, 336.
73. H. A. Boger, L. D. Middaugh, K. S. Patrick, S. Ramamoorthy, E. D. Denehy, H. Zhu, A. M. Pacchioni, A. C. Granholm and J. F. McGinty, *J. Neurosci.*, 2007, **27**, 8816.

74. E. R. Kramer, L. Aron, G. M. Ramakers, S. Seitz, X. Zhuang, K. Beyer, M. P. Smidt and R. Klein, *PLoS Biol.*, 2007, **5**, e39.
75. J. G. Nutt, K. J. Burchiel, C. L. Comella, J. Jankovic, A. E. Lang, E. R. Laws Jr, A. M. Lozano, R. D. Penn, R. K. Simpson Jr, M. Stacy and G. F. Wooten, ICV GDNF Study Group, *Neurology*, 2003, **60**, 69.
76. S. S. Gill, N. K. Patel, G. R. Hotton, K. O'Sullivan, R. McCarter, M. Bunnage, D. J. Brooks, C. N. Svendsen and P. Heywood, *Nat. Med.*, 2003, **9**, 589.
77. J. T. Slevin, G. A. Gerhardt, C. D. Smith, D. M. Gash, R. Kryscio and B. Young, *J. Neurosurg.*, 2005, **102**, 216.
78. A. E. Lang, S. Gill, N. K. Patel, A. Lozano, J. G. Nutt, R. Penn, D. J. Brooks, G. Hotton, E. Moro, P. Heywood, M. A. Brodsky, K. Burchiel, P. Kelly, A. Dalvi, B. Scott, M. Stacy, D. Turner, V. G. Wooten, W. J. Elias, E. R. Laws, V. Dhawan, A. J. Stoessl, J. Matcham, R. J. Coffey and M. Traub, *Ann. Neurol.*, 2006, **59**, 459.

CHAPTER 9

Animal Models for ALS

RITSUKO FUJII[*a,b] AND TORU TAKUMI[a,c]

[a] Laboratory of Integrative Bioscience, Graduate School of Biomedical Sciences, Hiroshima University, Japan; [b] Laboratory for Anatomy and Physiology, Faculty of Human Science, Hiroshima Bunkyo Woman's University, Japan; [c] Japan Science and Technology Agent (JST), Core Research for Evolutional Science and Technology (CREST), Japan

9.1 Introduction

Amyotrophic lateral sclerosis (ALS) is a fatal neurodegenerative disease which results in paralysis and death within two to five years after disease onset.[1] It typically strikes adults in their mid-life and leads to progressive muscle weakness and atrophy in limbs, and eventually to death due to respiratory failure. These events are caused by a progressive and selective loss of upper motor neurons in the cerebral cortex and the lower motor neurons in the brainstem and spinal cord. The major histopathology of ALS-affected motor neurons is formation of abnormal protein aggregates/inclusions in the cytoplasm, which correlates well with the progression and severity of the clinical symptoms of the disease.[1,2]

A worldwide incidence of ALS is estimated at 2–8 new cases per 100 000 population each year, which makes the disease the most common adult-onset motor neuron disease.[1] Approximately 90% of ALS cases are sporadic ALS (SALS), while inherited forms of ALS or familial ALS (FALS) constitutes approximately 5–10% of all cases.[3] To date, more than 10 genetic loci have been linked to ALS.[3,4] Most of them were shown to contain genes linked to FALS, including *SOD1* (ALS1), *Alsin/ALS2* (ALS2), *senataxin/SETX* (ALS4), *spastic paraplegia 11/SPG11*(ALS5), *FUS/TLS* (ALS6), *VAPB*

RSC Drug Discovery Series No. 6
Animal Models for Neurodegenerative Disease
Edited by Jesús Avila, Jose J. Lucas and Félix Hernández
© Royal Society of Chemistry 2011
Published by the Royal Society of Chemistry, www.rsc.org

(ALS8), *Angigenin/ANG* (ALS9), *TARDBP/TDP-43*(ALS10), *FIG4* (ALS11), *Optineurin/OPTN* (ALS12).[3,4]

In 1993, Rosen *et al.* discovered that 20% of all FALS cases carry specific point mutations in a gene encoding for Cu/Zn superoxide dismutase 1 (SOD1).[5] This epoch-making discovery led to generation of transgenic rodents that express various types of ALS-related mutant SOD1 s. The mutant *SOD1* transgenic rodents display major ALS symptoms that lead to motor neuron degeneration and defects in neuromuscular junction.[6,7] Importantly, over-expression of wild-type SOD1 does not lead to overt motor neuron degenera-tion, but slight toxicity to motor neurons.[8] Ablation of the endogenous *SOD1* gene has also very mild toxicity.[9] These results support the 'gain of function' hypothesis proposing that SOD1 carrying an ALS-associated mutation can acquire a toxic function responsible for the selective motor neuron death. Although numbers of transgenic mice with known ALS-associated mutant genes have been generated, most of them could not provide evidence to show their pathological relevance to ALS.

The cause of selective motor neuron degeneration in ALS remains elusive. It is widely accepted that multiple mechanisms may contribute to the disease pathogenesis.[1,3,4] The possible molecular mechanisms include glutamate excitotoxicity, endoplasmic reticulum stress, mitochondrial dysfunction, cytos-keletal abnormalities, deficiency of neurotrophic growth factors, abnormal accumulation of protein aggregates (*i.e.* SOD1, TDP-43, neurofilament), defects in

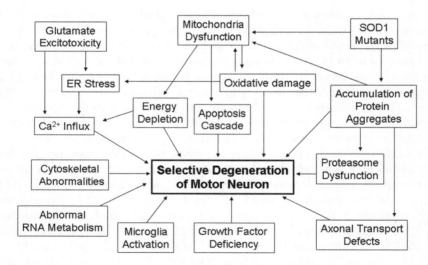

Figure 9.1 A model for selective motor neuron degeneration in ALS. A current working model proposes that ALS may be caused by a combination of ALS-causal genes (*i.e.* mutations in *SOD1*, defects in cytoskeletal proteins or motor proteins implicated in axonal transport), and environmental cues to trigger the disease (*i.e.* excessive glutamate, oxidative mitochondria damage, endoplasmic reticulum stress). Their complicated interplay still remains speculative and unsolved.

axonal transport, activation of apoptotic pathways, and so on (Figure 9.1). Recent studies also suggest that motor neuron death in ALS depends not only on abnormal intracellular events in motor neurons but also malfunctions of surrounding non-neuronal cells.

In general, animal disease models have been used in developing and testing new drugs to evaluate their effectiveness to the disease. Established first in 1994, the SOD1 ALS mouse model remains a robust ALS model for preclinical research.[10] Some of the recent breakthrough in ALS research are discoveries of ALS-associated novel mutations in *TARDBP/TDP*-43,[11,12] *FUS/TLS*[13,14] and *optineurin*[15] which encode proteins implicated in RNA processing or intra-cellular trafficking. Future studies on these genes should hopefully expand our understanding of the disease pathogenesis and lead to generation of new animal models alternative for the mutant SOD1 mice.

In this chapter, we describe important studies of ALS animal models, most of which have been demonstrated in mutant SOD1 transgenic mice, and discuss the applicability of ALS rodent models for better understanding and preclinical studies of the human disease.

9.2 Genetics of ALS

9.2.1 Genes Related to Familial ALS

To date, 16 familial ALS (FALS) including classic FALS (ALS1–ALS12), FALS with frontotemporal lobar dementia (FTD) (ALS-FTD1 and ALS-FTD2), X chromosome-linked ALS (X-linked ALS) and dynactin-related FALS have been identified.[3,4] Most of the classical FALS are inherited as an autosomal dominant disorder except *alsin*-linked ALS2 and *spastic paraplegia 11* (*SPG11*)-linked ALS5. However, ALS1 (*SOD1*), ALS6 (*FUS/TLS*), and ALS12 (*optineurin*) are also found to be an autosomal recessive disorder in some cases.[1,4,10–15] The aetiology of ALS could be even more complicated because mutations of some FALS-related genes such as *SOD1*, *angiogenin*, *TARDBP/TDP-43* and *FIG4* are also found in non-inherited SALS.[3,4]

9.2.2 Genes Related to Sporadic ALS

Sporadic ALS (SALS) that are not associated with a family history account for 90–95% of ALS cases.[3] SALS is clinically distinguishable from FALS, although some FALS-related genes are shared with SALS.[3,4]

While the genetic basis of FALS has been well explored, the molecular basis of SALS is less well understood. To date, SALS-related mutations are exclusively found in *charged multivesicular body protein 2B* (*CHMP2B*),[16] *neurofilament heavy chain* (*NF-H*)[17,18] and *peripherin* genes.[19,20]

9.2.2.1 Charged Multivesicular Body Protein 2B (CHMP2B) Gene

The mutation in *charged multivesicular body protein 2B* (*CHMP2B*) gene on chromosome 3 was first found in a large Danish pedigree with frontotemporal dementia (FTD), which is characterized by frontotemporal lobar degeneration (FTLD) with ubiquitin-immunoreactive inclusions (FTLD-U).[16] Subsequently *CHMP2B* mutants were found in patients with a non-inherited form of ALS with FTD symptoms designated as ALS-FTD3.

9.2.2.2 *Neurofilament Heavy Chain Subunit Gene*

Aberrant neurofilament accumulations are one of the major pathological features for both SALS and FALS.[21] However, no conclusive linkage between mutations in three neurofilament genes and either SALS or FALS patients have been reported, although in-frame insertions or deletions within the normal array of 44-45 lysine–serine–proline (KSP) repeats region in the tail domain of NF-H have been reported in ~1% of SALS cases.[22–24] In other investigations, dominant point mutations in the *NF-L* gene have been linked to a mild motor neuron generative disease, Charcot–Marie–Tooth (CMT) disease.[25,26]

9.2.2.3 *Peripherin Gene*

A frame-shift mutation in *peripherin* on chromosome 12q12 has been found in SALS patients[27] which produces a truncated peripherin that disrupts neurofilament assembly. Peripherin was found in intermediate filament (IF) inclusions in ALS patients.[27,28]

9.3 SOD1 Mouse Model of ALS

9.3.1 SOD1 Mutations

Cu/Zn superoxide dismutase 1 (SOD1) is a ubiquitously distributed cytoplasmic enzyme that catalyses conversion of superoxide into peroxide and oxygen. Ever since mutations in *SOD1* were linked to FALS,[4] SOD1 have been a hallmark of ALS research.[3,4] To date, more than 146 different mutations scattered throughout the SOD1 protein have been identified (www.alsod.org) in FALS (ALS1) cases, independent of exon, domains or motif. ALS-related SOD1 mutants (dismutase active G37R and G93A; dismutase inactive G85R, G86R, G127X, SOD1^{Quq}, H46R, D90A) have been overexpressed in mice and examined if they develop disease pathologies similar to that observed in human ALS.[7] These mutant SOD1 transgenic rodents develop comparable disease phenotypes such as mitochondria abnormalities and microglia activation in spinal cords, hindlimb weakness, progressive paralysis and muscular dystrophy but to different degrees between the transgenic strains.[7,29–36] It is interesting that SOD1 gene knock-out mice do not lead to motor neuron disease and expression of G85R mutant in SOD1-deficient mice does not affect the disease course. In addition, overexpression of WT hSOD1 in WT mice has no effect on SOD1 activity which is regulated by catalytic copper-binding catalysed by CCS (copper chaperone for SOD1).[37–40] Motor neurons of CCS-deleted mice have an increased sensitivity to axonomy-induced death.

Multiple molecular mechanisms have been proposed for SOD1-related ALS pathogenesis such as glutamate excitotoxicity, mitochondrial dysfunction, microglial dysfunction, axonal transport blockade, and inflammatory and apoptotic signals triggered by aberrant redox reaction or protein aggregation (Figure 9.1).

9.3.2 SOD1 Mutant Transgenic Mice

Studies of the mutant SOD1 transgenic rodents revealed that the levels and physical locations of mutant *SOD1* alleles are critical for promotion of the disease. For example, abnormally shaped mitochondria have been observed in mice transgenic for mutant SOD1 s and several mutant SOD1 proteins have been selectively recruited and accumulated in mitochondria in spinal cord.[39,41] Mice transgenic for SOD1$^{G[37R]}$ or SOD1^{D90A} do not develop ALS unless bred to homozygosity.[42,43] Similarly, ALS-like phenotype is absent in SOD1^{A4V} transgenic mice, but evident when the levels of total SOD1 protein are raised by co-expression of wild-type (WT) human SOD1,[39] suggesting that mutant SOD1 entraps WT SOD1 and form toxic aggregates. Supporting this hypothesis, it has been reported that overexpression of wild-type human SOD1 hastens the onset of disease without prolongation of the disease duration in the case of SOD1^{G93A}, SOD1^{L126Z} and SOD1^{G85R} transgenic mice, although the results may differ between the transgenic strains.[31,39,44] In addition, severer dose-effect phenotypes have been reported in the rare consanguineous human cases carrying G27ΔGP, L84F, N86S, or L126S homozygous SOD1 mutant alleles.[7]

Among SOD1 mutations identified in FALS cases, G93A mutant (SOD1^{G93A}) seems particularly vulnerable and gives rise to a shorter disease duration between the onset and death (2.2 ± 1.5 years) compared with an average duration (10 ± 5–6 years) in other SOD1 mutant causing FALS.[3,7,30,31] Mice hemizygous for human SOD1^{G93A} transgene are viable and fertile while expressing human SOD1^{G93A} proteins. This SOD1^{G93A} mouse, often referred to as G1 H founder line, contains the high transgene copy number up to 25 copies and recapitulates phenotypes similar to those observed in human ALS caused by the same mutant.[30] Mice carrying the high copy number SOD1^{G93A} transgenes show aggregate formation in motor neurons, muscle weakening and motor neuron loss, and die at 135 days of age where normal life span of WT mice is two years.[33] In contrast, a mouse subline SOD1^{G93AG1L} with low *SOD1*^{G93A} transgenic copies (~eight copies) delays onset and lengthens duration of the disease.[45] Moreover, mitochondrial abnormalities in spinal cords is minimal in a low-copy SOD1^{G93AG1L} subline,[45,46] indicating that higher expression levels of SOD1^{G93A} lead to an earlier disease onset and a shorter duration. Johnston *et al.* reported that soluble type of high molecular weight (HMW) SOD1-containing aggregates were detected in SOD1^{G93A} mice prior to the formation of insoluble heat- and reducing agent-resistant HMW SOD1 aggregates in motor neuron cells.[47] This is also observed in transgenic mice carrying FALS-linked mutations A4 V and G85 R in SOD1.[31,48] Collectively, overexpression of mutant SOD1 such as SOD1^{G93A}, SOD1^{L126Z} and SOD1^{G85R}

predominantly affect motor neurons and trigger ALS-like symptoms in mice, regardless of their ability to bind catalytic Cu^{2+} and dismutase activity.

Transgenic ALS mice are powerful tools to unravel the complicated *in vivo* biological events in the disease course, especially when crossbred with mice carrying other genes of interest. Functional interactions between ALS-causing mutant *SOD1* and other ALS-associated genetic modifiers are described below. Details of mouse ALS models and pathological features are summarized in Table 9.1.

9.4   Genetic Modifiers of ALS Mouse Model

9.4.1   Glutamate Excitotoxicity

9.4.1.1   *Glial Glutamate Transporter*

Glutamate is the major excitatory neurotransmitter in the mammalian central nervous system, exerting its effects through a wide variety of receptors that can be either ionotropic or metabotropic. Glutamate excitotoxicity in motor neurons is generally induced when excessive glutamate signalling *via* an ionotropic glutamate receptor AMPA receptor leads to intracellular calcium-overload and generation of reactive oxygen species.

In motor neurons, a glial glutamate transporter EAAT2 removes excess glutamate from the synapse and protects motor neurons from the glutamate-mediated excitotoxicity.[49,50] In a subset of ALS patients, expression of EAAT2 is significantly reduced and is accompanied by decreased glutamate uptake in the motor cortex and spinal cord.[50–52] Consistent with human ALS cases, reduction of EAAT2 expression in astrocytes is observed in spinal cords of symptomatic mutant SOD1[G85R] mice.[32] In SOD1[G93A] mice, however, the EAAT2 protein level is progressively decreased in the ventral horn, but not in the dorsal horn, of lumber spinal cords. This becomes prominent only at late-stage or not at all[53,54] and, basically, no change in EAAT2 level occurs before the SOD1[G93A]-related ALS symptoms appear. Despite this, oxidative or caspase-3-mediated inactivation of EAAT2 was observed in spinal cords of mutant SOD1 mice, suggesting a link between apoptotic signals and glutamate excitotoxicity in ALS pathogenesis.[55,56]

The diversity in glutamate handling in mutant SOD1 transgenic mouse models suggests that, in at least some cases, glutamate toxicity maybe a secondary event rather than a primary cause of disease. To further explore the glutamate uptake in astrocytes, SOD1[G93A] mice expressing human EAAT2 under the control of glial fibrillary acidic protein (GFAP) promoter were generated by crossing SOD1[G93A] mice with GFAP promoter-driven EAAT2 transgenic mice. These double transgenic EAAT2/SOD1[G93A] mice replenishing EAAT2 in astrocytes show a significant delay in motor neuron degeneration and associated motor functions at early-stage of the disease progression compared with observations in SOD1[G93A] mice, but no change in the disease duration. These results indicate that defects in astrocyte-mediated glutamate uptake may be critical for the disease progression from onset to early-stage,

Table 9.1 Mouse ALS models and pathological features.

Mouse	Inclusions in MN	Axonal transport	Development of MND	References
Mutant hSOD1				
SOD1^{G93A}	SOD1, IF	Reduced	Yes	10, 33, 66, 140
SOD1^{G37R}	SOD1, IF	Reduced	Yes	39, 42, 140
SOD1^{G85R}	SOD1, IF	Reduced	Yes	31, 66, 140, 152
SOD1^{D90A}	SOD1, IF	Reduced	Yes	41, 43
SOD1^{H46R}	SOD1, IF	Reduced	Yes, but moderate	115, 141, 195
NF deletion and mutant				
hNF-H	Perikaryal NF	ND	Moderate axonopathy	23, 155, 152
hNF-H (KSP mutation)	Yes	Blocked	Yes, severe phenotype	24
NF-L–/–	Perikaryal NF-M and NF-H	ND	Yes, no MN death	62, 118, 152
hNF-L	Perikaryal NF	ND	Yes, no MN death	151, 156
Peripherin	IF	Reduced	Slow MN death	20, 28, 157,158
Kinesin deletion				
KIF1B+/–	Yes	Blocked	Neuropathy	111
KIF1B–/–	Yes	Blocked	Neuropathy	111
KIF5A–/–	Yes	Reduced	Neuropathy	112
Dynein mutations				
Loa+/–	ND	ND	Yes, but normal lifespan	117, 123, 129
Cra1+/–	ND	No change	Yes, but normal lifespan	119, 122, 128
Others				
hVAPB^{P56S}	Aggregates in ER	ND	Yes	179, 180, 181
Alsin–/–	ND	ND	Yes, but moderate	184
hTDP-43^{M337V}	TDP-43, SOD1, htt	ND	Yes	204

SOD1, human Cu–Zn superoxide dismutase 1; NF-H, high molecular weight neurofilament; NF-L, low molecular weight neurofilament; KIF, kinesin family; *Loa, legs at odd angle*, vesicle-associated membrane protein B; *Cra1 (cramping 1)*, dynein heavy chain 1 gene mutants; VAPB, vesicle associated membrane protein-associated protein B; TDP-43, transactive response DNA-binding protein; MND, motor neuron disease; MN, motor neuron; htt, huntingtin; ER, endoplasmic reticulum; ND, not determined.

although it is still unclear whether loss of functional EAAT2 is the primary cause for neuronal degeneration or is the secondary.

9.4.1.2 AMPA Receptor

Motor neurons expressing SOD1^{G93A} mutant are more susceptible to glutamate toxicity than wild type cells and show changes in expression of glutamate receptor subunits,[57,58] suggesting mutant SOD1 affects glutamate transmitter signalling pathways as well as glutamate levels. Glutamate toxicity in motor neurons is primarily mediated through AMPA receptors and in fact, administration of AMPA receptor antagonists (*i.e.* NBQX) to SOD1^{G93A} mice significantly reduced symptoms and increase survival.[59,60] AMPA receptors can exert functional diversity as composed of four subunits, designated GluR1-4.[61] For example, permeability for Ca^{2+} ions of a particular AMPA receptor is determined by its subunit composition. The GluR2 subunit is widely expressed and included in most of Ca^{2+}-impermeable AMPA receptors.[61] This property is rendered by RNA editing of GluR2 transcripts. Investigations into changes in AMPA receptor subunit expression in ALS have identified reduced RNA editing and expression of GluR2 subunit in motor neurons compared to other neuronal populations,[62–64] but failed to find clear differences between subunit expression in motor neurons from ALS and that from control subjects.[62,64] These differences in glutamate receptor profile between motor neurons and other neuronal populations may partly explain the selective vulnerability of motor neurons to glutamate toxicity, but do not suggest changes in subunit expression lead to ALS. A study on AMPA receptor properties has shown no changes in Ca^{2+} permeability of AMPA receptor in SOD1^{G93A} expressing motor neurons, but an increase in Na^+ and K^+ permeability.[58] In this way, Ca^{2+} influx is triggered through voltage-gated calcium channels.

9.4.1.3   Calcium Binding Proteins

Calcium signalling is usually attenuated by rapid buffering of free Ca^{2+}, either by binding to cytosolic calcium binding proteins (CaBP) or by uptake into organelles. Motor neurons express only low levels of CaBP and thus there appears to be correlation between low CaBP expression and susceptibility to ALS whereas GABAergic neurons have marked expression of seven different CaBP such as calbindin 28K, calrectin and paralbumin.[65] Conversely, the effects of glutamate toxicity in SOD1^{G93A} mutant-expressing motor neurons can be reduced by increased expression of CaBP or calbindin.[66] Selective motor neuron degeneration in ALS may be due to the low CaBP level in motor neurons. Therefore, in human ALS cases and ALS-causing rodent models, AMPA receptor activation causes greater mitochondrial calcium uptake than in GABAergic neurons and results in mitochondrial ROS generation leading to neuronal degeneration.[67] Taken together, calcium loading and handling should be a key modifier for the ALS phenotypes.

9.4.2   Mitochondrial Dysfunction and Apoptosis

Mitochondria are not only responsible for production of ATP, but also the major site for the generation of reactive oxygen species (ROS) as a by-product in the consequence of the energy production. Mitochondria also have roles in calcium buffering and initiation of apoptotic cell death, all of which are considered to be important in ALS pathogenesis.[39] Morphologically abnormal mitochondria have been described in motor neurons from ALS patients,[68,69] mutant SOD1 transgenic mice (SOD1^{G93A}, SOD1^{G37R} and SOD1^{G85R}) and a motor neuron cell expressing mutant SOD1 s.[38,42,45,65,68–70] In SOD1^{G93A} mice, mitochondrial abnormalities such as mitochondrial swelling, membrane fragmentation and dilation of cristae appear before onset of muscle weakness.[70] Massive mitochondrial vacuolation is also occurred with the onset of symptoms, suggesting that mitochondrial dysfunction may be an important early event in disease pathogenesis and constitute part of the toxic gain of function by ALS-causing SOD1 mutants (*i.e.* SOD1^{G93A}). Biochemical studies of mitochondria from ALS patients and ALS cell culture models have consistently shown reduced electron transport chain activity,[71–73] decreased mitochondrial membrane potential,[74] disrupted calcium homeostasis and altered mitochondrial proteome.[75–77]

The elevated levels of reactive oxygen species have been proposed for causing the increased frequency of mitochondrial DNA mutations in the motor cortex and spinal cord of ALS patients.[73,77,78] Several studies have shown mitochondria damages in mutant SOD1 transgenic ALS mouse models. It has been reported that translocation of cytochrome C from mitochondria into cytoplasm, which is considered as a trigger of apoptosis, occurs and increases during the disease progression in SOD1^{G93A} mice. The translocation of mitochondrial cytochrome C is not observed in age-matched non-transgenic littermates.[79–82]

Age-dependent increase in mutant mitochondrial SOD1 accumulation has been shown to precede onset of human ALS. In mutant SOD1^{D90A} and SOD1^{G93A} transgenic mice, exogenously expressed mutant SOD1^{D90A} and SOD1^{G93A} also enter mitochondria and competes out intracellular cupper availability over endogenous WT SOD1 by forcing disulfide reduction.[83] Age-dependent increase in mutant mitochondrial SOD1 accumulation has been shown to precede onset of symptoms. The accumulation of mutant SOD1 in transgenic mice coincides with increased oxidative damages, decreased respiratory activity of mitochondria,[72] and appearance of mitochondrial swelling and vacuolization.[8,84] However, the role for mutant mitochondrial SOD1 in ALS pathogenesis is still equivocal as the mutant SOD1 accumulation is not selectively occurred in spinal cords.

Pasinelli *et al.* have first shown a link between mitochondria dysfunction and apoptosis in ALS-affected motor neurons.[85] In SOD1^{G93A} mice, both dimeric and high molecular weight SOD1-containing aggregates, both of which are thought to be toxic to mitochondria, directly bind an anti-apoptotic protein Bcl-2 and trigger apoptosis.[85] Piles of evidence have indicated that neuronal apoptosis in ALS patients and FALS mouse models is mediated by deregulation of Bcl-2,[86–89] death signalling p75 neurotrophin receptor (p75NTR), Fas and ER stress,[90–97] however none of them is conclusive.

9.4.3   Effects of Non-neuronal Cell Activation in ALS

9.4.3.1   *Effects of Microglia Activation*

It is generally accepted that motor neuron neurodegeneration in FALS is non-cell autonomous, with ALS-causing mutant SOD1 affects multiple cell types. Non-neuronal cells in the spinal cord may contribute to motor neuron protection, rather than degeneration, in mutant SOD1-associated ALS (ALS1). This concept is based on numerous studies of mutant SOD1 transgenic mice in which selective overexpression of mutant SOD1 in neurons, microglia and astrocytes was achieved by cell type-specific promoters (Table 9.2).

Selective expression of mutant SOD1 in astrocytes induced reactive gliosis without ALS-symptoms in mice, indicating that astrocytes do not contribute to the disease.[98,99] Expression of mutant SOD1 in neurons, astrocytes and muscle but not in microglia driven by the prion promoter could cause motor neuron degeneration in mice.[99] Effects of neuron-specific expression of SOD1^{G93A} were investigated in Thy1 promoter-driven $SOD1^{G93A}$ transgenic mice (Thy1-$SOD1^{G93A}$).[100] Thy1-$SOD1^{G93A}$ mice did not develop motor neuron degeneration unless bred to homozygosity.[100] Another targeted mutant SOD1 expression using the Cre-loxP system was able to show that *Cre*-mediated

Table 9.2 Non-cell autonomous effects in SOD1 transgenic ALS mouse models.

Genotype	*SOD1 transgenic*	*Target*	*Onset*	*Survival*	*Reference*
Chimeric dilution					
SOD1G37R		motor neuron	↑	↑	30, 31
SOD1G93A		motor neuron	↑	↑	30, 31, 38, 39
Cell-specific promoter					
Prp-SOD1G37R		neuron, astrocyte, muscle	↑	ND	99
Thy1-SOD1G85R		neuron postnatal	↑	↓	100
Thy1-SOD1G93A		neuron postnatal	↑	↓	100
GFAP-SOD1G86R		astrocyte	↑	ns	98
Cre-Lox transgene excision					
Islet1-Cre	Lox-SOD1^{G37R}	inmature motor & sensory neuron	↑	↑	102
CD11b-Cre	Lox-SOD1^{G37R}	myeloid cells	ND	↑	102
VAChT-Cre	Lox-SOD1^{G37R}	motor neuron	↑	↑	205
GFAP-Cre	Lox-SOD1^{G37R}	astrocyte	ns	↑	98
P$_0$-Cre	Lox-SOD1^{G37R}	Schwann cells	ns	↑	107
Bone marrow graft transplant					
PU.1 −/−	SOD1^{G93A} + SOD1^{G93A} BMT	microglia (+SOD1^{G93A}-microglia)	ns	ns	103
PU.1 −/−	SOD1^{G93A} + WT BMT	microglia (+WT-microglia)	ns	↑	103
	SOD1^{G37R} + MyD88 −/− BMT	microglia	↓	↓	104
	SOD1^{G93A} + MyD88 −/− BMT	microglia	ns	ns	104

ND, not determined; ns, not sensitive; BMT, bone marrow transplant activation of mutant $SOD1^{G85R}$ transgene in subpopulations of spinal motor neurons and interneurons could trigger selective degeneration of motor neurons.[101] These results indicate that motor neuron degeneration is an intrinsic event in SOD1-associated ALS.

Selective expression of mutant SOD1 within motor neurons is a determinant of the disease initiation, whereas mutant expression within neighbouring non-neural cells (*i.e.* microglia, astrocyte) may accelerate disease progression. To test the beneficial effect of microglia on ALS, novel mice lines carrying *Cre* gene driven by the myeloid CD11b promoter were generated and bred with floxed $SOD1^{G37R}$ mice.[102] The specific transgene inactivation in CD11b-positive cells (*i.e.* myeloid-derived microglia) dramatically prolonged lifespan by approximately three months in Cre-excised mice without causing microgliosis in the spinal cord.[102] $SOD1^{G93A}$ mice, in which myeloid cells were ablated by deletion of PU.1 and restored by transplant of wild-type bone marrow grafts giving rise to microglia with normal functions, showed disease onset but delayed progression of the disease.[103] In support of this, it has been demonstrated that deletion of myeloid-differentiation factor 88 (MyD88) in microglia derived from bone marrow worsens disease outcomes in $SOD1^{G37R}$ ALS mice.[104] These results strongly suggest a neuroprotective role for microglia during ALS progression rather than initiation of the disease.

9.4.3.2 Effects of Schwann Cells in Peripheral Nervous System

Schwann cells, the myelinating glia of the peripheral nervous system, could be damaged by SOD1 mutant. Schwann cells are associated with almost the entire surface of peripheral axons and are essential for the survival of motor neurons during neuronal development, and again become important during neuronal regeneration. Lobsieger *et al.* have assessed the non-cell-autonomous contribution of Schwann cell-expressed mutant SOD1($SOD1^{G37R}$) to ALS.[105–107] They mated mice heterozygous for human $SOD1^{G37R}$ transgene flanked by loxP sites (Lox$SOD1^{G37R}$ mice) with Cre-mice in which Cre recombinase was expressed under the control of mouse myelin-protein-zero (P_0) regulatory sequences.[102] In this system, Cre-mediated excision of Lox$SOD1^{G37R}$ can remove $SOD1^{G37R}$ specifically from Schwann cells.[107] Unexpectedly, reduction of a fully dismutase-active $SOD1^{G37R}$ within Schwann cells significantly accelerates disease progression.[107]

9.4.4 Axonal Transport Blockade

9.4.4.1 Cytoskeleton Networks in Motor Neuron Axons

Motor neurons are highly specialized cells with extensive dendrite arbors and axonal processes that can extend up to one metre from the cell body. This particular morphology is maintained by cytoskeletal networks comprised of microtubules, actin filaments and neurofilaments, and continuous transport of proteins and organelles to and from the cell body. Many studies have shown that defects of axonal transport are one of the causes for motor neuron degeneration in ALS.[1,3,108] Here we summarize recent data showing a close link between axonal transport defects in motor neurons and ALS aetiology.

9.4.4.2 Microtubule-dependent Retrograde Transport

Microtubules provide stability and polarity to the axonal compartment of the neuron as they are polarized with a slow-growing minus end directed toward the cell body and a fast-growing plus end directed peripherally in the axon, while actin contributes mainly to the integrity of the cell periphery. In addition to their structural roles, the microtubules provide mainly two long-range transport paths, the fast anterograde movement (away from cell body) and the retrograde movement (toward cell body) mediated by kinesin motor proteins and dynein motor complex, respectively.[109,110]

Mice with heterozygous disruption of a kinesin family member KIF1B gene show defects in the anterograde axonal transport and provoke neurodegeneration similar to human neuropathies.[111] In accordance with this mice phenotype, a loss-of-function mutation in the motor domain of the KIF1B gene has been found in patients with an inherited form of peripheral neuropathy, Charcot–Marie–Tooth disease type 2A.[111] Missense mutations in a conventional kinesin KIF5A have also been found responsible for an inherited form of spastic paraplegia and disruption of KIF5A gene in mice has been known to impair the transport of neurofilament proteins.[112,113]

In SOD1^{G93A} transgenic mice, a considerable inhibition of retrograde axonal transport was observed at a very early stage of disease before animals became symptomatic.[8] However, the mechanisms by which mutant SOD1 affects axonal transport have not been uncovered. Compared to WT SOD1, several SOD1 mutants have been shown to interact more stably with the dynactin–dynein complex.[114,115] Moreover, the dynein has been reported to be colocalized with protein inclusions formed by mutant SOD1 in ALS.[115] Mutant SOD1 may influence neuronal survival by impairing the dynein-dependent retrograde transport of neurotrophic factors and mitochondria.[108,114,116,117]

Both anterograde and retrograde axonal transport have been shown to be disrupted in mutant SOD1 transgenic mice.[118] Different motor proteins are believed to be involved in controlling anterograde and retrograde transport. While ALS-causing mutations in proteins engaged in anterograde transport have not been identified, a mutation in dynactin, a protein involved in dynein-mediated retrograde transport, caused a progressive lower motor neuron disorder.[108,111] Two point mutations in a dynein subunit caused progressive motor neuron degeneration in heterozygous mice.[119] Similarly, disruption of the dynactin complex (thereby reducing activation of cytoplasmic dynein) inhibited retrograde transport and triggered late-onset motor neuron degeneration in genetically engineered mice.[120]

9.4.4.3 Deficits in Dynein-mediated Axonal Transport

As described above, decreased dynein-mediated retrograde axonal transport affects the disease course both in ALS patients and in transgenic ALS animal models. To investigate the genetic modifying effects of dynein, a genetic crossing between ALS-causing transgenic and non-lethal dynein mutant mice may provide informative outcomes.

There are several mouse lines with point mutations in cytoplasmic dynein heavy chain (DHC) 1 available, including *Legs at odd angles* (*Loa*) and *Cramping* 1(*Cra1*) mice.[120,121] *Loa* and *Cra1* mice not only develop late-onset, non-fatal motor neuron degeneration with impaired retrograde axonal transport, but also suffer from sensory neuropathy that occurs prior to the onset of motor symptoms. Interestingly, crossing heterozygous *Loa* and *Cra1* mice with SOD1[G93A] ALS mice ameliorated the transport defect and delayed disease onset, and slowed disease progression instead of showing an additive phenotype.[122–124] The precise mechanisms underlying this effect remain unclear, though several explanations have been proposed: First, the dynein mutations alter intracellular transport and thereby change the subcellular localization of SOD1 with other proteins or organelles. For example, it is possible that decreased interaction of mutant SOD1 with mitochondria could improve cell survival by reducing apoptosis or other downstream consequences. SOD1-positive aggregates have been observed in homozygous *Loa* mice suggesting that the dynein mutation can affect WT SOD1 distribution in the cell. Secondly, the decreased rates of retrograde transport by the *Loa* or *Cra1* mutants might counterbalance an inhibition of anterograde transport caused by SOD1[G93A], thereby restoring the balance between anterograde and retrograde transport.

Lai *et al.* reported that mice with a G59S amino acid substitution in a dynactin 150 kDa subunit (p150[Glued]) showed signs of reduced motor neuron axonal transport and developed motor neuron disease in a similar way to phenotypes of *Loa* and *Cra1* mice.[125] However, when heterozygous G59S-p150[Glued] knock-in mice were crossed with SOD1[G93A] mice, no improvement of the SOD1[G93A] phenotype was observed[126,127] *Sprawling* (*Swl*) heterozygous mice harbouring another DHC mutation display proprioceptive sensory neuropathy without causing motor neuron deficits; however, they do not show late-onset motor neurons loss unlike *Loa* or *Cra1*.[128] Crossing *Swl* mice with SOD1[G93A] mice did not affect ALS disease onset or progression.[129]

9.4.4.4 Effects of Mutant SOD1 Aggregation on Axonal Transport

In ALS transgenic mice, dynein colocalizes with mutant SOD1 and forms aggregates in motor neuron axons.[116,121] It is possible that mutant SOD1 aggregates could directly hinder dynein transport on microtubules in axons.

Recent studies have demonstrated that interaction of mutant SOD1 with dynein is more stable than that of WT SOD1 in both ALS cell culture and animal models.[115] Despite no changes in the dynein–dynactin subunit interactions, the amount of mutant SOD1 that interacted with dynein increased as the disease progressed. Knowing the increased association of mutant SOD1 with dynein in mouse ALS models, the transport capacity of dynein might be saturated by mutant SOD1 over time. Supporting this, a study in monkeys showed that the dynein–dynactin complex undergoes age-related changes including increased amounts of dynein in nerve endings and a decrease in the dynein–dynactin interaction.[130] These changes suggest that less functional dynein–dynactin complexes may be available during aging. Either the increase in dynein–mutant SOD1 association or the decrease in functional dynein–dynactin while aging could contribute to reduced retrograde transport to levels that are no longer able to sustain neuronal survival, thereby producing a phenotype of adult-onset motor neuron degeneration.

9.4.5   Depletion of Neurotrophic Factors

A homozygous mutation in a splice site of ciliary neurotrophic factor (CNTF) is present among ~2% of European and Japanese populations.[131,132] The mutation resulting in production of an inactive protein causes a 15–20% reduction in motor neuron number, but does not by itself cause neurodegenerative disease.[132] One individual in a family carrying SOD1^{V148G} mutation developed the disease at an early age; subsequent analysis showed the disease is sporadic and that the individual has the homozygous mutation in CNTF.[133] Genetic reduction of CNTF in ALS was investigated by crossing SOD1^{G93A} mice with CNTF−/− mice, which resulted in significantly earlier onset and increased severity of the disease compared to SOD1^{G93A} (CNTF+/+) mice. This result supports a hypothesis that homozygous CNTF-null mutation may be a risk factor for early onset of disease.[133]

The role of angiogenic factor vascular endothelial growth factor (VEGF) is implicated in ALS since hypoxic induction of VEGF increased disease severity of SOD1^{G93A} mice.[134] Importantly, subsequent overexpression of VEGF ligand and receptor prolonged survival in SOD1^{G93A} mice, implicating VEGF as a modifier of SOD1-related ALS and a potential neuroprotective factor.[134–136]

9.4.6   Protein Misfolding and Aggregation

Protein aggregates formed from misfolded mutant proteins are a common feature in many neurodegenerative diseases, but it has been largely unknown whether or not they are a primary cause of disease pathogenesis, a harmless by-product or even a cellular defence mechanism to sequester potentially toxic proteins. Protein aggregates that are immunoreactive to antibodies against ubiquitin, a protein tag that targets proteins for proteolytic degradation, are present in all ALS cases tested.[137,138] SOD1 forms protein aggregates in human FALS linked to SOD1 mutations and also in a population of sporadic cases.[37,139] Analogous cytoplasmic aggregates consisted of ubiquitinated mutant SOD1 have also been observed in spinal cords of mouse ALS models during the symptomatic disease.[30,32,37] The number of aggregates increased with age in SOD1^{G93A} and SOD1^{G85R} mice consistent with proteasome-mediated turnover.[140] Most aggregates were positive for SOD1 in SOD1^{G85R}

mice, whereas there was more variability in the presence of SOD1-positive aggregates in SOD1^{G93A} mice. Protein aggregates were not detected in control mice and rarely seen in SOD1^{G37R} mice.[99,141]

Protein aggregation has been shown to inhibit the ubiquitin–proteasome system. In support of this hypothesis, reduced chaperone and proteasome activity was reported in mutant SOD1 transgenic mouse.[140,142–145] Over-expression of a molecular chaperone HSP-70 and E3 ubiquitin ligase partly protected against mutant SOD1 toxicity in cell culture models and HSP-70 was chronically upregulated in several mutant SOD1 models.[146] However, elevated levels of HSP-70 in mutant SOD1 models did not change disease onset or survival.[147] In genetic cross experiments to assess the role of the ubiquitin–proteasome system, SOD1^{G93A} mice were bred with mice lacking the LMP2 subunit of 20S core proteasome;[148] these mice were essentially not affected, although overall proteolysis in spinal cords was reduced. Therefore, genetic induction of chaperone proteins and proteasomes in transgenic mouse ALS models fails to change the disease onset, suggesting that the primary insult in the disease is upstream of refolding and degradation of SOD1 mutants.

9.4.7 Neurofilament Defects

Neurofilaments (NFs), which assemble from three subunits NF-L, NF-M and NF-H, are particularly abundant in motor neurons and regulate myelinated large calibre axons that conduct electrical impulses. One of the major components of neurofilaments, heavy chain form NF-H subunit, was identified as a causal gene for SALS.[22,24] In transgenic mouse overexpressing human NF-H protein, both transcript and protein levels of NF-L and NF-M are increased.[149–151] In contrast, expression of NF-L is reduced in cases of both SALS and FALS, and in SOD1^{G93A} ALS mice,[152] causing changes in stoichiometry of NF subunits. These model mice with neurofilament deficit progressively developed ALS-like symptoms and defects in axonal transport, but no significant loss of motor neurons.

Neurofilaments are particularly important in motor neurons with their large size and long axons. There is much evidence supporting a role for abnormal NF assembly in ALS pathogenesis: aggregation of neurofilaments is a pathological feature of many neurodegenerative disorders including ALS and is also seen in SOD1 transgenic mice.[2,30] Transport along neurofilaments was also shown to be disrupted in mutant SOD1 transgenic mice.

Although it is not known how mutant SOD1 causes neurofilament changes, it has been shown that mutant SOD1, but not wild-type SOD1, binds directly to 3′ UTR of NF-L mRNA, thereby leading to destabilization and subsequent degradation of the mRNA.[153,154] Since NF-L is required for NF-M and NF-H to assemble into filaments, changes in expression may disrupt neurofilament assembly and trigger neurofilament aggregation. In fact, deletion of NF-L in SOD1^{G85R} mice significantly delays onset and progression of the motor neuron disease, but accelerates degeneration of other neurons and reduces the selective toxicity of the SOD1 mutant to motor neurons.[152] Another study has demonstrated that SOD1^{G93A} mice overexpressing either NF-H or NF-L show late onset of the disease and survive longer than SOD1^{G93A} mice.[155] Overexpression of human NF-H in SOD1^{G85R} mice also has beneficial effects resulting in increased survival, with lifespan extended by up to 65% compared with SOD1^{G85R} mice.[150] In these transgenic models, an increase in perikaryal NF and a decrease in axonal NF inclusions are observed. However, there is minor controversy about whether extra neurofilaments can rescue the motor neuron disease caused by mutant SOD1. Couillard-Despres *et al.* have reported that overexpression of human NF-L proteins resulting in extra axonal filaments does not shorten the lifespan of transgenic mice expressing mutant SOD1^{G37R} despite its very modest neuroprotective effects on spinal cord and ventral roots.[156] Their results indicate that axonal neurofilaments may not be an exacerbating factor in motor neuron disease caused by mutant SOD1 but that an increase in perikaryal NF may play a neuroprotective role. In either model, neurofilaments are determinants for the selective vulnerability of motor neurons to ALS-causing SOD1 mutants.

Another intermediate filament protein, peripherin, has been identified in intermediate filament inclusions in ALS cases and SOD1 transgenic mice.[27,28] Overexpression of peripherin in mice causes a late-onset selective motor neuron disease with intermediate filament aggregates containing peripherin. Deletion of *peripherin* gene has no effect on the disease progression in SOD1^{G93A} mice.[157,158] It was subsequently shown that one of the known splice form variants of peripherin, the 61 kDa isoform (peripherin-61) was responsible for this toxicity.[159] Peripherin-61 has been reported to be detected in motor neurons from SOD1^{G37R} mice and human sporadic ALS patients.[159] Although neither upregulation nor suppression of *peripherin* gene had any effect on disease onset, severity and progression of SOD1^{G37R}-ALS mice, peripherin may be responsible for the pathogenesis in a small population of ALS cases.

9.4.8 Inflammatory Cytokines

Several inflammatory cytokines and enzymes including various interleukins, tumor necrosis factor-α (TNF-α), cyclooxygenase 2 (COX2) and prostaglandin E2 (PGE2) have been shown to be upregulated in the spinal cord of ALS patients and mutant SOD1 mice.[160–163] Microglia, the resident macrophage population in the central nervous system, become activated by these inflammatory cytokines and reactive oxygen species.

There is growing evidence to suggest that inhibition of microglia activation is protective in ALS. For example, microglia cultured from adult pre-symptomatic mutant SOD1 mice showed increased TNFα release upon stimulation compared to wild-type microglia and TNFα antagonists significantly increased survival in mutant SOD1 mice.[162–164] Similarly, reduction of PGE2 levels by inhibition of COX2 protected against motor neuron loss and increased survival in mutant SOD1 mice.[165,166]

9.5 Protein Degradation in ALS Transgenic Mouse Model

Dynein-mediated transport is a key mechanism in protein degradation and accumulation of misfolded proteins in cells. Interaction between mutant SOD1 and dynein may directly influence SOD1 aggregation and degradation. Proteins are mainly degraded by two pathways—the ubiquitin–proteasome system (UPS) and autophagy.[167,168]

Proteins destined to be degraded by the UPS are marked by the covalent attachment of ubiquitin and then transported to the proteasome for proteolytic cleavage. It has been widely accepted that dynein-mediated retrograde transport could be involved in shuttling ubiquitin-tagged proteins to the proteasome.

Autophagy is another intracellular process for the degradation of proteins, organelles and protein aggregates.[168] Autophagy is important, particularly for neurons, as mice lacking the autophagy-related gene Atg7 develops axonal dystrophy characterized by distal accumulation of membrane structures and swelling of axonal terminals.[169] Dynein-mediated transport is involved in autophagic clearance of aggregated proteins. In the autophagy process, the dynein–protein transporting complex collects misfolded proteins from the cell periphery and transports them to perinuclear region of the cells where they form intracellular inclusions called aggresomes. Dynein has also been implicated in lysosomal transport and in mediating the fusion of the autophagosome and the lysosome.

Mutant SOD1 is believed to be degraded by both UPS and autophagy.[170] It has been shown that mutant SOD1 interacts strongly with dynein while WT SOD1 does not, or to a very little extent.[108,114] Consistent with this view, mutant but not WT SOD1 interacts with p62/sequestosome,[171] a protein that has been linked to autophagy.[169] Moreover, it has been demonstrated that overexpression of p62/sequestosome 1 increased the formation of large mutant SOD1 inclusions resembling aggresomes.[171] These data suggest that cells utilize dynein-mediated transport to collect mutant SOD1 as aggresomes and target them for autophagic degradation.[108,167] In this context, the interaction between mutant SOD1 and dynein might be of beneficial effect to the cell. Supporting the importance of dynein in autophagic clearance of aggregates in preventing neuronal toxicity, *Loa* mutation in fact decreased autophagic clearance of huntingtin protein resulting in increased inclusions and toxicity in Huntington's disease model mice.[172] This result may or may not be relevant to SOD1-mediated ALS as the negative effect of *Loa* is not observed in SOD1^{G93A} mice.[117] Both *Loa* and *Cra1* ameliorate the ALS symptoms of SOD1^{G93A} mice and thus impaired autophagy can be a positive effect in ALS.[122]

9.6 SOD1 Transgenic Rat Model of ALS

Howland *et al.* reported that they developed transgenic rats expressing mutated human SOD1^{G93A} displaying ALS-like phenotypes, including motor neuron degeneration in spinal cord.[33] Despite the intense studies on genetic ALS mouse models,[1,7] rats transgenic for ALS-causing SOD1 mutations render great advantages to biochemical analysis and surgical manipulation due to their size.

To date, rat transgenic strains overexpressing two different human SOD1 mutants have been established as ALS rat models. One is a rat transgenic for SOD1[H46R] with prominent protein cytopathology in the spinal cord.[36] Another is transgenic for ALS-causing SOD1[G93A] with predominant vacuolar neurodegeneration.[173] The main difference between mouse and rat ALS models lies in the prominence of the forelimb onset, which can predict rapid progression of ALS in the rat model but which may contrast to human situations.

In SOD1[G93A] rats, a pronounced loss of a predominant glutamate transporter EAAT2 (GLT-1) protein is observed in the ventral horn at the end stage of the disease,[33] reminiscent of a selective loss of GLT in spinal cord and motor cortex from ALS patients. Consistently, the SOD1[G93A] rat ALS model shows marked reduction of glutamate uptake in the spinal cord but not in the brainstem or other brain areas in which levels of a neuronal glutamate transporter EAAC1 are not affected by the disease.[173] The reduced glutamate uptake capacity in spinal cord is accompanied by decreased expression of astrocytic GLT-1 (or EAAT2) and glial high affinity glutamate transporter (GLAST); thus significant deficits in glutamate uptake are most likely mediated by glial cells in the SOD1[G93A] transgenic rat.[173]

Rat ALS models have several advantages over mice ALS models. Generally speaking, rat ALS models allow easier experimental manipulation for a large-scale biochemical analysis such as protein extraction, detection of neurotransmitters, synaptosome preparation or intrathecal injection of new drugs.[31,36,173–177] The large rat spinal cord confers much easier access for microsurgery, which is required for testing of the gene or stem cell therapies.

9.7 VAPB Transgenic Mouse

Vesicle-associated membrane protein-associated protein A (VAPA) and vesicle-associated membrane protein-associated protein B (VAPB) interact with lipid-binding proteins carrying a short motif containing two phenylalanines in an acidic tract (FFAT motifs) and target them to the cytosolic surface of the endoplasmic reticulum (ER). A genetic mutation by which one of the phenylalanines in the conserved major sperm protein homology domain of VAPB is substituted with serine (P56S) has been linked to motor neuron degeneration in affected amyotrophic lateral sclerosis 8 (ALS8) patients.[178–181] It has been reported that VAPB is abundant in motor neurons and that the P56S substitution causes aggregation of mutant VAPB in immobile tubular ER clusters, perturbs FFAT-motif binding and traps endogenous vesicle-associated membrane protein-associated protein (VAP) in mutant aggregates; thereby VAPB with P56S substitution (P56S-VAPB) causes motor neuron degeneration *via* a dominant negative mechanism.[179,181] Expression of mutant VAPB or reduction of VAP by short hairpin RNA causes Golgi dispersion and neuronal cell death *in vitro*.

Both VAPA and VAPB are reduced in human ALS patients and mouse ALS models,[180,181] suggesting that VAP family proteins may be involved in the pathogenesis of sporadic and SOD1-linked ALS. Enforced expression of wild-type VAPB, mainly localized in the endoplasmic reticulum, induced activation of one of the main UPR pathways, the IRE1/XBP1 pathway. P56S-VAPB mutant forms cytosolic aggregates and loses the ability to be involved in the activation of the IRE1/XBP1pathway[179] in which deficient VAP family protein levels result in decreased ER anchoring of lipid-binding proteins.

Animal models carrying VAPB mutations will provide additional insight into the cellular mechanisms by which a reduced level of functional VAP proteins in ER can result in specific degeneration of motor neurons *in vivo*. Loss of activity of IRE1/XBP1 pathway by mutant VAPB would lead to the accumulation of unfolded proteins in motor neurons which eventually increase sensitivity to ER stress-induced death. Supporting this idea, several reports have demonstrated that that ER stress contributes to the development of sporadic ALS-related motor neuron cell death.[182,183]

9.8 Alsin Transgenic Mouse

9.8.1 Alsin

The recessive form of juvenile-onset familial ALS designated as ALS2 has been linked to chromosome 2q33.[1,4] This locus was overlapped with a familial juvenile primary lateral sclerosis (JPLS) locus.[184,185] ALS2 is an early onset and slowly progressive disease: onset of ALS2 varies from one to 10 years of age where the duration of the disease may be as long as 50 years. This ALS2 gene encodes short and long forms of alsin protein, which is a novel guanine nucleotide exchanging factor (GEF) for the small GTPase Rab5 involved in a macropinocytosis-associated endosome fusion, and trafficking and neurite growth.[184,185]

Alsin contains a pleckstrin homology domain that targets host proteins to membrane by binding to phosphoinositides as well as seven MORN (Membrane Occupation and Recognition Nexus) repeats that are usually found in phosphatidylinositol signalling proteins.[184,185] Mutations in *alsin* gene cause chronic juvenile ALS (ALS2), juvenile primary lateral sclerosis (JPLS) and infantile-onset ascending spastic paralysis—all of which are neurodegenerative conditions—depending on the location of the mutations.[186] Among 12 mutations reported for ALS2, two missense mutations (A46fsX50 and T185fsX189) have been revealed to cause familial ALS2, while the others seem to result in upper motor neuron syndromes.[184,185] There is a report that the long form of alsin protects motor neurons from the toxicity induced by mutant SOD1 through the Rac1/PI3K/Akt3 pathway.[187] This study has raised a possibility that loss of alsin is a upstream signalling pathway of motor neuron degeneration occurring both in ALS1 and ALS2, or that loss of alsin may contribute to development of ALS1.

9.8.2   Alsin Knock-out Mice

The major pathology in ALS2 patients is prominent in the upper motor neuron system, including upper motor neurons in the cortex, corticobulbar and corticospinal tracts (CST), with lesser involvement of the lower motor neuron system. Alsin-deficient mice show variable phenotypes of motor impairment and degenerative pathology in CST without apparent motor neuron pathology.[186–188] A moderate impairment in motor coordination, a higher level of anxiety response and increased susceptibility to oxidative stress were reported by Cai *et al*.[186] Slow degeneration of the corticospinal axons (CSA) in the dorsolateral column was observed in other *Alsin*−/− mouse lines. Motor behavioural abnormalities and disturbances in endosome trafficking were also observed.[189,190] Alsin knock-out mice show predominantly a distal axonopathy rather than a neuropathy in the murine central nervous system, which is similar to the predominant upper motor neuron symptoms in human ALS2 cases. ALS2 is characterized by bilateral pyramidal syndrome, weakness with atrophy and fasciculation of the hands and/or legs, but without sensory disturbance. *Alsin*−/− mice do not display any apparent neurological phenotype even in mice over 400 days old.[186,191] In a Rota-Rod test, both male and female *Alsin*−/− mice showed a significantly reduced motor activity compared with age-matched controls.[186,191]

9.8.3   Phenotypes of Alsin Knock-out Mice

Alsin has two transcriptional forms with two distinct poly (A) signals and encodes one short and one long form of protein product. The short form of *alsin* gene has four exons while the long form of *alsin* gene has 34 exons sharing the first four exons. A 1-bp deletion in exon3 causes frame shift and premature stop codon (A46fsX50), which is verified in the lymphoblasts of human ALS2 patients.[185] A homozygous deletion mutation A46fsX50 that interrupts both forms of alsin proteins leads to ALS2, whereas a homozygous L623fsX647 which only interrupts the long form, leaving the short form intact, leads to JPLS.[184]

The major pathology in ALS2 patients is prominent in the upper motor neuron system, including upper motor neurons in the cortex, corticobulbar and CST, with lesser involvement of the lower motor neuron system. It has been well-documented that transgenic mice overexpressing SOD1^{G93A} show massive ubiquitinated SOD1-containing aggregates in neurons and neurite processes in the spinal cord sections.[8,192] In contrast, *Alsin*−/− mice do not show substantial neuron loss in cortex and spinal cord despite apparent SOD1-positive ubiquitinated aggregates.[191] These data indicates that alsin may not be indispensable for motor neuron survival. In fact, alsin-mediated ALS2 is primarily a distal axonopathy rather than a neuropathy, because the CST around and above the pyramidal decussation is more affected than the CST in the spinal cord. Recently increased susceptibility to glutamate excitotoxicity was observed in the cultured spinal cord slice of alsin-deficient mice.[190,193]

The transgenic lines with high expression of mutant SOD1^{G93A} develop ALS around 100 days of the age.[3,9] However, *Alsin* $-/-$ mice do not display any motor abnormality by the age of one year. This suggests that alsin deficiency is not responsible for the development of ALS1, although alsin levels are decreased in ALS1 cases and SOD1^{G93A} mice. In addition, SOD1^{G93A} mice on the *Alsin* $-/-$ background did not change progression or severity of the disease, suggesting that loss of alsin does not trigger motor neuron degeneration in ALS1.[194] Therefore, it seems likely that the signalling pathways triggering motor neuron degeneration in ALS1 and ALS2 are independent or that the extremely rapid progression of motor dysfunction observed in the high copy number of SOD1^{G93A} could overwhelm the modest symptoms by loss of *Alsin*.

In contrast, alsin-deficiency in SOD1^{H46R} mice has been shown to cause widespread axonal degeneration with slowly progressive motor neuron degeneration in the spinal cord.[195] Alsin-deficient SOD1^{H46R} mice show enhanced accumulation of SOD1 and polyubiquitinated proteins, and macro-autophagy-associated proteins such as polyubiquitin-binding protein p62/SQSTM1 and a lapidated form of light chain 3 (LC3-II).[195] Alsin is colocalized with LC3 and p62 and partly with SOD1 on autophapsome/endosome hybrid compartments, and loss of alsin significantly lowered the lysosome-dependent clearance of LC3 and p62 *in vitro*.[195] Alsin-deficiency impairs the endosomal system and may exacerbate SOD1^{H46R} mediated neurotoxicity by accelerating the accumulation of immature vesicles and misfolded proteins in the spinal cord.

Human ALS2 pathology is not always replicated in mouse models. For example, *Alsin* $-/-$ mice do not show apparent locomotion deficits in their lifetime, while human ALS2 patients are forced to become bedridden before the age of ~ 59 years. This may be due to the anatomical and functional differences between humans and mice. In humans, a substantial proportion of the CST axons are located in the middle portion of the lateral column. In contrast, the main contingent of the CST axons in mice is located in the ventral part of the dorsal column and only a small portion of the CST descends in the dorsal portion of the lateral column.[8,192] Unlike humans, the input from the CST seems not to be essential for normal overground locomotion in rodents, possibly due to the lack of direct cortico-motoneuronal synaptic connections between corticospinal axon boutons and motor neurons in anterior horns.

9.9 Transgenic Mutant TDP-43 Rodent

9.9.1 TDP-43 in ALS

TDP-43 (TAR DNA binding protein 43 kDa) is encoded by *TARDBP*, a highly-conserved gene on human chromosome 1. TDP-43 was initially identified as a transcriptional repressor of HIV-1 gene expression and later found to be a multifunctional protein involved in transcription, splicing and mRNA stabilization. Recent studies show that TDP-43 is a major protein component of neuronal inclusion bodies in the affected tissues in a range of neurodegenerative disorders, including ALS, FTLD, Alzheimer's disease (AD) and other types of dementia.[196] Neuropathology related TDP-43 has currently been identified in a wide spectrum of neurodegenerative diseases collectively termed as TDP-43 proteinopathy, including ALS and FTLD. Decreased protein solubility, hyperphosphorylation, abnormal cleavage and cytoplasmic mis-localization of TDP-43 have been associated with TDP-43 proteinopathy.[196] It is not clear whether TDP-43 proteinopathy is caused by 'loss-of function' or 'gain-of-function' of TDP-43.

In ALS and FTLD patients, TDP-43-immunoreactive inclusions are observed in the cytoplasm and nucleus of both neurons and glial cells, suggesting that the two disorders share the common underlying mechanism[11,12,196] TDP-43 neuropathy induces characteristic abnormal hyperphosphorylation and ubiquitination of TDP-43 and production of $\sim 25\,kDa$ C-terminus fragments lacking the nuclear targeting domains in the brains and spinal cords of patients. TDP-43 is also partly cleared from the nuclei of neurons containing cytoplasmic aggregates, supporting a view that pathogenesis of ALS in these cases may be driven, at least in part, by loss of normal TDP-43 function in the nucleus. Combined with subsequent reports, TDP-43 inclusions are now considered a common characteristic of most ALS patients except those with FALS caused by SOD1 mutations (ALS1).[1,4] However, the pathology alone leaves it unclear as to whether aggregation of TDP-43 is a primary event in ALS pathogenesis or whether it is a by-product of the disease process.

Dominant mutations in the *TARDBP* gene were reported by several groups as a primary cause of ALS (now designated ALS10).[11,12,197–200] A total of 30 different mutations are now known in 22 unrelated families ($\sim 3\%$ of FALS cases) and in 29 sporadic cases of ALS ($\sim 1.5\%$ of sporadic cases).[201] Interestingly, all but one of the mutations identified so far are localized in the C-terminal region encoded by exon 6 of *TARDBP*.[199,201] All these mutations are dominantly inherited missense mutations except a truncating mutation (Y374X) at the extreme C-terminus of the protein.[200] These missense mutations affect amino acids that are highly conserved during evolution.

FTLD and ALS are neurodegenerative diseases that show considerable clinical and pathologic overlaps. Approximately 20% of patients with ALS develop FTLD. FTLD is a relatively common cause of a dementia among patients with onset before 65 years of age, typically manifesting with behavioural changes or language impairment due to degeneration of sub-populations of cortical neurons in the frontal, temporal and insular regions.

A direct role for TDP-43 in neurodegeneration has been supported by recent findings of dominant missense mutations in TDP-43 in FALS patients. FALS-related mutations in TDP-43 were found in the C-terminal glycine-rich region, which is involved in protein–protein interactions between TDP-43 and other heterogeneous nuclear ribonuclear proteins (hnRNPs).[201] Furthermore, C-terminal fragments of TDP-43 are observed selectively in ALS and FTLD tissues, suggesting that proteolytic cleavage of TDP-43 may cause protein aggregation or another toxic property.[202]

9.9.2 Mutant TDP-43 Transgenic Mice

Transgenic mice expressing a human TDP-43 construct containing the A315T mutation, which was identified in FALS patients, were generated under the control of the mouse prion promoter (Prp-TDP-43^{A315T}).[203] Prp-TDP-43^{A315T} transgenic mice were born at normal Mendelian ratios, weighed the same as non-transgenic littermates and appeared normal up to three months of age. In Prp-TDP-43^{A315T} mice, the exogenous TDP-43^{A315T} was expressed highest in the brain and spinal cord, but also expressed at lower level in most other tissues, which is a typical pattern of PrP promoter-driven expression. TDP-43^{A315T} showed the nuclear localization in both neuron and glia throughout brain and spinal cord, similar to the endogenous TDP-43.

Although Prp-TDP-43^{A315T} mice initially appeared normal and weighed the same as their wild-type littermates, Prp-TDP-43^{A315T} mice developed a gait abnormality by 3–4 months of age. By approximately 4–5 months of age, TDP-43^{A315T} mice began losing weight and developed a swimming gait. At this stage, they were unable to hold their body off from the ground, but could use their limbs for propulsion to slide on their stomachs. During this end-stage, they either died spontaneously or had to be euthanized if they were unable to obtain food and water. Average survival of the TDP-43^{A315T} mice is 154 ± 19 days. Despite the universal expression of TDP-43^{A315T} protein in all layers of the cortex, cytoplasmic accumulation of ubiquitinated proteins was detected exclusively in layer 5 in cortex. Increased ubiquitination levels appeared in pyramidal cells and were prominent in motor cortex. Although *Prp-TDP-43^{A315T}* transgene was expressed in the nervous system including caudate/putamen, substantia nigra, thalamus and other structures, no ubiquitin aggregates were observed even at the late/end stage of TDP-43^{A315T} mice.[203] In TDP-43^{A315T} transgenic mice, glial fibrillary acidic protein (GFAP) was also selectively increased in cortical layer 5, suggesting that neuronal degeneration led to local activation of astrocytes and microglia. Tau and α-synuclein were not present as aggregates, resembling FTLD brain pathology.[202] TDP-43^{A315T} is quite selectively involved in certain neuronal subpopulations, including cortical upper motor neurons. Taken together with these results, TDP-43^{A315T} transgenic mice recapitulate key features of human ALS, including ubiquitin-positive aggregates or selective vulnerability of cortical projection neurons and spinal motor neurons, but without the presence of TDP-43-positive aggregates.[202,203]

9.9.3 TDP-43 and FUS/TLS

The identification of TDP-43 mutations in ALS pathogenesis fuelled the discovery of ALS mutations in another DNA/RNA binding protein-encoding gene called FUS (*fused in sarcoma*) or TLS (*translocated in liposarcoma*).[13,14] Vance et al.[14] prioritized sequencing of genes within the linkage region identified in a large British family with familial ALS so as to target genes encoding DNA/RNA proteins like TDP-43. This strategy led to identification of a dominant missense mutation in the FUS/TLS gene on chromosome 16.

Subsequent survey of 197 familial ALS cases identified the same point mutation in four additional families, as well as two additional missense mutations in another four families.[14] Kwiatkowski *et al.* conducted a linkage study in an ALS family originating from the Cape Verde islands in which disease transmission was compatible with an autosomal recessive inheritance pattern.[13] A region of homozygosity shared by all affected members of this family was overlapped with previously reported FALS (ALS6) locus on chromosome 16.[4]

Interestingly, TDP-43 mutant is partially cleared from the nucleus of either neuronal or glial cells when TDP-43 proteins aggregate in the cytoplasm.[11,12] In a minority of neurons from ALS patients with *FUS/TLS* mutations or cells overexpressing symptomatic FUS/TLS mutants, FUS/TLS aggregates are observed in the cytoplasm.[13,14] Cytoplasmic inclusions containing TLS/FUS protein are absent in normal unaffected individuals, in FALS patients with SOD1 mutations, and in SALS patients who presumably are positive for TDP-43 aggregates.[200,202] In contrast, TDP-43-positive inclusions are absent in ALS patients with *FUS/TLS* mutations,[13,14] implying that neurodegenerative processes driven by *FUS/TLS* mutations are independent of TDP-43 aggregation. It will necessary to assess FUS/TLS accumulation and localization in ALS patients with TDP-43 mutations, as well as in patients with other neurodegenerative diseases, especially those with mislocalized TDP-43.

9.9.4 Mutant TDP-43 Transgenic Rats

Transgenic rats expressing normal human TDP-43 or mutant form of humanTDP-43 with a M337 V amino acid substitution (TDP-43^{M337V}) have recently been developed.[204] TDP-43^{M337V} transgenic rats manifest ALS phenotypes such as progressive degeneration of motor neurons and denervation atrophy of skeletal muscles.[204] TDP-43^{M337V} transgenic rats also recapitulate major pathological features of ALS and TDP-43 proteinopathies such as formation of TDP-43 inclusions, cytoplasmic accumulation of phosphorylated TDP-43, and fragmentation of TDP-43 protein.[202]

To fully understand how TDP-43 mutants cause motor neuron degeneration, more sophisticated transgenic models will be required, although rats transgenic for TDP-43^{M337V} have provided evidence that TDP-43 mutation is neurotoxic and develops ALS in rodent models.

9.10   Conclusions

The numerous studies carried out to understand the mechanisms of motor neuron degeneration in ALS by genetic and pathohistological approaches have revealed that ALS can be a multifactorial disease caused by various neurotoxic insults such as mutations in a gene encoding ubiquitously expressed enzymes or defects in the cytoskeleton-dependent intracellular trafficking, all of which lead the selective degeneration of motor neurons.[1,3,4]

ALS is a progressive neurodegenerative disease with limited survival, though promising therapeutic trials are underway. Nonetheless, animal models play a key role in ALS research, though their preclinical application requires extra caution. One of the long-standing arguments is how and when the disease begins and progresses. This problem often recurs even in the use of well-established mutant SOD1 transgenic rodent models. There cannot be a single standard ALS animal model for human ALS because of intrinsic differences between the two organisms. Knowing the various neurodegenerative mechanisms implicated in ALS pathogenesis, researchers must carefully select the disease model based on its applicability to what they are aiming to examine and clarify.

Enormous efforts have been made over the past 15 years, mainly in the study of ALS-related mutations in SOD1, to understand ALS pathophysiology.[3] No consensus has yet been reached as to how SOD1 mutations lead to selective and premature death of motor neurons.[205] Identification of genetic modifiers in ALS has provided potential therapeutic targets in ALS. Nonetheless, the transgenic approach is a robust tool for the identification and evaluation of the effects of genetic modifiers. A recent discovery of ALS-associated mutations in TDP-43 and FUS/TLS, both of which are DNA/RNA-binding proteins implicated in RNA processing,[201,206,207] has shed light on RNA metabolism in ALS pathogenesis.[11-14] Furthermore, a new causal gene product optineurin (OPTN), an adaptor protein which was already known to bind with multiple proteins, has recently been discovered.[15] OPTN may share commonality with TLS/FUS or TDP-43 because, for example, TLS/FUS binds to actin motor protein, myosin Va and VI,[208,209] whereas OPTN also binds to myosin VI.[210] Various animal models of TLS/FUS and OPTN will provide further insights to understanding the pathophysiology of ALS.

There are limitations in the application of the animal models to human ALS due to intrinsic differences in genetics and anatomy. To compensate for these, multiple ALS models should be incorporated into preclinical trials. Translational research has made great progress with the development of ALS rodent models for the past 15 years. However, none of them has been successful yet. Additional breakthroughs in technologies and therapeutic approaches are still demanded to successfully translate animal ALS models to human patients, or *vice versa*.

Acknowledgements

This work was supported by grants from Ministry of Education, Sports and Technology in Japan (RF) and CREST of Japan Science and Technology Agency (TT).

References

1. F. Gros-Louis, C. Gaspar and G. A. Rouleau, *Biochim. Biophys. Acta*, 2006, **1762**, 956–972.
2. A. Hirano, *Adv. Neurol.*, 1991, **56**, 91–101.
3. S. Boillee, C. Vande Velde and D. W. Cleveland, *Neuron*, 2006, **52**, 39–59.
4. P. N. Valdmanis, H. Daoud, P. A. Dion and G. A. Rouleau, *Curr. Neurol. Neurosci. Rep.*, 2009, **9**, 198–205.
5. D. R. Rosen, T. Siddique, D. Patterson, D. A. Figlewicz, P. Sapp, A. Hentati, D. Donaldson, J. Goto, J. P. O'Regan, H.-X. Deng, Z. Rahmani, A. Krizus, D. McKenna-Yasek, A. Cayabyab, S. Gaston, R. Tanzi, J. J. Halperin, B. Herzfeldt, R. Van den Berg, W.-Y. Hung, T. Bird, G. Deng, D. W. Mulder, C. Smith, N. G. Laing, E. Soriano, M. A. Pericak-Vance, J. Haines, G. A. Rouleau, J. Gusella, H. R. Horvitz and R. H. Brown Jr., *Nature*, 1993, **362**, 59–62.
6. L. Dupuis and J. -P. Loeffler, *Curr. Opin.Phrmacol.*, 2009, **9**, 341–346.
7. J. P. Jullien and J. Kriz, *Biochim. Biophys. Acta*, 2006, **1762**, 1013–1124.
8. D. Jaarsma, E. D. Haasdijk, J. A. Grashorn, R. Hawkins, W. Van Duijn, H. W. Verspaget, J. London and J. C. Holstege, *Neurobiol. Dis.*, 2000, **7**, 623–643.
9. A. G. Reaume, J. L. Elliott, E. K. Hoffman, N. W. Kowall, R. J. Ferrante, D. F. Siwek, H. M. Wilcox, D. G. Flood, M. F. Beal, R. H. Brown Jr., R. W. Scott and W. D. Snider, *Nat. Genet.*, 1996, **13**, 43–47.
10. M. E. Gurney, H. Pu, A. Y. Chiu, M. C. Dal Canto, C. Y. Polchow, D. D. Alexander, J. Caliendo, A. Hentati, Y. W. Kwon, H. X. Deng, W. Chen, P. Zhai, R. L. Sufit and T. Siddique, *Science*, 1994, **264**, 1772–1775.
11. E. Kabashi, P. N. Valdmanis, P. Dion, D. Spiegelman, B. J. McConkey, C. Vande Velde, J. P. Bouchard, L. Lacomblez, K. Pochigaeva, F. Salachas, P. F. Pradat, W. Camu, V. Meininger, N. Dupre and G. A. Rouleau, *Nat. Genet.*, 2008, **40**, 572–574.
12. J. Sreedharan, I. P. Blair, V. B. Tripathi, X. Hu, C. Vance, B. Rogelj, S. Ackerley, J. C. Durnall, K. L. Williams, E. Buratti, F. Baralle, J. de Belleroche, J. D. Mitchell, P. N. Leigh, A. Al-Chalabi, C. C. Miller, G. Nicholson and C. E. Shaw, *Science*, 2008, **319**, 1668–1672.
13. T. J. Kwiatkowski Jr, D. A. Bosco, A. L. Leclerc, E. Tamrazian, C. R. Vanderburg, C. Russ, A. Davis, J. Gilchrist, E. J. Kasarskis, T. Munsat, P. Valdmanis, G. A. Rouleau, B. A. Hosler, P. Cortelli, P. J. de Jong, Y. Yoshinaga, J. L. Haines, M. A. Pericak-Vance, J. Yan, N. Ticozzi, T. Siddique, D. McKenna-Yasek, P. C. Sapp, H. R. Horvitz, J. E. Landers and R. H. Brown Jr., *Science*, 2009, **323**, 1205–1208.
14. C. Vance, B. Rogelj, T. Hortobágyi, K. J. De Vos, A. L. Nishimura, J. Sreedharan, X. Hu, B. Smith, D. Ruddy, P. Wright, J. Ganesalingam, K. L. Williams, V. Tripathi, S. Al-Saraj, A. Al-Chalabi, P. N. Leigh, I. P. Blair, G. Nicholson, J. de Belleroche, J. M. Gallo, C. C. Miller and C. E. Shaw, *Science*, 2009, **323**, 1208–1211.

15. H. Maruyama, H. Morino, H. Ito, Y. Izumi, H. Kato, Y. Watanabe, Y. Kinoshita, M. Kamada, H. Nodera, H. Suzuki, O. Komure, S. Matsuura, K. Kobatake, N. Morimoto, K. Abe, N. Suzuki, M. Aoki, A. Kawata, T. Hirai, T. Kato, K. Ogasawara, A. Hirano, T. Takumi, H. Kusaka, K. Hagiwara, R. Kaji and H. Kawakami, *Nature*, 2010, **465**, 223–226.

16. G. Skibinski, N. J. Parkinson, J. M. Brown, L. Chakrabarti, S. L. Lloyd, H. Hummerich, J. E. Nielsen, J. R. Hodges, M. G. Spillantini, T. Thusgaard, S. Brandner, A. Brun, M. N. Rossor, A. Gade, P. Johannsen, S. A. Sørensen, S. Gydesen, E. M. C. Fisher and J. Collinge, *Nat. Genet.*, 2005, **37**, 806–808.

17. S. G. Lindquist, H. Braedgaard, K. Svenstrup, A. M. Isaacs and J. E. Nielsen, FReJA Consortium, *Eur. J. Neurol.*, 2008, **15**, 667–670.

18. J. D. Vechio, L. I. Bruijn, Z. Xu, R. H. Brown Jr. and D. W. Cleveland, *Ann. Neurol.*, 1996, **40**, 603–610.

19. M. Corbo and A. P. Hays, *J. Neuropathol. Exp. Neurol.*, 1992, **51**, 531–537.

20. R. C. Lariviere, J. M. Beaulieu, M. D. Nguyen and J. P. Julien, *Neurobiol. Dis.*, 2003, **13**, 158–166.

21. A. Hirano, H. Donnenfeld, S. Sasaki and I. Nakano, *J. Neuropathol. Exp. Neurol.*, 1984, **43**, 461–470.

22. J. Tomkins, P. Usher, J. Y. Slade, P. G. Ince, A. Curtis, K. Bushby and P. J. Shaw, *Neuroreport*, 1998, **9**, 39670–39670.

23. A. Al-Chalabi, P. M. Andersen, P. Nilsson, B. Chioza, J. L. Andersson, C. Russ, C. E. Shaw, J. F. Powell and P. N. Leigh, *Hum. Mol. Genet.*, 1999, **8**, 157–164.

24. D. A. Figlewicz, A. Krizus, M. G. Martinoli, V. Meininger, M. Dib, G. A. Rouleau and J. P. Julien, *Hum. Mol. Genet.*, 1994, **3**, 1757–1761.

25. P. De Jonghe, I. Mersivanova, E. Nelis, J. Del Favero, J. J. Martin, C. Van Broeckhoven, O. Evgrafov and V. Timmerman, *Ann. Neurol.*, 2001, **49**, 245–249.

26. A. Jordanova, P. De Jonghe, C. F. Boerkoel, H. Takashima, E. De Vriendt, C. Ceuterick, J. J. Martin, I. J. Butler, P. Mancias, S. Papasozomenos, D. Terespolsky, L. Potocki, C. W. Brown, M. Shy, D. A. Rita, I. Tournev, I. Kremensky, J. R. Lupski and V. Timmerman, *Brain*, 2003, **126**, 590–597.

27. F. Gros-Louis, R. Lariviere, G. Gowing, S. Laurent, W. Camu, J. P. Bouchard, V. Meininger, G. A. Rouleau and J. P. Julien, *J. Biol. Chem.*, 2004, **279**, 45951–45956.

28. N. K. Wong, B. P. He and M. J. Strong, *J. Neuropathol. Exp. Neurol.*, 2000, **59**, 972–982.

29. M. E. Ripps, G. W. Huntley, P. R. Hof, J. H. Morrison and J. W. Gordon, *Proc. Natl. Acad. Sci. U. S. A.*, 1995, **92**, 689–693.

30. L. I. Bruijn, M. K. Houseweart, S. Kato, K. A. Anderson, S. D. Anderson, E. Ohama, A. G. Reaume, R. W. Scott and D. W. Cleveland, *Science*, **281**, 1851–1853.

31. L. I. Bruijn, M. W. Becher, M. K. Lee, K. L. Anderson, N. A. Jenkins, N. G. Copeland, S. S. Sisodia, J. D. Rothstein, D. R. Borchelt, D. L. Price and D. W. Cleveland, *Neuron*, 1997, **18**, 327–338.

32. M. E. Gurney, *J. Neurol.*, 1997, **244**, Suppl. 2, S15.

33. D. S. Howland, J. Liu, Y. She, B. Goad, N. J. Maragakis, B. Kim, J. Erickson, J. Kulik, L. DeVito, G. Psaltis, L. J. DeGennaro, D. W. Cleveland and J. D. Rothstein, *Proc. Natl. Acad. Sci. U. S. A.*, 2002, **99**, 1604–1609.

34. M. Nagai, M. Aoki, I. Miyoshi, M. Kato, P. Pasinelli, N. Kasai, R. H. Brown Jr. and Y. Itoyama, *J. Neurosci.*, 2001, **21**, 9246–9254.

35. J. Wang, H. Slunt, V. Gonzales, D. Fromholt, M. Coonfield, N. G. Copeland, N. A. Jenkins and D. R. Borchelt, *Hum. Mol. Genet.*, 2003, **12**, 2753–2764.

36. P. A. Jonsson, K. Ernhill, P. M. Andersen, D. Bergemalm, T. Brannstrom, O. Gredal, P. Nilsson and S. L. Marklund, *Brain*, 2004, **127**, 73–88.

37. L. I. Bruijn, M. K. Houseweart, S. Kato, K. A. Anderson, S. D. Anderson, E. Ohama, A. G. Reaume, R. W. Scott and D. W. Cleveland, *Science*, 1998, **281**, 1851–1854.

38. D. Jaarsma, F. Rognoni, W. van Duijn, H. W. Verspaget, E. D. Haasdijk and J. C. Holstege, *Acta Neuropathol.*, 2001, **102**, 293–305.

39. H. X. Deng, Y. Shi, Y. Furukawa, H. Zhai, R. Fu, E. Liu, G. H. Gorrie, M. S. Khan, W. Y. Hung, E. H. Bigio, T. Lukas, M. C. Dal Canto, T. V. O'Halloran and T. Siddique, *Proc. Natl. Acad. Sci. U. S. A.*, 2006, **103**, 7142–7147.

40. L. B. Corson, J. J. Strain, V. C. Culotta and D. W. Cleveland, *Proc. Natl. Acad. Sci. U. S. A.*, 1998, **95**, 6361–6366.

41. P. A. Jonsson, K. S. Graffmo, P. M. Andersen, T. Brannstrom, M. Lindberg, M. Oliveberg and S. L. Marklund, *Brain*, 2006, **129**, 451–464.

42. P. C. Wong, C. A. Pardo, D. R. Borchelt, M. K. Lee, N. G. Copeland, N. A. Jenkins, S. S. Sisodia, D. W. Cleveland and D. L. Price, *Neuron*, 1997, **14**, 1105–1116.

43. W. J. Broom, C. Russ, P. C. Sapp, D. McKenna-Yasek, B. A. Hosler, P. M. Andersen and R. H. Brown Jr., *Neurosci. Lett.*, 2006, **392**, 52–57.

44. J. Wang, G. Xu, H. Li, V. Gonzales, D. Fromholt, C. Karch, N. G. Copeland, N. A. Jenkins and D. R. Borchelt, *Hum. Mol. Genet.*, 2005, **14**, 2335–2764.

45. M. C. Dal Canto and M. E. Gurney, *Acta. Neuropathol.*, 1997, **93**, 537–550.

46. S. Sasaki, H. Warita, T. Murakami, K. Abe and M. Iwata, *Acta. Neuropathol.*, 2004, **107**, 461–474.

47. J. A. Johnston, M. J. Dalton, M. E. Gurney and R. R. Kopito, *Proc. Natl. Acad. Sci. U. S. A.*, 2000, **97**, 12571–12576.

48. T. Siddique T and H. X. Deng, *Hum. Mol. Genet.*, 1996, **5**, 1465–1470.

49. J. D. Rothstein, L. J. Martin and R. W. Kuncl, *N. Engl. J. Med.*, 1992, **326**, 1464–1468.

50. J. D. Rothstein, M. Van Kammen, A. I. Levey, L. J. Martin and R. W. Kuncl, *Ann. Neurol.*, 1995, **38**, 73–84.
51. J. D. Rothstein, M. Dykes-Hoberg, L. B. Corson, M. Becker, D. W. Cleveland, D. L. Price, V. C. Culotta and P. C. Wong, *J. Neurochem.*, 1999, **72**, 422–429.
52. P. R. Heath and P. J. Shaw, *Muscle Nerve*, 2002, **26**, 438–458.
53. C. Bendotti, M. Tortarolo, S. K. Suchak, N. Calvaresi, L. Carvelli, A. Bastone, M. Rizzi, M. Rattray and T. Mennini, *J. Neurochem.*, 2001, **79**, 737–746.
54. J. S. Deitch, G. M. Alexander, L. Del Valle and T. D. Heiman-Patterson, *J. Neurol. Sci.*, 2002, **193**, 117–126.
55. D. Trotti, A. Rolfs, N. C. Danbolt, R. H. Brown Jr. and M. A. Hediger, *Nat. Neurosci.*, 1999, **2**, 427–433.
56. W. Boston-Howes, S. L. Gibb, E. O. Williams, P. Pasinelli, R. H. Brown Jr. and D. Trotti, *J. Biol. Chem.*, 2006, **281**, 14076–14084.
57. I. I. Kruman, W. A. Pedersen, J. E. Springer and M. P. Mattson, *Exp. Neurol.*, 1999, **160**, 28–39.
58. A. Spalloni, F. Albo, F. Ferrari, N. Mercuri, G. Bernardi, C. Zona and P. Longone, *Neurobiol. Dis.*, 2004, **15**, 340–350.
59. P. Van Damme, M. Leyssen, G. Callewaert, W. Robberecht and L. Van Den Bosch, *Neurosci. Lett.*, 2003, **343**, 81–84.
60. M. Tortarolo, G. Grignaschi, N. Calvaresi, E. Zennaro, G. Spaltro, M. Colovic, C. Fracasso, G. Guiso, B. Elger, H. Schneider, B. Seilheimer, S. Caccia and C. Bendotti, *J. Neurosci. Res.*, 2006, **83**, 134–146.
61. M. Hollmann and S. Heinemann, *Ann. Rev. Neurosci.*, 1994, **17**, 31–108.
62. T. L. Williams, N. C. Day, P. G. Ince, R. K. Kamboj and P. J. Shaw, *Ann. Rev. Neurol.*, 1997, **42**, 200–207.
63. Y. Kawahara, S. Kwak, H. Sun, K. Ito, H. Hashida, H. Aizawa, S. Y. Jeong and I. Kanazawa, *J. Neurochem.*, 2003, **85**, 680–689.
64. Y. Kawahara, K. Ito, H. Sun, H. Aizawa, I. Kanazawa and S. Kwak, *Nature*, 2004, **427**, 801.
65. S. H. Hendry and E. G. Jones, *Brain Res.*, 1991, **543**, 45–55.
66. C. B. Kunst, E. Mezey, M. J. Brownstein and D. Patterson, *Nat. Genet.*, 1997, **15**, 91–94.
67. S. G. Carriedo, S. L. Sensi, H. Z. Yin and J. H. Weiss, *J. Neurosci.*, 2000, **20**, 240–250.
68. S. Sasaki and M. Iwata, *Neurosci. Lett.*, 1999, **268**, 29–32.
69. S. Vielhaber, K. Winkler, E. Kirches, D. Kunz, M. Buchner, H. Feistner, C. E. Elger, A. C. Ludolph, M. W. Riepe and W. S. Kunz, *J. Neurol. Sci.*, 1999, **169**, 133–139.
70. J. Kong and Z. Xu, *J. Neurosci.*, 1998, **18**, 3241–3250.
71. C. Jung, C. M. Higgins and Z. Xu, *J. Neurochem.*, 2002, **83**, 535–545.
72. M. Mattiazzi, M. D'Aurelio, C. D. Gajewski, K. Martushova, M. Kiaei, M. F. Beal and G. Manfredi, *J. Biol. Chem.*, 2002, **277**, 29626–29633.

73. F. R. Wiedemann, G. Manfredi, C. Mawrin, M. F. Beal and E. A. Schon, *J. Neurochem.*, 2002, **80**, 616–625.

74. M. T. Carri, A. Battistoni, F. Polizio, A. Desideri and G. Rotilio, *FEBS Lett.*, 1994, **356**, 314–316.

75. R. H. Swerdlow, J. K. Parks, D. S. Cassarino, P. A. Trimmer, S. W. Miller, D. J. Maguire, J. P. Sheehan, R. S. Maguire, G. Pattee, V. C. Juel, L. H. Phillips, J. B. Tuttle, J. P. Bennett Jr, R. E. Davis and W. D. Parker Jr, *Exp. Neurol.*, 1998, **153**, 135–142.

76. M. Damiano, A. A. Starkov, S. Petri, K. Kipiani, M. Kiaei, M. Mattiazzi, M. Flint Beal and G. Manfredi, *J. Neurochem.*, 2006, **96**, 1349–1361.

77. K. Fukada, F. Zhang, A. Vien, N. R. Cashman and H. Zhu, *Mol. Cell Proteomics*, 2004, **3**, 1211–1223.

78. G. K. Dhaliwal and R. P. Grewal, *Neuroreport*, 2000, **11**, 2507–2509.

79. R. Liu, B. Li, S. W. Flanagan, L. W. Oberley, D. Gozal and M. Qiu, *J. Neurochem.*, 2002, **80**, 488–500.

80. S. Zhu, I. G. Stavrovskaya, M. Drozda, B. Y. Kim, V. Ona, M. Li, S. Sarang, A. S. Liu, D. M. Hartley, D. C. Wu, S. Gullans, R. J. Ferrante, S. Przedborski, B. S. Kristal and R. M. Friedlander, *Nature*, 2002, **417**, 74–78.

81. C. M. Higgins, C. Jung and Z. Xu, *BMC Neurosci.*, 2003, **4**, 16.

82. I. G. Kirkinezos, S. R. Bacman, D. Hernandez, J. Oca-Cossio, L. J. Arias, M. A. Perez-Pinzon, W. G. Bradley and C. T. Moraes, *J. Neurosci.*, 2005, **25**, 164–172.

83. J. Liu, C. Lillo, P. A. Jonsson, C. V. Velde, C. M. Ward, T. M. Miller, J. R. Subramaniam, J. D. Rothstein, S. Marklund, P. M. Andersen, T. Brannstrom, O. Gredal, P. C. Wong, D. S. Williams and D. W. Cleveland, *Neuron*, 2004, **43**, 5–17.

84. D. Jaarsma, F. Rognoni, W. van Duijn, H. W. Verspaget, E. D. Haasdijk and J. C. Holstege, *Acta Neuropathol.*, 2001, **102**, 293–305.

85. P. Pasinelli, M. E. Belford, N. Lennon, B. J. Bacskai, B. T. Hyman, D. Trotti and R. H. Brown Jr., *Neuron*, 2004, **43**, 19–30.

86. X. Mu, J. He, D. W. Anderson, J. Q. Trojanowski and J. E. Springer, *Ann. Neurol.*, 1996, **40**, 379–386.

87. V. Kostic, V. Jackson-Lewis, F. de Bilbao, M. Dubois-Dauphin and S. Przedborski, *Science*, 1997, **277**, 559–562.

88. M. Azzouz, A. Hottinger, J. C. Paterna, A. D. Zurn, P. Aebischer and H. Bueler, *Hum. Mol. Genet.*, 2000, **9**, 803–811.

89. S. Vukosavic, L. Stefanis, V. Jackson-Lewis, C. Guegan, N. Romero, C. Chen, M. Dubois-Dauphin and S. Przedborski, *J. Neurosci.*, 2000, **20**, 9119–9125.

90. K. S. Lowry, S. S. Murray, C. A. McLean, P. Talman, S. Mathers, E. C. Lopes and S. S. Cheema, *Amyotroph. Lateral Scler. Other Motor Neuron Disord.*, 2001, **2**, 127–134.

91. B. J. Turner, I. K. Cheah, K. J. Macfarlane, E. C. Lopes, S. Petratos, S. J. Langford and S. S. Cheema, *J. Neurochem.*, 2003, **87**, 752–763.

92. B. J. Turner, S. S. Murray, L. G. Piccenna, E. C. Lopes, T. J. Kilpatrick and S. S. Cheema, *J. Neurosci. Res.*, 2004, **78**, 193–199.
93. S. Petri, M. Kiaei, E. Wille, N. Y. Calingasan and M. Flint Beal, *J. Neurol. Sci.*, 2006, **251**, 44–49.
94. C. Raoul, A. G. Estevez, H. Nishimune, D. W. Cleveland, O. deLapeyriere, C. E. Henderson, G. Haase and B. Pettmann, *Neuron*, 2002, **35**, 1067–1083.
95. C. Raoul, E. Buhler, C. Sadeghi, A. Jacquier, P. Aebischer, B. Pettmann, C. E. Henderson and G. Haase, *Proc. Natl. Acad. Sci. U. S. A.*, 2006, **103**, 6007–6012.
96. H. Kikuchi, G. Almer, S. Yamashita, C. Guegan, M. Nagai, Z. Xu, A. A. Sosunov, G. M. McKhann 2nd and S. Przedborski, *Proc. Natl. Acad. Sci. U. S. A.*, 2006, **103**, 6025–6030.
97. B. J. Turner and J. D. Atkin, *Curr. Mol. Med.*, 2006, **6**, 79–86.
98. Y. H. Gong, A. S. Parsadanian, A. Andreeva, W. D. Snider and J. L. Elliott, *J. Neurosci.*, 2000, **20**, 660–665.
99. J. Wang, G. Xu, H. H. Slunt, V. Gonzales, M. Coonfield, D. Fromholt, N. G. Copeland, N. A. Jenkins and D. R. Borchelt, *Neurobiol. Dis.*, 2005, **20**, 943–952.
100. M. M. Lino, C. Schneider and P. Caroni, *J. Neurosci.*, 2002, **22**, 4825–4832.
101. L. Wang, K. Sharma, H. X. Deng, T. Siddique, G. Grisotti, E. Liu and R. P. Roos, *Neurobiol. Dis.*, 2008, **29**, 400–408.
102. S. Boillée, K. Yamanaka, C. S. Lobsiger, N. G. Copeland, N. A. Jenkins, G. Kassiotis, G. Kollias and D. W. Cleveland, *Science*, 2006, **312**, 1389–1392.
103. D. R. Beers, J. S. Henkel, J. Wang, A. A. Yen, L. Siklos, S. R. McKercher and S. H. Appel, *Proc. Natl. Acad. Sci. U. S. A.*, 2006, **103**, 16021–16026.
104. J. Kang and S. Rivest, *J. Cell. Biol.*, 2007, **179**, 1219–1230.
105. C. S. Lobsiger, S. Boillée and D. W. Cleveland, *Proc. Natl. Acad. Sci. U. S. A.*, 2007, **104**, 7319–7326.
106. C. S. Lobsieger and D. W. Cleveland, *Nat. Neurosci.*, 2007, **10**, 1355–1360.
107. C. S. Lobsiger, S. Boillée, M. McAlonis-Downes, A. M. Khan, M. L. Feltri, K. Yamanaka and D. W. Cleveland, *Proc. Natl. Acad. Sci. U. S. A.*, 2009, **106**, 4465–4470.
108. A. L. Ström, J. Gal, P. Shi, E. J. Kasarskis, L. J. Hayward and H. Zhu, *J. Neurochem.*, 2008, **106**, 495–505.
109. N. Hirokawa, *Science*, 1998, **279**, 519–526.
110. L. S. Goldstein and Z. Yang, *Annu. Rev. Neurosci.*, 2000, **23**, 39–71.
111. C. Zhao, J. Takita, Y. Tanaka, M. Setou, T. Nakagawa, S. Takeda, H. W. Yang, S. Terada, T. Nakata, Y. Takei, M. Saito, S. Tsuji, Y. Hayashi and N. Hirokawa, *Cell*, 2001, **105**, 587–597.
112. E. Reid, M. Kloos, A. Ashley-Koch, L. Hughes, S. Bevan, I. K. Svenson, F. L. Graham, P. C. Gaskell, A. Dearlove, M. A. Pericak-Vance,

D. C. Rubinstein and D. A. Marchuk, *Am. J. Hum. Genet.*, 2002, **71**, 1189–1194.

113. C. H. Xia, E. A. Roberts, L. S. Her, X. Liu, D. S. Williams, D. W. Cleveland and L. S. Goldstein, *J. Cell Biol.*, 2003, **161**, 55–66.

114. F. Zhang, A. L. Ström, K. Fukada, S. Lee, L. J. Hayward and H. Zhu, *J. Biol. Chem.*, 2007, **282**, 16691–16699.

115. A. L. Ström, P. Shi, F. Zhang, J. Gal, R. Kilty, L. J. Hayward and H. Zhu, *J. Biol. Chem.*, 2008, **283**, 22795–22805.

116. A. J. Reynolds, S. E. Bartlett and I. A. Hendry, *Brain Res. Brain Res. Rev.*, 2000, **33**, 169–178.

117. D. Kieran, M. Hafezparast, S. Bohnert, J. R. Dick, J. Martin, G. Schiavo, E. M. Fisher and L. Greensmith, *J. Cell Biol.*, 2005, **169**, 561–567.

118. T. L. Williamson and D. W. Cleveland, *Nat. Neurosci.*, 1999, **2**, 50–56.

119. M. Hafezparast, R. Klocke, C. Ruhrberg, A. Marquardt, A. Ahmad-Annuar, S. Bowen, G. Lalli, A. S. Witherden, H. Hummerich, S. Nicholson, P. J. Morgan, R. Oozageer, J. V. Priestley, S. Averill, V. R. King, S. Ball, J. Peters, T. Toda, A. Yamamoto, Y. Hiraoka, M. Augustin, D. Korthaus, S. Wattler, P. Wabnitz, C. Dickneite, S. Lampel, F. Boehme, G. Peraus, A. Popp, M. Rudelius, J. Schlegel, H. Fuchs, M. Hrabe de Angelis, G. Schiavo, D. T. Shima, A. P. Russ, G. Stumm, J. E. Martin and E. M. Fisher, *Science*, 2003, **300**, 808–812.

120. B. H. LaMonte, K. E. Wallace, B. A. Holloway, S. S. Shelly, J. Ascano, M. Tokito, T. Van Winkle, D. S. Howland and E. L. Holzbauer, *Neuron*, 2002, **34**, 715–727.

121. L. I. Bruijn, T. M. Miller and D. W. Cleveland, *Ann. Rev. Neurosci.*, 2004, **27**, 723–749.

122. M. Teuchert, D. Fischer, B. Schwalenstoecker, H. J. Habisch, T. M. Bockers and A. C. Ludolph, *Exp. Neurol.*, 2006, **198**, 271–274.

123. H. S. Ilieva, K. Yamanaka, S. Malkmus, O. Kakinohana, T. Yaksh, M. Marsala and D. W. Cleveland, *Proc. Natl. Acad. Sci. U. S. A.*, 2008, **105**, 12599–12604.

124. E. Perlson, G. B. Jeong, J. L. Ross, R. Dixit, K. E. Wallace, R. G. Kalb and E. L. Holzbaur, *J. Neurosci.*, 2009, **29**, 9903–9917.

125. C. Lai, X. Lin, J. Chandran, H. Shim, W. J. Wang and H. Cai, *J. Neurosci.*, 2007, **27**, 13982–13990.

126. L. J. Corcoran, T. J. Mitchison and Q. Liu, *Curr. Biol.*, 2004, **14**, 488–492.

127. F. M. Laird, M. H. Farah, S. Ackerley, A. Hoke, N. Maragakis, J. D. Rothstein, J. Griffin, D. L. Price, L. J. Martin and P. C. Wong, *J. Neurosci.*, 2008, **28**, 1997–2005.

128. M. Hafezparast, A. Ahmad-Annuar, H. Hummerich, P. Shah, M. Ford, C. Baker, S. Bowen, J. E. Martin and E. M. Fisher, *Amyotroph. Lateral Scler. Other Motor Neuron Disord.*, 2003, **4**, 249–257.

129. X. J. Chen, E. N. Levedakou, K. J. Millen, R. L. Wollmann, B. Soliven and B. Popko, *J. Neurosci.*, 2007, **27**, 14515–15524.

130. N. Kimura, K. Tanemura, S. Nakamura, A. Takashima, F. Ono, I. Y. Ishii, S. Kyuwa and Yoshikawa, *Biochim. Biophys Res. Commun.*, 2003, **310**, 303–311.
131. R. Takahashi, H. Yokoji, H. Misawa, M. Hayashi, J. Hu and T. Deguchi, *Nat. Genet.*, 1994, **7**, 79–84.
132. R. W. Orrell, A. W. King, R. J. Lane and J. S. de Bellero, *J. Neurol. Sci.*, 1995, **132**, 123–128.
133. R. Giess, B. Holtmann, M. Braga, T. Grimm, B. Muller-Myhsok, K. V. Toyka and M. Sendtner, *Am. J. Hum. Genet.*, 2002, **70**, 1277–1286.
134. M. Azzouz, G. S. Ralph, E. Storkebaum, L. E. Walmsley, K. A. Mitrophanous, S. M. Kingsman, P. Carmeliet and N. D. Mazarakis, *Nature*, 2004, **429**, 413–417.
135. L. Van Den Bosch, E. Storkebaum, V. Vleminckx, L. Moons, L. Vanopdenbosch, W. Scheveneels, P. Carmeliet and W. Robberecht, *Neurobiol. Dis.*, 2004, **17**, 21–28.
136. A. Brockington, S. B. Wharton, M. Fernando, C. H. Gelsthorpe, L. Baxter, P. G. Ince, C. E. Lewis and P. J. Shaw, *J. Neuropathol. Exp. Neurol.*, 2006, **65**, 26–36.
137. P. G. Ince, J. Tomkins, J. Y. Slade, N. M. Thatcher and P. J. Shaw, *J. Neuropathol. Exp. Neurol.*, 1998, **57**, 895–904.
138. C. A. Ross and M. A. Poirier, *Nat. Med.*, 2004, **10**, Suppl., S10–17.
139. S. M. Chou, H. S. Wang and A. Taniguchi, *J. Neurol. Sci.*, 1996, **139**(Suppl), 16–26.
140. M. Watanabe, M. Dykes-Hoberg, V. C. Culotta, D. L. Price, P. C. Wong and J. D. Rothstein, *Neurobiol. Dis.*, 2001, **8**, 933–941.
141. C. M. Karch, M. Prudencio, D. D. Winkler, P. J. Hart and D. R. Borchelt, *Proc. Natl. Acad. Sci. U. S. A.*, 2009, **106**, 7774–7779.
142. H. Tummala, C. Jung, A. Tiwari, C. M. Higgins, L. J. Hayward and Z. Xu, *J. Biol. Chem.*, 2005, **280**, 17725–17731.
143. M. Urushitani, J. Kurisu, K. Tsukita and R. Takahashi, *J. Neurochem.*, 2002, **83**, 1030–1042.
144. E. Kabashi and H. D. Durham, *Biochim. Biophys. Acta.*, 2006, **1762**, 1038–1050.
145. M. Basso, T. Massignan, G. Samengo, C. Cheroni, S. De Biasi, M. Salmona, C. Bendotti and V. Bonetto, *J. Biol. Chem.*, 2006, **281**, 33325–33335.
146. J. Niwa, S. Ishigaki, N. Hishikawa, M. Yamamoto, M. Doyu, S. Murata, K. Tanaka, N. Taniguchi and G. Sobue, *J. Biol. Chem.*, 2002, **277**, 36793–36798.
147. D. Kieran, B. Kalmar, J. R. Dick, J. Riddoch-Contreras, G. Burnstock and L. Greensmith, *Nat. Med.*, 2004, **10**, 402–405.
148. K. Puttaparthi, L. Van Kaer and J. L. Elliott, *Exp. Neurol.*, 2007, **206**, 53–58.
149. F. Cote, J. F. Collard and J. P. Julien, *Cell*, 1993, **73**, 35–46.

150. S. Couillard-Després, Q. Zhu, P. C. Wong, D. L. Price, D. W. Cleveland and J. P. Julien, *Proc. Natl. Acad. Sci. U. S. A.*, 1998, **95**, 9626–9630.
151. J. Meier, S. Couillard-Després, H. Jacomy, C. Gravel and J. P. Julien, *J. Neuropathol. Exp. Neurol.*, 1999, **58**, 1099–1110.
152. T. L. Williamson, L. I. Bruijn, Q. Zhu, Q. K. L. Anderson, S. D. Anderson, J. P. Julien and D. W. Cleveland, *Proc. Natl. Acad. Sci. U. S. A.*, 1998, **95**, 9631–9636.
153. H. Lin, J. Zhai, R. Cañete-Soler and W. W. Schlaepfer, *J. Neurosci.*, 2004, **24**, 2716–2726.
154. H. Lin, J. Zhai and W. W. Schlaepfer, *Hum. Mol. Genet.*, 2005, **14**, 3643–3659.
155. J. Kong and Z. Xu, *J. Neurosci. Lett.*, 2000, **281**, 72–74.
156. S. Couillard-Despres, J. Meier and J. P. Julien, *Neurobiol. Dis.*, 2000, **7**, 462–470.
157. P. H. Tu, P. Raju, K. A. Robinson, M. E. Gurney, J. Q. Trojanowski and V. M. Lee, *Proc. Natl. Acad. Sci. U. S. A.*, 1996, **93**, 3155–3160.
158. J. P. Julien and J. M. Beaulieu, *J. Neurol. Sci.*, 2000, **180**, 7–14.
159. J. Robertson, M. M. Doroudchi, M. D. Nguyen, H. D. Durham, M. J. Strong, G. Shaw, J. P. Julien and W. E. Mushynski, *J. Cell Biol.*, 2003, **160**, 939–949.
160. J. L. Elliott, *Brain Res. Mol. Brain Res.*, 2001, **95**, 172–178.
161. K. Yasojima, W. W. Tourtellotte, E. G. McGeer and P. L. McGeer, *Neurology*, 2001, **57**, 952–956.
162. K. Hensley, J. Fedynyshyn, S. Ferrell, R. A. Floyd, B. Gordon, P. Grammas, L. Hamdheydari, M. Mhatre, S. Mou, Q. N. Pye, C. Stewart, M. West, S. West and K. S. Williamson, *Neurobiol. Dis.*, 2003, **14**, 74–80.
163. T. Yoshihara, S. Ishigaki, M. Yamamoto, Y. Liang, J. Niwa, H. Takeuchi, M. Doyu and G. Sobue, *J. Neurochem.*, 2002, **80**, 158–167.
164. M. West, M. Mhatre, A. Ceballos, R. A. Floyd, P. Grammas, S. P. Gabbita, L. Hamdheydari, T. Mai, S. Mou, Q. N. Pye, C. Stewart, S. West, K. S. Williamson, F. Zemlan and K. Hensley, *J. Neurochem.*, 2004, **91**, 133–143.
165. P. N. Pompl, L. Ho, M. Bianchi, T. McManus, W. Qin and G. M. Pasinetti, *FASEB J.*, 2003, **17**, 725–727.
166. P. Klivenyi, M. Kiaei, G. Gardian, N. Y. Calingasan and M. F. Beal, *J. Neurochem.*, 2004, **88**, 576–582.
167. N. Mizushima, B. Levine, A. M. Cuervo and D. J. Klionsky, *Nature*, 2008, **451**, 1069–1075.
168. K. Kanekura, H. Suzuki, S. Aiso and M. Matsuoka, *Mol. Neurobiol.*, 2009, **39**, 81–89.
169. M. Komatsu, S. Waguri, M. Koike, Y. S. Sou, T. Ueno, T. Hara, N. Mizushima, J. Iwata, J. Ezaki, S. Murata, J. Hamazaki, Y. Nishito, S. Iemura, T. Natsume, T. Yanagawa, J. Uwayama, E. Warabi, H. Yoshida, T. Ishii, A. Kobayashi, M. Yamamoto, Z. Yue, Y. Uchiyama, E. Kominami and K. Tanaka, *Cell*, 2007, **131**, 1149–1163.

170. T. Kabuta, Y. Suzuki and K. Wada, *J. Biol. Chem.*, 2006, **281**, 30524–30533.
171. J. Gal, A. L. Storm, R. Kilty, F. Zhang and H. Zhu, *J. Biol. Chem.*, 2007, **282**, 11068–11077.
172. Ravikumar, A. Acevedo-Arozena, S. Imarisio, Z. Berger, C. Vacher, C. J. O'Kane, S. D. Browm and D. C. Rubinsztein, *Nat. Genet.*, 2005, **37**, 771–776.
173. J. Dunlop, H. Beal McIlvain, Y. She and D. S. Howland, *J. Neurosci.*, 2003, **23**, 1688–1696.
174. H. Z. Yin, D. T. Tang and J. H. Weiss, *Exp. Neurol.*, 2007, **207**, 177–185.
175. C. Boucherie, A. S. Caumont, J. M. Maloteaux and E. Hermans, *Exp. Neurol.*, 2008, **212**, 557–561.
176. W. Tang, U. Tasch, N. K. Neerchal, L. Zhu and P. J. Yarowsky, *Neurosci. Methods*, 2009, **176**, 254–262.
177. M. Suzuki, S. Klein, E. A. Wetzel, M. Meyer, J. McHugh, C. Tork, A. Hayes and C. N. Svendsen, *Exp. Neurol.*, 2010, **221**, 346–352.
178. J. E. Landers, A. L. Leclerc, L. Shi, A. Virkud, T. Cho, M. M. Maxwell, A. F. Henry, M. Polak, J. D. Glass, T. J. Kwiatkowski, A. Al-Chalabi, C. E. Shaw, P. N. Leigh, I. Rodriguez-Leyza, D. McKenna-Yasek, P. C. Sapp and R. H. Brown Jr., *Neurology*, 2008, **70**, 1179–1185.
179. K. Kanekura, I. Nishimoto, S. Aiso and M. Matsuoka, *J. Biol. Chem.*, 2006, **281**, 30223–30233.
180. E. Teuling, S. Ahmed, E. Haasdijk, J. Demmers, M. O. Steinmetz, A. Akhmanova, D. Jaarsma and C. C. Hoogenraad, *J. Neurosci.*, 2007, **27**, 9801–9815.
181. H. Suzuki, K. Kanekura, T. P. Levine, K. Kohno, V. M. Olkkonen, S. Aiso and M. Matsuoka, *J. Neurochem.*, 2009, **108**, 973–985.
182. D. Kieran, I. Woods, A. Villunger, A. Strasser and J. H. Prehn, *Proc. Natl. Acad. Sci. U. S. A.*, 2007, **104**, 20606–20611.
183. J. D. Atkin, M. A. Farg, B. J. Turner, D. Tomas, J. A. Lysaght, J. Nunan, A. Rembach, P. Nagley, P. M. Beart, S. S. Cheema and M. K. Horne, *J. Biol. Chem.*, 2006, **281**, 30152–30165.
184. S. Hadano, C. K. Hand, H. Osuga, Y. Yanagisawa, A. Otomo, R. S. Devon, N. Miyamoto, J. Showguchi-Miyata, Y. Okada, R. Singaraja, D. A. Figlewicz, T. Kwiatkowski, B. A. Hosler, T. Sagie, J. Skaug, J. Nasir, R. H. Brown Jr, S. W. Scherer, G. A. Rouleau, M. R. Hayden and J. E. Ikeda, *Nat. Genet.*, 2001, **29**, 166–173.
185. Y. Yang, A. Hentati, H. X. Deng, O. Dabbagh, T. Sasaki, M. Hirano, W. Y. Hung, K. Ouahchi, J. Yan, A. C. Azim, N. Cole, G. Gascon, A. Yagmour, M. Ben-Hamida, M. Pericak-Vance, F. Hentati and T. Siddique, *Nat. Genet.*, 2001, **2**, 160–165.
186. H. Cai, X. Lin, C. Xie, F. M. Laird, C. Lai, H. Wen, H. C. Chiang, H. Shim, M. H. Farah, A. Hoke, D. L. Price and P. C. Wong, *J. Neurosci.*, 2005, **25**, 7567–7574.
187. S. Hadano, S. C. Benn, S. Kakuta, A. Otomo, K. Sudo, R. Kunita, K. Suzuki-Utsunomiya, H. Mizumura, J. M. Shefner, G. A. Cox,

Y. Iwakura, R. H. Brown Jr and J. E. Ikeda, *Hum. Mol. Genet.*, 2006, **15**, 233–250.

188. K. Yamanaka, T. M. Miller, M. McAlonis-Downes, S. J. Chun and D. W. Cleveland, *Ann. Neurol.*, 2006, **60**, 95–104.

189. R. S. Devon, P. C. Orban, K. Gerrow, M. A. Barbieri, C. Schwab, L. P. Cao, J. R. Helm, N. Bissada, R. Cruz-Aguado, T. L. Davidson, J. Witmer, M. Metzler, C. K. Lam, W. Tetzlaff, E. M. Simpson, J. M. McCaffery, A. E. El-Husseini, B. R. Leavitt and M. R. Hayden, *Proc. Natl. Acad. Sci. U. S. A.*, 2006, **103**, 9595–9600.

190. C. Lai, C. Xie, H. Shim, J. Chandran, B. W. Howell and H. Cai, *Mol. Brain*, 2009, **2**, 23.

191. H. X. Deng, H. Zhai, R. Fu, Y. Shi, G. H. Gorrie, Y. Yang, E. Liu, M. C. Dal Canto, E. Mugnaini and T. Siddique, *Hum. Mol. Genet.*, 2007, **16**, 2911–2920.

192. J. Niwa, S. Yamada, S. Ishigaki, J. Sone, M. Takahashi, M. Katsuno, F. Tanaka, M. Doyu and G. Sobue, *J. Biol. Chem.*, 2007, **282**, 28087–28095.

193. C. Lai, C. Xie, S. G. McCormack, H. C. Chiang, M. K. Michalak, X. Lin, J. Chandran, H. Shim, M. Shimoji, M. R. Cookson, R. L. Huganir, J. D. Rothstein, D. L. Price, P. C. Wong, L. J. Martin, J. J. Zhu and H. Cai, *J. Neurosci.*, 2006, **26**, 11798–11806.

194. X. Lin, H. Shim and H. Cai, *Neurobiol. Aging*, 2007, **28**, 1628–1630.

195. S. Hadano, A. Otomo, R. Kunita, K. Suzuki-Utsunomiya, A. Akatsuka, M. Koike, M. Aoki, Y. Uchiyama, Y. Itoyama and J. E. Ikeda, *PLoS One*, 2010, **5**, e9805.

196. M. Neumann, D. M. Sampathu, L. K. Kwong, A. C. Truax, M. C. Micsenyi, T. T. Chou, J. Bruce, T. Schuck, M. Grossman, C. M. Clark, L. F. McCluskey, B. L. Miller, E. Masliah, I. R. Mackenzie, H. Feldman, W. Feiden, H. A. Kretzschmar, J. Q. Trojanowski and V. M. Lee, *Science*, 2006, **314**, 130–133.

197. G. T. Banks, A. Kuta, A. M. Isaacs and E. M. Fisher, *Mamm. Genome*, 2008, **19**, 299–305.

198. V. M. Van Deerlin, J. B. Leverenz, L. M. Bekris, T. D. Bird, W. Yuan, L. B. Elman, D. Clay, E. M. Wood, A. S. Chen-Plotkin, M. Martinez-Lage, E. Steinbart, L. McCluskey, M. Grossman, M. Neumann, I. L. Wu, W. S. Yang, R. Kalb, D. R. Galasko, T. J. Montine, J. Q. Trojanowski, V. M. Lee, G. D. Schellenberg and C. E. Yu, *Lancet Neurol.*, **7**, 409–416.

199. L. Corrado, A. Ratti, C. Gellera, E. Buratti, B. Castellotti, Y. Carlomagno, N. Ticozzi, L. Mazzini, L. Testa, F. Taroni, F. E. Baralle, V. Silani and S. D'Alfonso, *Hum. Mutat.*, 2009, **30**, 688–694.

200. H. Daoud, P. N. Valdmanis, E. Kabashi, P. Dion, N. Dupré, W. Camu, V. Meininger and G. A. Rouleau, *J. Med. Genet.*, 2009, **46**, 112–114.

201. C. Lagier-Tourenne and D. W. Cleveland, *Cell*, 2009, **136**, 1001–1004.

202. T. Arai, M. Hasegawa, T. Nonoka, F. Kametani, M. Yamashita, M. Hosokawa, K. Niizato, K. Tsuchiya, Z. Kobayashi, K. Ikeda,

M. Yoshida, M. Onaya, H. Fujishiro and H. Akiyama, *Neuropathology*, 2010, **30**, 170–181.

203. I. Wegorzewska, S. Bell, N. J. Cairns, T. M. Miller and R. H. Baloh, *Proc. Natl. Acad. Sci. U. S. A.*, 2009, **106**, 18809–18814.

204. H. Zhou, C. Huang, H. Chen, D. Wang, C. P. Landel, P. Y. Xia, R. Bowser, Y. J. Liu and X. G. Xia, *PLoS Genet.*, 2010, **6**, e1000887.

205. K. Yamanaka, S. Boillee, E. A. Roberts, M. L. Garcia, M. McAlonis-Downes, O. R. Mikse, D. W. Cleveland and L. S. Goldstein, *Proc. Natl. Acad. Sci. U. S. A.*, 2008, **105**, 7594–7599.

206. R. Fujii, S. Okabe, T. Urushido, K. Inoue, A. Yoshimura, T. Tachibana, T. Nishikawa, G. G. Hicks and T. Takumi, *Curr. Biol.*, 2005, **15**, 587–593.

207. R. Fujii and T. Takumi, *J. Cell Sci.*, 2005, **118**, 5755–5765.

208. A. Yoshimura, R. Fujii, Y. Watanabe, S. Okabe, K. Fukui and T. Takumi, *Curr. Biol.*, 2006, **16**, 2345–2351.

209. T. Takarada, K. Tamaki, T. Takumi, M. Ogura, Y. Ito, N. Nakamichi and Y. Yoneda, *J. Neurochem.*, 2009, **110**, 1457–1468.

210. D. A. Sahlender, R. C. Roberts, S. D. Arden, G. Spudich, M. J. Taylor, J. P. Luzio, J. Kendrick-Jones and F. Buss, *J. Cell Biol.*, 2005, **169**, 285–295.

CHAPTER 10

Animal Models for Huntington's Disease

ZAIRA ORTEGA AND JOSÉ J. LUCAS[*]

Centro de Biología Molecular 'Severo Ochoa' (CBMSO), Consejo
Superior de Investigaciones Científicas (CSIC), Universidad Autonoma
de Madrid and CiberNed, Madrid, Spain

10.1 Huntington's Disease

Huntington's disease (HD) is a genetic autosomal dominant neurodegenerative disease in which the symptoms begin in adulthood (although it also exists in a juvenile form in which the symptoms start earlier at the age of two, and a senile form in which symptoms start at the age of 70 or 80[1]) and last over 10–15 years until death.[2] The disease is characterized by motor dysfunction, cognitive decline and psychological dysfunction, and there is currently no effective treatment to prevent or delay disease progression.[3]

All the studies on HD succeed in identifying the age of onset of the symptoms,[1,4–6] but since previous to the obvious symptoms of extrapyramidal dysfunction (three years before) thee are other minor symptoms (tremor, involuntary eye movements, hyperreflexia, etc.) that may remain unnoticed, identifying the age at which the disease starts is more difficult to establish.[7] Motor symptoms include, among others, rigidity, dystonia and oculomotor dysfunction.[8] Cognitive dysfunction involves subcortical dementia including affective and personality changes, problems in acquiring new knowledge, depression suicide, and maniac and psychotic symptoms.[9,10]

RSC Drug Discovery Series No. 6
Animal Models for Neurodegenerative Disease
Edited by Jesús Avila, Jose J. Lucas and Félix Hernández
© Royal Society of Chemistry 2011
Published by the Royal Society of Chemistry, www.rsc.org

The neuropathology of the disease involves striatal (caudate and putamen) atrophy and gliosis due to the death of the GABAergic medium size spiny neurons of the striatum (St) that project to the substantia nigra (SN) and globus pallidus (GP).[11] The pyramidal neurons present in layers III, V, VI of the cortex (Cx) are also affected in HD, but to a lesser extent.[12,13] This degeneration spreads to other brain regions as the disease progresses, spanning the GP, subthalamic nucleus (STN), SN, hippocampus (Hipp), spinal cord and others.[3,13]

The mutation responsible for HD consists of an expansion of a CAG repeat in the huntingtin gene (*htt*).[14] This triplet expansion encodes a polyglutamine stretch (polyQ) in the N' terminus of the protein huntingtin (htt).[15,16] Normal individuals have between six and 35 CAG triplets, while expansions longer than 40 repeats lead to HD.[14,17] The onset and the severity of the disease depend directly on the length of the polyQ tract; the longer the polyQ, the earlier the disease begins and the more severe the symptoms are.[17,18] The main histopathological hallmark of HD is the presence of neuronal inclusion bodies (IBs) containing the polyQ-harbouring protein,[19,20] mainly localized in the layers V and VI of the Cx—even in stages of the disease in which degeneration is not yet detected.[21,22] Nevertheless, in spite of the St being the most affected brain structure in HD, IBs are rarely localized in it.[21,22]

10.2   Invertebrate Models of Huntington's Disease

The mechanisms that are affected in HD are still unknown. But because HD is the consequence of a very well identified and characterized single mutation, this makes it a good candidate about which to generate animal models.

Rodents are the most widely used animal models, although lately non-human primates are being used quite often. Other invertebrate species such as *Caenorhabditis elegans* and *Drosophila melanogaster* are used too in the study of HD, though to a lesser degree. These invertebrate models allow a quick test of the hypothesis and the study of new therapeutic strategies.

10.2.1   *Caenorhabditis elegans*

C. elegans is a nematode widely used in developmental studies. It has neither the homolog for the human htt gene nor polyQ sequences in any of its proteins, both of which are very useful characteristics for an HD model since the phenotype will be due only to the inserted transgen.[23] However, it has orthologous proteins that interact with htt, which could interact with the inserted transgenic htt[24] and allow us to perform protein–protein interaction studies.[25]

The main HD model based on *C. elegans* contains the first 57 amino acids of the human htt with a polyQ under the control of a promoter expressed only in mecanosensorial neurons.[26] Its phenotype consists of mecanosensorial deficit in the tail, axonal abnormal morphologies, and neuronal dysfunction directly proportional to the polyQ length. All this takes place in absence of neuronal death. Apart from these features it also presents perinuclear proteinaceous aggregates or IBs, though these are not a consequence of the polyQ length because they are also present when the length of the expressed polyQ is not pathogenic.[27]

10.2.2   *Drosophila melanogaster*

At least 50% of the genes of *Drosophila melanogaster* (fruit fly) are similar to human genes[28] and it has orthologous genes for up to 75% of the genes related to human diseases.[29] Transgenic models based on *Drosophila* are obtained from microinjection of fly embryos with a conditional promoter regulated by the transactivator GAL4 (UAS) followed by the exon 1 of the human htt or the full length htt followed by a pathogenic polyQ (37 CAG or more). This fly has to be crossed with strains expressing the transactivator GAL4, which will bind to the promoter UAS, to express the transgenic construction. Several tissue-specific GAL4 strains are already developed, thus restricting the expression of mutant htt (htt*) to specific tissue.[30] Htt* expression in *Drosophila* triggers late neuronal degeneration similar to the human degeneration, and unleashes motor deficits and early neuronal death.[31] The severity of the disease, as in humans, is progressive and depends on polyQ length. This model has been used to study the affected mechanisms in HD as well as possible treatments for the disease.

10.3   Vertebrate Models of Huntington's Disease

10.3.1   Mouse Models

Rodents have been used in laboratories for over three decades, with the common house mouse (*Mus musculus*) is the most used in laboratories. Their genome is very similar to the human genome and they have orthologous genes for almost every protein involved in human disease.

HD mouse models can be classified into two different categories depending on the procedure used to generate them:

- chemically induced animal models;
- genetic mouse models.

10.3.1.1   Chemically Induced Mouse Models

10.3.1.1.1   Direct Excitotoxic Mouse Models.   The word 'excitotoxic' refers to the harmful effect that high concentrations of glutamate or agonists produce in neurons (by intrastriatal injection or systemic administration). Glutamate or glutamic acid is a non-essential amino acid and the main excitatory neurotransmitter released by the corticostriatal afferents neurons in the St; this is why glutamate agonists are so commonly used to obtain HD mouse models.[32] Glutamate selectively activates ionotropic (ion channels) and metabotropic receptors (with seven transmembrane domains or G-protein coupled receptors).

This mouse model reproduces some of the HD features such as hyperkinesia, motor deficit, and spatial and executive learning deficit. However, it does not reproduce the progressive worsening of the disease.[33] Moreover, the characteristic IBs made of htt* observed in human brains and in other mouse models are not present in this model because it does not express the mutation; these mice develop the disease by exogenously adding the compounds that trigger the degeneration.

10.3.1.1.1.1 Quinolinic Acid. Quinolinic acid (QA) is the most commonly used protocol to obtain mouse and non-human primates models of HD. QA is one of two products derived from the metabolism of tryptophan in the kynurenine pathway and acts as selective glutamate agonists for *N*-methyl-D-aspartate (NMDA) receptors. Normal levels of QA do not cause damage, but only small increases in QA levels cause toxicity. QA is not able to cross the blood–brain–barrier and is therefore administered directly to the St.[34] QA administration produces cellular depolarization due to intracellular calcium influx; as a consequence of this influx, free radicals are produced which carry out the death of the GABAergic striatal neurons,[35–37] leaving the neurons expressing NADPH diaphorase and the cholinergic neurons intact.[38–40] Other brain regions such as GP and SN are also affected by QA administration, although the damage present in these areas is not due to the QA toxicity but to the striatal release of excitotoxic substances—nitric oxide, glial fibrillar acidic protein (GFAP)—in these areas. Both unilateral and bilateral QA injections produce behavioural and neuroanatomic subtle phenotypes, similar to the early symptoms of HD.

10.3.1.1.1.2 Kainic Acid. Kainic acid (KA) was the first toxin used to produce a HD model.[41] It is an acidic pyrrolidine isolated from *Digenea simplex* seaweed and acts as a glutamate agonist to ionotropic receptors. It is not as selective for the striatal GABAergic neurons as QA as it also affect NADPH diaphorase positive interneurons and projective neurons.[42,43] KA injections also damage brain areas far from the injury reason, which is why it was replaced by QA as the main HD model used.

Excitotoxic mouse models triggered the use of glutamate antagonists as HD treatments. NMDA receptor antagonists such as riluzole or amantidine are frequently used in HD patients.[44]

10.3.1.1.2 Indirect Excitotoxic Mouse Models. These mouse models are the result of the administration of mitochondrial toxins. These toxins produce an increase in the generation of lactate and an adenosine triphosphate (ATP) depletion that trigger a disruption in the energetic metabolism of the mitochondria, directly responsible for the specific degeneration of the GABAergic neurons and indirectly responsible for the neuronal excitotoxicity. As in direct excitotoxic mouse models, these indirect models neither reproduce some of the motor symptoms of HD (mainly the progressive factor of the disease) nor present IBs due to the absence of the mutated protein.

Several mitochondrial toxins are used to obtain these animal models (malonate, Mn^{2+}, MPP^+, aminooxyacetate, rotenone, 3-acetyl-pyridine), although the most widely used is the 3-nitropropionic acid.

3-Nitropropionic acid (3-NP) acts as a chronic and irreversible inhibitor of the mitochondrial succinate dehydrogenase enzyme, responsible for the oxidation of succinate to fumarate, present in the complex II of the Krebs cycle and localized in the internal membrane of the mitochondria.[42,45,46] The disruption of the mitochondrial activity is associated with the formation of reactive oxygen species (ROS) such as superoxide, hydroxyl and hydrogen peroxide radicals that damage the cellular membrane and the DNA.[47] Moreover, a deficiency in some enzymes involved in the tricarboxylic acid cycle and the electron transport chain, and a reduction in the activity of aconitase and complexes II, III and IV have been described in the caudate and putamen of the brains of HD patients.[47,48]

Intraperitoneal administration of 3-NP causes selective bilateral degeneration in the dorsolateral St[49,50] similar to the most severe putamen dorsolateral phenotype in HD brains.[13] High doses of 3-NP administered over short periods of time (1–5 days) do not reproduce HD pathology because they produce total depletion of the neurons present in the injection area. Chronic administration (approximately one month) of low doses of 3-NP (10–12 mg/kg per day) do reproduce hypokinetic movements as well as working memory and attention deficits similar to the early symptoms of HD.[51,52] Diffuse neuronal death gradually increases from non-affected striatal areas to the injury. Around the injury loss of GABAergic neurons can be observed, however NADPH-diaforase positive neurons and afferent dopaminergic neurons remain intact.[49,53,54]

Although this mouse model reproduces some of the main features of the disease, the abnormal movements displayed by this model are not similar to those present in HD patients due to the different organization of the basal ganglia in rodents with respect to primates.[51,54] Moreover, injuries produced by the administration of 3-NP are very heterogeneous and just half of them produce striatal injuries. But despite the limitations of this model, it is highly recommended for the study of motor treatments and mitochondrial physiology in HD.[55]

10.3.1.2   Genetic Mouse Models

Since the mutation responsible for HD was discovered in 1993,[14] several genetic mouse models of HD have been produced. These models carry modifications in the htt gene that trigger neurological, electrophysiological and behavioural degeneration.

10.3.1.2.1 Modification of the Endogenous htt Protein

10.3.1.2.1.1 Knock-out Genetic Models. Knock-out genetic models are obtained by removing the murine orthologous gene of htt protein (Figure 10.1). This mouse model showed the relevance of htt during embryonic development since its absence led to embryonic lethality at day 8.5;[56,57] however, the heterozygous knock-out mice developed normally. The same happened when these mice were crossed with transgenic mice not carrying normal htt but one allele of htt*.[58] The model has also been studied with mice lacking the seven CAG normal repeats present in murine htt.[59] These mice were born under Mendelian frequency and did not show the HD phenotype either at an early age or at later ages. These results prove that the polyQ is not involved in htt function during embryonic development.

There is a conditional knock-out model of HD obtained using the Cre/loxP approach[60] where htt is under the control of the CamKIIα promoter in order to postnatally inhibit htt expression in brain and testis. This mouse model displays behavioural phenotype and progressive neuronal degeneration similar to other transgenic models of HD. Using this model it has been shown that htt is vital for neuronal function and survival, and that htt loss of function is not responsible for HD although it contributes to the pathogenesis.

10.3.1.2.1.2 Knock-in Genetic Models. Knock-in genetic models are the best for reproducing human pathogenesis. They are obtained by inserting polyQ expansions in the same location along murine orthologous htt gene where it should naturally be (Figure 10.1). As a result, these models carry either both copies (homozygous) of the htt gene, or just one of them (heterozygous), mutated and under the control of endogenous htt promoter. There are several knock-in models carrying from 50[58] to 150[61] CAG repeats. Homozygous knock-in models develop behavioural deficits at a very early age, even previous to the neuropathology. Moreover, they present striatal pathology in absence of striatal degeneration. From two months (depending on the model), htt translocates to the nucleus promoting the appearance of nuclear IBs.[61–64] Some models show molecular (decreased enkephalin mRNA levels in the St)[65] and cellular (increased glutamate sensibility)[66] alterations, though none of them show neuronal death or gliosis even in very old age.[61] These results prove that neuronal dysfunction precedes neuronal death, which is in good agreement with the presence of subtle motor deficits several years before the appearance of other mayor motor symptoms or atrophy in HD patients.[67]

10.3.1.2.2 Transgenic Genetic Mouse Models.

These models are obtained by hazardously inserting full human htt* gene, or part of it, into the mouse genome under the control of different promoters that restrict the expression to specific brain areas (Figure 10.1). These models express two copies of normal murine htt and one copy of human htt*. They can be classified into

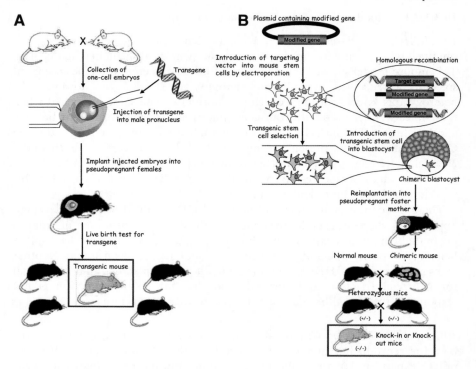

Figure 10.1 Diagram explaining transgenic, knock-in and knock-out mice generation. (**A**) Development of transgenic mice by implanting injected embryos
carrying the modified transgene into pseudopregnant foster mothers.
(**B**) Developed of knock-in or knock-out mice by implanting in
pseudopregnant foster mothers blastocists carrying the modified gene.

two different groups depending on the length of the htt* fragment inserted in
the mouse genome. Table 10.1 summarizes the main features of selected
transgenic genetic mouse models.

10.3.1.2.2.1 Mouse Carrying a Fragment of Human htt Protein. Usually,
the inserted region corresponds to exon 1 or exons 1 and 2 of human htt
carrying a polyQ expansion in its *N*-terminus. Some of these models are
described briefly below.

- R6/1 and R6/2 are the most used models. They ubiquitously express the
  exon 1 of human htt carrying 116 or 144 CAG repeats, respectively, under
  the control of human htt promoter.[68] The long polyQ tract triggers
  an early and severe phenotype, similar to juvenile HD, with a very
  fast progression. These mouse models present progressive body weight
  loss (their final weight is 70% lower than that of the non-transgenic
  mouse[68]), cerebral volume and weight loss (starting at 30 and 60 postnatal

Table 10.1 Main characteristics of the genetic mouse models. This table summarizes genetic background, htt amount considering 1× normal levels of htt in mice, the promoter that controls the transgene expression, htt length and number of CAG repeats, age of onset of symptoms expressed in days (d) and months (m), cerebral structures harbouring IBs, and brain areas displaying neuronal death and atrophy. aa: amino acids; Cb: cerebellum; Cx: cortex; Hipp: hippocampus; St: striatum.

Mouse model	Genetic background	Amount of htt	Promoter	Transgene length/CAG repeats	Phenotype	Inclusion bodies	Neuronal death	Brain atrophy
R6/1-R6/2	CBA/C57Bl6	RNA <1×	Human huntingtin	Exon1 / 116Q-144Q	Rotarod deficit (40 d) Clasping, tremor, chorea, dystonia, swallow impairment, hypokinesia, narcolepsy, weight loss (2 m)	In neuropil, dendrites and spines in Cx, Hipp and St	Frontal Cx Dorsal St and Cb	Whole brain
N171-82Q	C3H/Bl6	From 0.1–0.2×	Murine prion protein	171 aa / 82Q, 44Q, 18Q	Clasping and rotarod deficit (2.5 m) Tremor, swallow impairment, hypokinesia, weight loss, early death	In St, Cx, Hipp, amygdala, Cb Diffuse nuclear staining	Lateral St	Whole brain
HD94	CBA/C57Bl6	>endogenous	CamKIIα-tTA	Exon1/ 94Q	Clasping from 2.5 m Rotarod deficit from 2.5 m Tremor, swallow impairment	In St, septum, Cx, Hipp	St (17 m)	Whole brain, progressive in the St
YAC	FVB/N	2×	Human huntingtin	Full htt/ 72Q, 128Q	Clasping (3 m) Ataxia, hyperactivity, rotatory behaviour (3 m) and hypokinesia (6 m) Weight loss, rotatory behaviour, swallow impairment and ataxia (12 m)	In St	In St	

days, respectively), decrease in the striatal volume due to the loss of striatal neurons expressing enkephalin from 90 days,[69] and motor and cognitive decline (tremor, chorea, stereotypical movements and dystonia).[68,69] In R6/2 mice, motor symptoms start at 9–11 weeks of age and they dye when they are 10–13 weeks old. In R6/1 mice motor symptoms start later, at 22 weeks of age, and they live longer than R6/2 mice. Approximately one month before the symptoms start, the loss of different neurotransmitter receptors (such as type 1 glutamate meta-botropic receptor, D1 and D2dopamine receptors, and cholinergic muscarinic receptors) takes place, similar to what happens in adult human HD patients.[70] The R6/1 and R6/2 mice also present the charac-teristic IBs, appearing first in the cortex and hippocampus, and progressing later to the St.

- N-171-82Q mice express exons 1 and 2 (first 171 amino acids) of human htt with 82 CAG repeats under the control of the murine prion protein promoter which restricts their expression to neurons.[71] The mice express a shorter polyQ than R6 mice so their phenotype is less severe and starts later. Motor phenotype starts at 2.5 months of age and displays tremor, hypokinesia and clasping in both fore and hind limbs. They also develop cognitive decline represented by working memory deficits at 14 weeks,[72] striatal neuronal loss (25% at 16 weeks of age) and IBs in St, Cx and Hipp.

- Inducible HD94 mice carry the bidirectional tetO responsive promoter followed by both a chimeric mouse/human exon 1 with a polyQ expansion of 94 repeats in one direction and β-galactosidase (lacZ) reporter sequences in the other.[90] Regulation of the system is achieved through the tetracycline-regulated transactivator (tTA), a fusion protein between the tet-repressor binding domain and a VP16 activation domain.[73] This protein binds specifically to the tetO operator and induces transcription from an adjacent cytomegalovirus (CMV) minimal promoter. The tTA transgene used is under the control of the calcium/calmodulin kinase IIa promoter (CamKIIα-tTA).[74] When the BiTetO mice are crossed with CamKIIα-tTA mice, the resulting double transgenic progeny con-stitutively express both transgenes. This expression, however, can be abolished in the presence of doxycycline. These mice live longer than R6 mice (up to two years) and develop a milder phenotype consistent with HD patients: an abnormal motor phenotype, a decrease in striatal size, intra- and extranuclear aggregates, selective reactive astrocytosis in the St, and a decrease in D1 receptor levels. Also, as in HD patients, most of the pathological characteristics increase in severity over time. The phenotype of these mice confirms previous findings that full-length htt is not necessary to recreate the key characteristics of HD.[3,68,71]

10.3.1.2.2.2 *Mouse Carrying Full Length Human htt.* These mice express two copies of normal htt and one copy of htt*. The most used are as follows.

- YAC mice were obtained using yeast artificial chromosomes (YAC) as vectors where the complete human htt gene with a polyQ expansion under the control of human htt promoter was inserted.[75] There are two different models, one carrying 72 CAG repeats and the other carrying 128CAG repeats. Mice carrying 72 CAG repeats develop rotator behaviour, ataxia and clasping in their hindlimbs. Mice carrying 128 CAG repeats develop hyperkinesia starting at three months old that progressively decreases until it turns into hypokinesia at six months of age and a deficit in the learning of new tasks.[76] Both models present striatal neuronal loss, but only those carrying 128 CAG repeats show IBs and at very old ages (18 months old). YAC mice live longer than R6 mice, making them good models for long-lasting therapy studies.
- Inducible model mice express full-length human htt with 148 CAG repeats under the control of the tetracycline-regulated promoter.[77] Tetracycline expression is under the control of the prion protein (PrP) promoter so it expresses in the brain. The mice live for 8–11 months and show motor deficits, involuntary movements, ataxia, tremor and body weight loss. They also present IBs in the Cx, St, Hipp and Cb. These mice are commonly used to study therapeutic targets in HD.

The main difference between mouse models expressing partial human htt and full-length human htt is the absence of striatal cell death in the former as well as an increase in astrogliosis and in the ventricular size. The neuropathology developed by mice models expressing full length htt is more similar to that of human. But since the phenotype in mice models expressing partial human htt is more obvious, they are more useful in HD studies.

10.3.1.3 Viral Genetic Mouse Models

These models are obtained by intrastriatally injecting viruses carrying full-length or partial human htt. It is a very selective approach which achieves specific death of the striatal GABAergic neurons.[78,79] The most used viruses are adenovirus and lentivirus. The main difference between them is that the lentiviruses allow longer genomic sequences and for longer time periods than adenoviruses.[80] Nuclear and cytoplasmic IBs can be detected five days after injection,[79] and the longer the polyQ, the more IBs there are.

10.3.2 Sheep Models

Sheep models of HD have recently been developed. These transgenic sheep express htt*.[81] Although their histopathology has not been published in the scientific literature, it has been described and it is very similar to human. The reasons for using sheep to study HD include:

- sheep have a similar brain structure to humans';
- sheep have a lifespan of 10–15 years (longer than mouse models);
- sheep are cheaper than primates.

Experiments performed in sheep have focused on the analysis of the pre-symptomatic stages of the disease and in testing potential therapeutic drugs that could modify its onset. This model was developed very recently and so there are no conclusive results yet.

10.3.3 Primate Models

While the mouse's St is not divided into caudate and putamen like the human St, that of a primate is. This means that their motor and neurological symptoms are more similar to those of humans, making them the best model for the analysis of HD. As in mouse models, there are several models and they are classified depending on the approach used to obtain them.

10.3.3.1 Chemically Induced Models

As in mouse models, these non-human primate models are obtained by intra-cerebrally injecting QA or 3-NP. These models are commonly used in experiments that monitor the evolution of the disease using neuroimaging approaches. Due to the resemblance of their brains to human brain, they are very useful in the study of drugs in preclinical phases.[82]

10.3.3.1.1 Direct Excitotoxic Model (QA). QA is either unilaterally or bilaterally injected in the St of non-human primates, although the most used is unilateral injection. These animals show motor dysfunction in their fore contralateral limbs and rotatory movements induced by apomorphine.[83] When the injury takes place in the putamen, it triggers symptoms similar to chorea 48 hours after the injection of QA. Bilateral injections produce memory deficits,[84] deficits in the planification of tasks[85] and visuospatial deficits[86] when they take place in the caudate as well as the putamen—all of them similar to a human patient's symptoms.

As in the direct excitotoxic mouse model, the main limitation of this model is that it does not present IBs due to the absence of the mutated protein. This model is widely used to study the effect of neuroprotective drugs.

10.3.3.1.2 Indirect Excitotoxic Model (3-NP). The administration process, the cellular mechanism involved and the symptoms displayed are the same as those present in indirect excitotoxic mouse models, although non-human primate models do reproduce the progressive component of the disease as well as other characteristic movements of the disease (dyskinesia, dystonia and brady-kinesia) that mouse model does not display.[87] The administration of low doses of 3-NP during 3–6 weeks triggers chorea.[88] Longer treatment (four months) triggers spontaneous dyskinesia and dystonia, and memory deficits.[88]

10.3.3.2 Genetic Models

A genetic non-human primate model of HD[89] was recently obtained by injecting a lentivirus containing the exon 1 of human htt with 84 CAG repeats in its *N*-terminus under the control of the human polyubiquitin-C promoter in Rhesus macaque (*Macaca mulatta,* Rhesus monkey) oocytes. Animals obtained by this protocol show high variation of the polyQ tract (from 29 to 88 CAG repeats), htt* amount (from one to four copies) and their life expectancy (from one day to six months). They do not show striatal degeneration but they present nuclear, axonal and dendritic IBs in the St and Cx. They develop motor symptoms such as involuntary movements, severe chorea and swallowing problems—all of which worsen with longer CAG repeats.

This model is expected to be very useful in studies of the pathogenesis of HD and in developing new drugs.

References

1. D. Craufurd and A. Dodge, *J. Med. Genet.*, 1993, **30**, 1008–1011.
2. C. M. Ambrose, M. P. Duyao, G. Barnes, G. P. Bates, C. S. Lin, J. Srinidhi, S. Baxendale, H. Hummerich, H. Lehrach, M. Altherr, J. Wasmuth, A. Buckler, D. Church, D. Housman, M. Berks, G. Micklem, R. Durbin, A. Dodge, A. Read, J. Gusella and M. MacDonald, *Somat. Cell Mol. Genet.*, 1994, **20**, 27–38.
3. J. P. Vonsattel and M. DiFiglia, *J. Neuropathol. Exp. Neurol.*, 1998, **57**, 369–384.
4. R. D. Currier, J. F. Jackson and E. F. Meydrech, *Neurology*, 1982, **32**, 907–909.
5. T. Foroud, J. Gray, J. Ivashina and P. M. Conneally, *J. Neurol. Neurosurg. Psychiatry*, 1999, **66**, 52–56.
6. R. A. Roos, J. Hermans, M. Vegter-van der Vlis, G. J. van Ommen and G. W. Bruyn, *J. Neurol. Neurosurg. Psychiatry*, 1993, **56**, 98–100.
7. G. M. de Boo, A. Tibben, J. B. Lanser, A. Jennekens-Schinkel, J. Hermans, A. Maat-Kievit and R. A. Roos, *Arch. Neurol.*, 1997, **54**, 1353–1357.
8. J. P. van Vugt, S. Siesling, K. K. Piet, A. H. Zwinderman, H. A. Middelkoop, J. J. van Hilten and R. A. Roos, *Mov. Disord.*, 2001, **16**, 481–488.
9. M. S. Haddad and J. L. Cummings, *Psychiatr. Clin. North Am.*, 1997, **20**, 791–807.
10. A. Rosenblatt and I. Leroi, *Psychosomatics*, 2000, **41**, 24–30.
11. G. A. Graveland, R. S. Williams and M. DiFiglia, *Science*, 1985, **227**, 770–773.
12. H. Heinsen, M. Strik, M. Bauer, K. Luther, G. Ulmar, D. Gangnus, G. Jungkunz, W. Eisenmenger and M. Gotz, *Acta Neuropathol.*, 1994, **88**, 320–333.
13. J. P. Vonsattel, R. H. Myers, T. J. Stevens, R. J. Ferrante, E. D. Bird and E. P. Richardson, Jr., *J. Neuropathol. Exp. Neurol.*, 1985, **44**, 559–577.
14. Huntingdon's Disease Collaborative Research Group, *Cell*, 1993, **72**, 971–983.
15. P. A. Locke, M. E. MacDonald, J. Srinidhi, T. C. Gilliam, R. E. Tanzi, P. M. Conneally, N. S. Wexler, J. L. Haines and J. F. Gusella, *Somat. Cell Mol. Genet.*, 1993, **19**, 95–101.

16. J. F. Gusella, M. E. MacDonald, C. M. Ambrose and M. P. Duyao, *Arch. Neurol.*, 1993, **50**, 1157–1163.

17. S. E. Andrew, Y. P. Goldberg, B. Kremer, H. Telenius, J. Theilmann, S. Adam, E. Starr, F. Squitieri, B. Lin, M. A. Kalchman, R. K. Graham and M. R. Hayden, *Nat. Genet.*, 1993, **4**, 398–403.

18. R. G. Snell, J. C. MacMillan, J. P. Cheadle, I. Fenton, L. P. Lazarou, P. Davies, M. E. MacDonald, J. F. Gusella, P. S. Harper and D. J. Shaw, *Nat. Genet.*, 1993, **4**, 393–397.

19. C. A. Ross, *Neuron*, 1997, **19**, 1147–1150.

20. C. A. Ross and M. A. Poirier, *Nat. Med.*, 2004, **10**(Suppl), S10–17.

21. M. DiFiglia, E. Sapp, K. O. Chase, S. W. Davies, G. P. Bates, J. P. Vonsattel and N. Aronin, *Science*, 1997, **277**, 1990–1993.

22. C. A. Gutekunst, S. H. Li, H. Yi, J. S. Mulroy, S. Kuemmerle, R. Jones, D. Rye, R. J. Ferrante, S. M. Hersch and X. J. Li, *J. Neurosci*, 1999, **19**, 2522–2534.

23. *C. elegans* Sequencing Consortium, *Science*, 1998, 282, 2012–2018.

24. S. Holbert, I. Denghien, T. Kiechle, A. Rosenblatt, C. Wellington, M. R. Hayden, R. L. Margolis, C. A. Ross, J. Dausset, R. J. Ferrante and C. Neri, *Proc. Natl. Acad. Sci. U. S. A.*, 2001, **98**, 1811–1816.

25. L. Segalat and C. Neri, *Med. Sci.*, 2003, **19**, 1218–1225.

26. P. W. Faber, J. R. Alter, M. E. MacDonald and A. C. Hart, *Proc. Natl. Acad. Sci. U. S. A.*, 1999, **96**, 179–184.

27. J. A. Parker, J. B. Connolly, C. Wellington, M. Hayden, J. Dausset and C. Neri, *Proc. Natl. Acad. Sci. U. S. A.*, 2001, **98**, 13318–13323.

28. G. M. Rubin, M. D. Yandell, J. R. Wortman, G. L. Gabor Miklos, C. R. Nelson, I. K. Hariharan, M. E. Fortini, P. W. Li, R. Apweiler, W. Fleischmann, J. M. Cherry, S. Henikoff, M. P. Skupski, S. Misra, M. Ashburner, E. Birney, M. S. Boguski, T. Brody, P. Brokstein, S. E. Celniker, S. A. Chervitz, D. Coates, A. Cravchik, A. Gabrielian, R. F. Galle, W. M. Gelbart, R. A. George, L. S. Goldstein, F. Gong, P. Guan, N. L. Harris, B. A. Hay, R. A. Hoskins, J. Li, Z. Li, R. O. Hynes, S. J. Jones, P. M. Kuehl, B. Lemaitre, J. T. Littleton, D. K. Morrison, C. Mungall, P. H. O'Farrell, O. K. Pickeral, C. Shue, L. B. Vosshall, J. Zhang, Q. Zhao, X. H. Zheng and S. Lewis, *Science*, 2000, **287**, 2204–2215.

29. L. T. Reiter, L. Potocki, S. Chien, M. Gribskov and E. Bier, *Genome Res*, 2001, **11**, 1114–1125.

30. A. H. Brand and N. Perrimon, *Development*, 1993, **118**, 401–415.

31. J. M. Warrick, H. L. Paulson, G. L. Gray-Board, Q. T. Bui, K. H. Fischbeck, R. N. Pittman and N. M. Bonini, *Cell*, 1998, **93**, 939–949.

32. E. G. McGeer and P. L. McGeer, *Nature*, 1976, **263**, 517–519.

33. P. R. Sanberg and J. T. Coyle, *CRC Crit. Rev. Clin. Neurobiol.*, 1984, **1**, 1–44.

34. A. C. Foster, L. P. Miller, W. H. Oldendorf and R. Schwarcz, *Exp. Neurol.*, 1984, **84**, 428–440.

35. Y. M. Bordelon, M. F. Chesselet, D. Nelson, F. Welsh and M. Erecinska, *J. Neurochem.*, 1997, **69**, 1629–1639.
36. J. P. McLin, L. M. Thompson and O. Steward, *Eur. J. Neurosci*, 2006, **24**, 3134–3140.
37. C. A. Ribeiro, V. Grando, C. S. Dutra Filho, C. M. Wannmacher and M. Wajner, *J. Neurochem.*, 2006, **99**, 1531–1542.
38. M. F. Beal, R. J. Ferrante, K. J. Swartz and N. W. Kowall, *J. Neurosci*, 1991, **11**, 1649–1659.
39. R. J. Ferrante, N. W. Kowall, P. B. Cipolloni, E. Storey and M. F. Beal, *Exp. Neurol.*, 1993, **119**, 46–71.
40. R. C. Roberts, A. Ahn, K. J. Swartz, M. F. Beal and M. DiFiglia, *Exp. Neurol.*, 1993, **124**, 274–282.
41. J. T. Coyle and R. Schwarcz, *Nature*, 1976, **263**, 244–246.
42. M. F. Beal, N. W. Kowall, D. W. Ellison, M. F. Mazurek, K. J. Swartz and J. B. Martin, *Nature*, 1986, **321**, 168–171.
43. M. F. Beal, P. E. Marshall, G. D. Burd, D. M. Landis and J. B. Martin, *Brain Res.*, 1985, **361**, 135–145.
44. Z. H. Qin, J. Wang and Z. L. Gu, *Acta Pharmacol. Sin.*, 2005, **26**, 129–142.
45. T. A. Alston, L. Mela and H. J. Bright, *Proc. Natl. Acad. Sci. U. S. A.*, 1977, **74**, 3767–3771.
46. C. J. Coles, D. E. Edmondson and T. P. Singer, *J. Biol. Chem.*, 1979, **254**, 5161–5167.
47. S. E. Browne, A. C. Bowling, U. MacGarvey, M. J. Baik, S. C. Berger, M. M. Muqit, E. D. Bird and M. F. Beal, *Ann. Neurol.*, 1997, **41**, 646–653.
48. S. J. Tabrizi, M. W. Cleeter, J. Xuereb, J. W. Taanman, J. M. Cooper and A. H. Schapira, *Ann. Neurol.*, 1999, **45**, 25–32.
49. M. F. Beal, E. Brouillet, B. G. Jenkins, R. J. Ferrante, N. W. Kowall, J. M. Miller, E. Storey, R. Srivastava, B. R. Rosen and B. T. Hyman, *J. Neurosci*, 1993, **13**, 4181–4192.
50. D. Blum, D. Gall, L. Cuvelier and S. N. Schiffmann, *Neuroreport*, 2001, **12**, 1769–1772.
51. C. V. Borlongan, T. K. Koutouzis, T. B. Freeman, D. W. Cahill and P. R. Sanberg, *Brain Res.*, 1995, **697**, 254–257.
52. D. A. Shear, J. Dong, C. D. Gundy, K. L. Haik-Creguer and G. L. Dunbar, *Prog. Neuropsychopharmacol. Biol. Psychiatry*, 1998, **22**, 1217–1240.
53. E. Brouillet, B. G. Jenkins, B. T. Hyman, R. J. Ferrante, N. W. Kowall, R. Srivastava, D. S. Roy, B. R. Rosen and M. F. Beal, *J. Neurochem.*, 1993, **60**, 356–359.
54. M. C. Guyot, P. Hantraye, R. Dolan, S. Palfi, M. Maziere and E. Brouillet, *Neuroscience*, 1997, **79**, 45–56.
55. E. Brouillet, F. Conde, M. F. Beal and P. Hantraye, *Prog. Neurobiol.*, 1999, **59**, 427–468.
56. M. P. Duyao, A. B. Auerbach, A. Ryan, F. Persichetti, G. T. Barnes, S. M. McNeil, P. Ge, J. P. Vonsattel, J. F. Gusella, A. L. Joyner and M. E. MacDonald, *Science*, 1995, **269**, 407–410.

57. S. Zeitlin, J. P. Liu, D. L. Chapman, V. E. Papaioannou and A. Efstratiadis, *Nat. Genet.*, 1995, **11**, 155–163.
58. J. K. White, W. Auerbach, M. P. Duyao, J. P. Vonsattel, J. F. Gusella, A. L. Joyner and M. E. MacDonald, *Nat. Genet.*, 1997, **17**, 404–410.
59. E. B. Clabough and S. O. Zeitlin, *Hum. Mol. Genet.*, 2006, **15**, 607–623.
60. I. Dragatsis, M. S. Levine and S. Zeitlin, *Nat. Genet.*, 2000, **26**, 300–306.
61. C. H. Lin, S. Tallaksen-Greene, W. M. Chien, J. A. Cearley, W. S. Jackson, A. B. Crouse, S. Ren, X. J. Li, R. L. Albin and P. J. Detloff, *Hum. Mol. Genet.*, 2001, **10**, 137–144.
62. H. Li, S. H. Li, H. Johnston, P. F. Shelbourne and X. J. Li, *Nat. Genet.*, 2000, **25**, 385–389.
63. L. B. Menalled, J. D. Sison, Y. Wu, M. Olivieri, X. J. Li, H. Li, S. Zeitlin and M. F. Chesselet, *J. Neurosci.*, 2002, **22**, 8266–8276.
64. V. C. Wheeler, J. K. White, C. A. Gutekunst, V. Vrbanac, M. Weaver, X. J. Li, S. H. Li, H. Yi, J. P. Vonsattel, J. F. Gusella, S. Hersch, W. Auerbach, A. L. Joyner and M. E. MacDonald, *Hum. Mol. Genet.*, 2000, **9**, 503–513.
65. L. Menalled, H. Zanjani, L. MacKenzie, A. Koppel, E. Carpenter, S. Zeitlin and M. F. Chesselet, *Exp. Neurol.*, 2000, **162**, 328–342.
66. M. S. Levine, G. J. Klapstein, A. Koppel, E. Gruen, C. Cepeda, M. E. Vargas, E. S. Jokel, E. M. Carpenter, H. Zanjani, R. S. Hurst, A. Efstratiadis, S. Zeitlin and M. F. Chesselet, *J. Neurosci. Res.*, 1999, **58**, 515–532.
67. M. A. Smith, J. Brandt and R. Shadmehr, *Nature*, 2000, **403**, 544–549.
68. L. Mangiarini, K. Sathasivam, M. Seller, B. Cozens, A. Harper, C. Hetherington, M. Lawton, Y. Trottier, H. Lehrach, S. W. Davies and G. P. Bates, *Cell*, 1996, **87**, 493–506.
69. E. C. Stack, J. K. Kubilus, K. Smith, K. Cormier, S. J. Del Signore, E. Guelin, H. Ryu, S. M. Hersch and R. J. Ferrante, *J. Comp. Neurol.*, 2005, **490**, 354–370.
70. J. H. Cha, C. M. Kosinski, J. A. Kerner, S. A. Alsdorf, L. Mangiarini, S. W. Davies, J. B. Penney, G. P. Bates and A. B. Young, *Proc. Natl. Acad. Sci. U. S. A.*, 1998, **95**, 6480–6485.
71. G. Schilling, M. W. Becher, A. H. Sharp, H. A. Jinnah, K. Duan, J. A. Kotzuk, H. H. Slunt, T. Ratovitski, J. K. Cooper, N. A. Jenkins, N. G. Copeland, D. L. Price, C. A. Ross and D. R. Borchelt, *Hum. Mol. Genet.*, 1999, **8**, 397–407.
72. S. Ramaswamy, J. L. McBride, L. Zhou, E. M. Berry-Kravis, E. P. Brandon, C. D. Herzog, M. Gasmi, R. T. Bartus and J. H. Kordower, *Cell Transplant*, 2004.
73. M. Gossen and H. Bujard, *Proc. Natl. Acad. Sci. U. S. A.*, 1992, **89**, 5547–5551.
74. M. Mayford, M. E. Bach, Y. Y. Huang, L. Wang, R. D. Hawkins and E. R. Kandel, *Science*, 1996, **274**, 1678–1683.
75. J. G. Hodgson, N. Agopyan, C. A. Gutekunst, B. R. Leavitt, F. LePiane, R. Singaraja, D. J. Smith, N. Bissada, K. McCutcheon, J. Nasir, L. Jamot,

X. J. Li, M. E. Stevens, E. Rosemond, J. C. Roder, A. G. Phillips, E. M. Rubin, S. M. Hersch and M. R. Hayden, *Neuron*, 1999, **23**, 181–192.

76. J. M. Van Raamsdonk, Z. Murphy, E. J. Slow, B. R. Leavitt and M. R. Hayden, *Hum. Mol. Genet.*, 2005, **14**, 3823–3835.

77. Y. Tanaka, S. Igarashi, M. Nakamura, J. Gafni, C. Torcassi, G. Schilling, D. Crippen, J. D. Wood, A. Sawa, N. A. Jenkins, N. G. Copeland, D. R. Borchelt, C. A. Ross and L. M. Ellerby, *Neurobiol. Dis.*, 2006, **21**, 381–391.

78. L. P. de Almeida, C. A. Ross, D. Zala, P. Aebischer and N. Deglon, *J. Neurosci.*, 2002, **22**, 3473–3483.

79. M. C. Senut, S. T. Suhr, B. Kaspar and F. H. Gage, *J. Neurosci.*, 2000, **20**, 219–229.

80. A. Pfeifer and I. M. Verma, *Annu. Rev. Genomics Hum. Genet.*, 2001, **2**, 177–211.

81. BBC News UK, http://news.bbc.co.uk/2/hi/science/nature/6222250.stm (accessed 8th June 2009).

82. D. Blum, M. C. Galas, D. Gall, L. Cuvelier and S. N. Schiffmann, *Neurobiol. Dis.*, 2002, **10**, 410–426.

83. A. L. Kendall, P. Hantraye and S. Palfi, *Prog. Brain Res.*, 2000, **127**, 381–404.

84. N. Butters, J. Wolfe, M. Martone, E. Granholm and L. S. Cermak, *Neuropsychologia*, 1985, **23**, 729–743.

85. B. Z. Roitberg, M. E. Emborg, J. G. Sramek, S. Palfi and J. H. Kordower, *Neurosurgery*, 2002, **50**, 137–145; discussion 145–136.

86. E. Mohr, P. Brouwers, J. J. Claus, U. M. Mann, P. Fedio and T. N. Chase, *Mov. Disord.*, 1991, **6**, 127–132.

87. P. Hantraye, D. Riche, M. Maziere and O. Isacson, *Exp. Neurol.*, 1990, **108**, 91–104.

88. S. Palfi, R. J. Ferrante, E. Brouillet, M. F. Beal, R. Dolan, M. C. Guyot, M. Peschanski and P. Hantraye, *J. Neurosci.*, 1996, **16**, 3019–3025.

89. S. H. Yang, P. H. Cheng, H. Banta, K. Piotrowska-Nitsche, J. J. Yang, E. C. Cheng, B. Snyder, K. Larkin, J. Liu, J. Orkin, Z. H. Fang, Y. Smith, J. Bachevalier, S. M. Zola, S. H. Li, X. J. Li and A. W. Chan, *Nature*, 2008, **453**, 921–924.

90. A. Yamamoto, J. J. Lucas and R. Hen, *Cell*, 2000, **101**(1), 57–66.

CHAPTER 11

Mouse Models of Prion Protein Related Diseases

MARÍA GASSET[*a] AND ADRIANO AGUZZI[b]

[a] Insto Química-Física Rocasolano, Consejo Superior de Investigaciones Científicas, Madrid, Spain; [b] Institute of Neuropathology, University Hospital of Zürich, Zürich, Switzerland

11.1 Introduction

Prion protein (PrP) related diseases are a group of fatal mammalian neurodegenerative diseases commonly referred to as transmissible spongiform encephalopathies (TSEs) or prion diseases which include: Creutzfeldt–Jakob disease (CJD), Gerstamann–Strausler–Scheinker disease (GSS) and fatal familial insomnia (FFI) in humans; scrapie in sheep and goats; bovine spongiform encephalopathies (BSE) in cattle; and chronic wasting disease (CWD) in deer and elk (Table 11.1). Histological signs include neuronal loss, spongiform degeneration, gliosis and often, but not always, PrP-positive amyloid deposits.[1,2] Clinical symptoms (involving dementia, ataxia and behavioural abnormalities) and the types and distribution of CNS lesions vary between different diseases.[1,2] These diseases are also diverse in aetiology, since they can be acquired by infection, can occur sporadically, or can be inherited.[1,2]

Despite the heterogeneity, a common characteristic of PrP-related diseases is the aberrant metabolism of the cellular prion protein (PrP^C). This protein, encoded by a single gene and transcribed from a single open reading frame (ORF), belongs to a growing family of related genes, each of which exhibits complex posttranslational maturation and processing.[1–3] Mistargetting, misprocessing and misfolding events

RSC Drug Discovery Series No. 6
Animal Models for Neurodegenerative Disease
Edited by Jesús Avila, Jose J. Lucas and Félix Hernández
© Royal Society of Chemistry 2011
Published by the Royal Society of Chemistry, www.rsc.org

Table 11.1 PrP-related diseases of mammalians. PrP-related diseases are often referred by the common name assigned on their report, which indicates either the hallmarks of the histological lesion or the authors who described it. In addition to the common name, these diseases can be also classified according to their aetiology.

Disease		Abbreviation	Host	Aetiology
Scrapie		Scrapie	Goat, sheep	acquired
				genetic
Bovine spongiform encephalopathy		BSE	Cattle	acquired
Chronic wasting disease		CWD	Deer, elk	acquired
Mink transmissible encephalopathy		TME	Mink	acquired
Feline spongiform encephalopathy		FSE	Cat	acquired
Creutzfeldt–Jakob Disease		CJD	Human	
	Sporadic	sCJD		sporadic
	Genetic	gCJD		genetic
	Iatrogenic	iCJD		acquired
	Variant	vCJD		acquired
Gerstmann–Sträussler–Scheinker disease		GSS	Human	genetic
Fatal familiar insomnia		FFI	Human	genetic
				sporadic
Kuru		–	Human	acquired

cause the formation and/or accumulation of alternate forms (disease-linked PrP forms) which display neurotoxic or infectious properties.[1–4]

Of the disease-linked forms of PrP, those exhibiting infectious properties or prions have attracted much attraction given their unique features and the social need for their inactivation.[1,2] Prions were originally defined as proteinaceous infectious particles and later found to consist primarily of a misfolded PrP.[1,5] Mounting evidence supports the hypothesis that infectivity can be specified singly by a protein conformational code, yet important elements remain controversial.[6–13] Regarding the neurotoxic conformers, recent findings support mislocalization on the basis of the gain of toxic functions.[4,7,14,15]

One good reason to model PrP-related diseases in mice is to generate mammalian organisms that undergo a similar pathological process as in humans with controlled and shortened kinetics, so that they can be used to uncover their molecular basis, to establish pre-asymptomatic markers and to develop prophylactic strategies. Similarly to other degenerative disorders, intervention before the onset of symptoms caused by irreversible damage may provide the best scenario for life prolongation with acceptable quality.

11.2 Notes on PrP Biology

11.2.1 Molecular and Structural Diversity of PrP^C

PrP is encoded by a single-copy gene (*PRNP*) located on chromosome 20 and 2 in humans and mouse, respectively.[1,2,16,17] The *PRNP* gene has multiple exons, with the PrP ORF contained entirely in the most 3' exon. This gene is expressed at highest levels in neurons and other cells of the central nervous system (CNS), but also in the lymphoreticular system and the skeletal muscle.[2,18]

The PrP chain is made of about 253 amino acids, depending on the species, and it is organized as follows (Figure 11.1).[1,2] The first *N*-terminal 22 amino acids encode a signal peptide sequence displaying the characteristic tripartite structure and which is cleaved off during co-translational translocation. Residues 51–91 contain a nonapeptide followed by four identical octapeptide repeats (OR), which function as a binding site for multiple ligands.[4,19] This repetitive region is flanked by two clusters of basic residues (CC1 and CC2) which account for polyanion binding properties.[20,21] The chain then moves into a hydrophobic segment (HC) with a complex structural behaviour. This segment appears as a disordered region in three-dimensional (3D) structural models, as a transmembrane helical segment in the integrated forms, and as a β-sheet structure based aggregate with amyloid staining properties when considered as an isolated peptide.[3,22–24] Following this segment, the 121–231 region hosts the globular domain which is folded with the arrangement of three α-helices (α1, α2, α3) and two short antiparallel β-strands (β1, β2).[24] This final structure is essentially conserved for all PrP chains studied and it is considered as the PrP^C-like structure, despite the use of *in vitro* folding protocols and the lack of an activity indicator. Furthermore, this region contains a polymorphic site (M129V in humans, M132L in elk) with significance in disease, two consensus sites for *N*-glycosylation (N181, N197) and a site for glycosylphosphatidylinositol (GPI) addition that will release the hydrophobic tail and anchor the protein to membranes at a specific domain.[25–28]

This simple chain design encrypts a very complex molecular diversity (Figure 11.1) Presence of a second M residue at either side of the η-region (M8 in humans, M15 in rodents) allows for the synthesis of a minor form that displays a nucleocytoplasmic location.[29] However, chains segregated into the secretory route as a result of a complex interaction with the translocation machinery will exist under three different topological forms:[3]

- SecPrP (totally translocated);
- NtmPrP (transmembrane form projecting the 22–111 region to the endoplasmic reticulum (ER) lumen and the *C*-terminal to the cytosol);
- CtmPrP (transmembrane form projecting the *C*-terminal domain to the ER lumen and with the *N*-terminal region cytosolically).

Moreover, the polypeptide chain contains two sites for proteolytic maturation (α- and β-cleavage sites at positions 100/111 and about 89/90 respectively) that causes the production of four different fragments: N1 and C1 resulting from α-secretase action and N2 and C2 from β-cleavage.[30–33] Lastly, proteases and phospholipase can cleave the membrane anchor producing soluble forms of the full length, C1, and eventually C2 in a process known as shedding.[34,35]

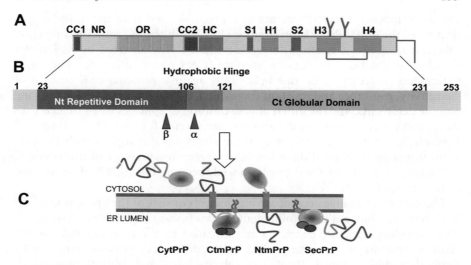

Figure 11.1 PrP polypeptide chain: its structural organization and molecular diversity. The mammalian PrP polypeptide chain is designed as a two domain (blue and orange boxes) protein with *N*- and *C*-terminal sequences signals (grey flanks) and an interdomain linker (red box) (**B**). This modular design is regulated by proteolytic processing at the α- and β-cleavage sites. (**A**) The primary structure of the chain contained between *N*- and *C*-signal sequences; motifs: shows a complex combination of motifs and structural elements: CC1 and CC2 define basic charged clusters; NR and OR are nona- and octapeptide repeats, respectively; HC stands for the hydrophobic region; S represents regions folded into β-strands; and H regions represent those with helical structure. The polypeptide chain contains the motifs that allow two glycosilations (N181 and N197), the formation of an intramolecular disulfide bond (C179 and C214) and the attachment of a GPI group. (**C**) Alternative translation start permits two polypeptide chains differing in the length of the *N*-terminus sequence. Chains with short *N*-terminal sequence are spilt into the cytosol and display a nucleocytoplasmic distribution; they are often referred to as CytPrP. Chains with long *N*-terminal sequences yield three different membrane-bound forms: SecPrP (fully translocated and membrane anchored by a GPI), NtmPrP (membrane integrated with the N-terminal domain projected to the ER lumen), and CtmPrP (membrane integrated with the N-terminal domain projected to the cytosol). Of these forms, on accumulation CytPrP and CtmPrP yield neurotoxic phenotypes whereas SecPrP through its conversion yields prion disorders.

11.2.2 PrP^C Conversion: Prions, Amyloids and Other Toxic Forms

The central molecular event in PrP-related diseases is the structural conversion of PrP^C into an ensemble of alternate conformers that display either infectious or neurotoxic properties and which are collectively referred to as PrP^{Sc}.[1] Infectivity of PrP^{Sc} forms is defined in terms of the capacity to seed and impose the conformation on a PrP^C acting as a precursor and resulting in the amplification of the initial PrP^{Sc} form (Figure 11.2).[1,36] When this molecular reaction takes place *in vivo* it causes a histological lesion characterized by its kinetics (incubation time) and its regional tropism (CNS region affected).[1] These two parameters together with the behaviour against proteinase K treatment (sensitivity, resistance and resistant fragments), the solubility in non-denaturing anionic detergents, and the differential exposition of epitopes when treated with increasing concentrations of denaturants are specific for each PrP^{Sc} strain.[37,38] Conversely, neurotoxic forms may share structural descriptors with the previous forms (as the formation of insoluble aggregates, presence of proteinase K resistance, *etc.*) but lack the capacity to seed their amplification both *in vivo* and *in vitro* (Figure 11.2).[39,40]

The structural change accompanying disease consists of the disruption of the PrP^C native α-fold and the refolding into a new β-sheet rich structural state.[22,41] The secondary structure change is accompanied by the rearrangement of the tertiary structure that can be probed by the differences in the recognition by antibodies and enzymes (proteases, glycosidases and phospholipases).[1,42] Together with these changes, on PrP^{Sc} formation the polypeptide chain acquires the capacity to aggregate into insoluble high-molecular weight forms with shapes varying from amorphous states, as in PrP^{Sc} 33-35, to rods, in

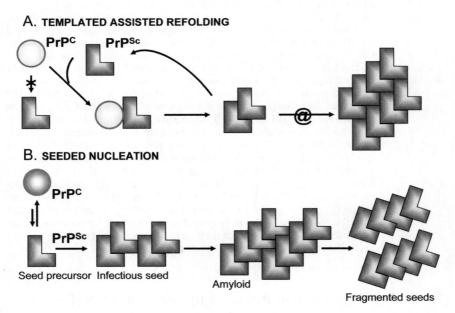

A. TEMPLATED ASSISTED REFOLDING

PrP^C PrP^{Sc}

B. SEEDED NUCLEATION

PrP^C

PrP^{Sc}

Seed precursor Infectious seed

Amyloid

Fragmented seeds

Figure 11.2 PrP^C in route to disease: kinetic models. To model prion conversion, two kinetic models have been proposed: (**A**) a template-assisted refolding and (**B**) a nucleation–polymerization. Models differ essentially in the permission for PrP^{Sc} existence.[2] A third model based on a branched chain polymerization mechanism has been recently proposed as a unifying approach.[135]

PrP27-30 when the *N*-terminal domain has been cleaved using proteinase K and anionic detergents.[43,44] However, it must be stressed that, in tissues, PrP^{Sc} exists in a heterogeneous and polydisperse ensemble of forms as, for instance, those engaged in extracellular amyloid plaques and those present in intracellular compartments.[45–48]

Recombinant chains modelling the mature polypeptide chain (lacking the *N*- and *C*-terminal hydrophobic sequences), containing an intramolecular disulfide bond and folded into β-sheet rich polymers displaying amyloid-staining properties have been used to address the structure–function relationship of the infectious activity. Of the different trials with proteins, only PrP truncated forms produced from heat-treated full-length chain polymers produced a neurodegenerative infectious disease in wild-type (wt) animals, despite differences in the clinical signs and lesions.[9,49–51] Inclusion of additives such as phospholipids and RNA with cycles of sonication and heating seems to improve the efficiency of synthetic infectivity generation.[52] Polymers formed from PrP chains mimicking the C2 fragments showed infectious properties but on transgenic hosts with increased susceptibility, whereas C1-like fragments that can oligomerize do not tolerate the infectious conversion.[9,53,54]

The differential competence for conversion displayed by the C1 fragment compared with the C2 fragment and the full-length precursor underlines PrP^C proteolytic processing as an essential metabolic process regulating the entity and concentration of disease-related precursors. In addition, the synthetic route accounting for the division of the nascent chain into the different members of the family plays a critical role. In this sense, overpopulation of CtmPrP and cytosolic forms, which can be identified as proteinase K resistant and detergent-insoluble forms, also feature some PrP-related diseases.[3]

11.3 Phenotyping PrP^C Function: Ablation and Overexpression

11.3.1 PrP^C Ablation: Hints from Losing PrP^C Function

The discovery of the PrP chain as a major component of the mammalian prion as well as the occurrence of mutations in the open reading frame of its gene in human diseases initiated the link between the PrP chain and mammalian prion disorders.[1,17] Sequencing studies provided shortly thereafter a large collection of PrP sequences that, upon theoretical comparison, underlined a high degree of identity among the different mammalian species and conservation in avians.[55,56] Such degree of conservation suggested that PrP plays an essential role for life as well as the involvement of its function as a basis of disease.

Contrary to some expectations, ablation of the *Prnp* gene concurred with no obvious developmental defects and the mice live normally.[57] It must be stressed, however, that some Prnp0/0 lines displayed an ataxic phenotype that correlated with deregulation of the expression of other protein Dpl resulting from the ablation strategy.[2] To account for possible adaptative mechanisms operative during embryonic life that would compensate the loss of PrP function in knock-out mice and explain the absence of phenotype, conditional PrP depletion mice were constructed using two different approaches.[58,59] In both cases, animals remained healthy with no apparent lesion. Notwithstanding this, subtle changes in the hippocampal function, the circadian rhythms and behaviour have been reported.[59–63] However, none of these phenotypes were explained in molecular terms and in some cases were not fully reproducible.[64]

A revisited study of both Ngsk and Zrch-I PrP^C deficient lines has allowed the detection of a late-onset peripheral neuropathy indicating a role in myelin maintenance.[65] Indeed, the chronic demyelinating polyneuropathy lesion correlated with the deregulation of the expression and processing of PrP^C specifically in neurons. The observed lesions were rescued by expression in neurons of C1-PrP^C, the product of the protein metabolic cleavage at the a-site, suggesting that PrP cleavage may be relevant to its function. Since mice overexpressing N-terminally truncated forms of the prion protein also suffer from neuropathy (A.A., unpublished results), we suspect that the phenotypes resulting from the absence of PrP and those emanating from the overexpression of truncated PrP isoforms may be molecularly related.

Despite the contribution of the PrP ablated mice to the histological phenotype establishment, these animal models have permitted the description of a large variety of PrP ligands, as for example Cu(II), and, extremely importantly, they have been crucial for the construction of the immunological reagent battery.[66–68]

11.3.2 PrP^C Overexpression also Leads to Pathology

Similarly to the myelin lesion observed in PrP^C knock-out mice, older mice harbouring high copy numbers of wt PrP transgenes develop truncal ataxia, hindlimb paralysis, and tremors.[69] These symptoms correlated with a profound necrotising myopathy involving skeletal muscle, a demyelinating polyneuropathy, and focal vacuolation of the central nervous system. Development of disease was dependent on transgene dosage indicating that, on ageing, the sustained overexpression of wt PrP^C is pathogenic. Restriction of the overexpression to skeletal muscle under a highly regulated fashion showed the apparition of a rapidly progressive primary myopathy that correlates with the preferential accumulation of the C1 fragment.[70] In these mice, the C1/full length PrP ratio is three-fold increased compared with the levels in the skeletal muscle of wild-type animals. These findings strongly suggested that the elevated levels of PrP found in the skeletal muscles of human primary myopathies, such as inclusion body myositis), and inflammatory myopathies (as polymyositis and dermatomyositis) may play an important role in the pathogenesis.[71–73]

11.3.3 Sensitivity to Prion Infection: Dose and Site of PrP^C Expression

Despite the fact that both depletion and overexpression of wt PrP^C cause degenerative diseases at advanced ages, the observed phenotypes are distinct from the symptoms and signs of PrP-related diseases. Indeed, the clearest phenotype of PrP knock-outs is the resistance to infection with prions and to their pathology.[74–77] These findings, together with the partial reversion of the lesions upon conditional abrogation of PrP^C expression, underline the essentiality of sustained PrP^C expression for the generation of disease-active species and for their damaging function.[59]

Mirror experiments using overexpression strategies have shown that increased levels of PrPC accelerate the course of disease upon infection with exogenous prions.[77,78] In this field, Tga20 mice overexpressing wt PrP from a randomly integrated transgene have been adopted as conventional reference model. Although overdosing PrP can itself lead to pathology, Tga20 mice have been extensively characterized, with no spontaneous pathologies reported, and with sensitivity to prions with heterologous sequences.[79,80]

11.4 Models for PrP^C Function–Structure Relationship: Spontaneous Generation of Disease or Increased Sensitivity to Acquisition

11.4.1 Chain Length, Motifs and Covalent Modifications

The PrP knock-out mouse enabled another important series of studies that defined some of the sequence elements required for PrP^{Sc} to form and to propagate exogenous prions. Table 11.2 summarizes the different truncated forms of the PrP chain used for transgenesis approaches as well as the phenotype outcome, defined in terms of observed lesion, capacity to generate prions (*transmission*) and capacity to amplified exogenous prions (*propagation*).

The OR region hosts the copper-binding activity; its expansion causes inherited diseases but it is absent in the infectious PrP27–30. Transgenic mice expressing its deletion are healthy and retain the convertibility of the PrP chain into PrP^{Sc}, helping this region to play, at best, a regulatory role on PrP conversion.[81–83] Deletion of CC$_1$ both singly (PrPΔ23–88) and in combination with partial deletions of CC$_2$ (PrPΔ23–88 Δ95–107 and PrPΔ23–88 Δ108–121) did not induce pathologies in transgenic mice, and the convertibility to PrPSc was retained.[84] The deletion combination of *N*-terminal and amino acids 141–176 (PrPΔ23–88Δ141–176) was also innocuous and restored susceptibility to prion infection.[84]

With the exception of a small deletion between the CC$_2$ and HC junction that favours all SecPrP (PrPΔ104–114), ablation of CC$_2$ in combination with a partial or complete deletion of HC elicits severe pathologies in mice.[85] PrPΔ32–121, PrPΔ32–134, PrPΔ94–134 and PrPΔ105–125 transgenic mice suffer from ataxia, cerebellar granule cell loss and a widespread white matter disease.[86–88] These pathologies are different from those seen in prion infections and importantly are counteracted by coexpression of wt PrP.[53,87,88] Moreover, mice with deletions in the CC$_2$ and HC region (PrPΔ32–121, PrPΔ32–134

Table 11.2   PrP chain length mutants for transgenesis. The murine PrP chain
has been taken as model for deletion and mutation of different
regions in transgenesis. The phenotype column indicates presence
or absence of phenotypic abnormalities in transgenic mice when
expressed on a PrP-deficient genetic background. The rescue column indicates the phenotypic response on wt PrP^C expression
restore. The propagation column indicates the susceptibility of
transgenic mice to prions after intracerebral inoculation.

MoPrP deletion	Phenotype	Rescue	Propagation	Reference
Δ1−253		+	no	57, 65
Δ23−88	none		yes	82
Δ32−80	none		yes	81
Δ32−93	none		yes	53, 83
Δ32−106	none		no	53
Δ23−88, Δ95−107	none		no	82
Δ23−88, Δ108−121	none		no	82
Δ32−121	cerebellar disorder	+	no	53
Δ32−134	cerebellar disorder	+	no	53
PG14	cerebellar disorder	−	yes	113
Δ94−134	cerebellar disorder	+	nd	87
Δ104−114	none		nd	85
Δ105−125	cerebellar disorder	+	nd	88
Δ114−121	none		no	87
Δ145−253	short-lived protein	−	−	82
Δ23−88, Δ122−140	short-lived protein	−	−	82
Δ23−88, Δ141−176	none		yes	82
Δ23−88, Δ141−221	cerebellar disorder	−	no	82
Δ23−88, Δ177−200	storage disease	−	no	82
Δ23−88, Δ201−217	storage disease	−	no	82
Δ1−22, Δ232−253	cerebellar disorder	−	no	90
Δ232−253	none	−	yes	14
Δ23−80, C176A	none	−	no	82

PrPΔ104–114, PrPΔ114–121) do not support prion propagation indicating the
importance of this region in conversion.[85,89]

Regarding the C-terminal domain, which contains the different elements
that upon folding permit the globular structure, the deletion of any of helical
scaffolds (H2: Δ177–200, H3: Δ201–217; H2–H3: Δ141–221) on a PrP lacking the
N-terminal domain (Δ23–88) produce ataxic phenotypes with reminiscences of
neuronal storage disease.[82] These phenotypes were refractory to the coexpression
of wt PrP and none of theme proved to be transmissible to normal wild-type mice.[82]

11.4.2   Sorting Signals: Hydrophobic Regions and their Messages

Deletion of the N-terminal signal peptide that determines the segregation of PrP into
the secretory route in combination with the removal of the C-terminal hydrophobic
tail that encodes for the GPI attachment permitted transgenic mice expressing a

soluble an intracellularly located PrP. These mice developed ataxia with cerebellar degeneration and gliosis that were refractory to the presence of wt PrP.[90]

However, single deletion of the *C*-terminal hydrophobic segment would model a system in which PrPC is fully secreted from the cell. Although this strategy did not provoke any physiological alteration in mouse, mice expressing PrPD231–254 when challenged with exogenous prions, propagated amyloids that cause a cerebral angiopathy and transmitted these amyloids as prions on wild-type animal.[14] These observations indicate that lesions and phenotype of prion diseases strictly depend on a cell signalling process.

In addition to the *N*- and *C*-terminal hydrophobic regions, PrP contains a central hydrophobic segment that can adopt a transmembrane configuration. Two sets of point substitutions, PrP3AV (involving A113V, A115V, and A118V replacements with and without L9R) and PrPKHII (with K109I and H110I double mutation) force PrP to adopt the CtmPrP topology.[1,3,85] Transgenic mice expressing these proteins developed a fatal neurological disorder, which is aggravated by coexpression of the wt PrP.[1,3,85]

Mice phenotypes from intracellular mislocating PrP, such as those promoting cytosolic distribution or the overpopulation of the Ctm topology using either signal sequence mutants or mutations in the HC region, result from a toxic PrP gain of function.[4] PrP through its OR region interacts with and sequesters mahogunin, a cytosolic ubiquitin ligase whose functional depletion causes spongiform neurodegeneration.[4]

11.4.3 Posttranslational Covalent Modifications: Glycosylations and Disulfide Bond

PrP contains two *N*-glycosylation sites at N180 and N196 (mouse sequence numbering) which are variably modified *in vivo* such that three glycotypes (un-, mono- and diglycosylated) are produced.[91–93] Both glycosylation sites are conserved in the PrP gene (*Prnp*) from all species, suggesting that *N*-glycans play an important role in the protein function. Pioneer studies focused on establishing the role of the modification degree and the structure of the glycan chains as a potential index of disease revealed differences in the structure of the chains isolated from PrPC and PrP^Sc.[93] However, since PrP^C was isolated from the brain of normal animal and PrP^Sc was purified from prion infected animals, the observed differences included metabolic factors difficult to uncouple.[91–93]

The role of glycosylation was then assessed using a loss of signature approach. Out of the different mutagenesis strategies directed to abrogate the modification with site specificity, the replacement of the *N*-sites by threonine (T), singly or in combination, provided the best set-up.[94,95] Transgenic mice carrying the PrP substitution N180T and N196T, singly or combined, were generated by double replacement gene targeting.[95] The presence of one glycan chain was sufficient for the trafficking to the cell membrane, whereas the unglycosylated PrP remained mainly intracellular. However, this altered cellular localization of PrP does not lead to any phenotype in the double transgenic mice. *In vivo* unglycosylated PrP does not acquire the characteristics of the aberrant pathogenic form, in contrast to culture cell studies.[95,96] Despite that partial or total abrogation of PrP, glycosylation does cause spontaneous disease—indeed the glycosylation state of host PrP^C dictates the targeting of the lesions. Using a peripheral mode of infection, mice expressing unglycosylated PrP did not develop clinical disease, and mice expressing mono-glycosylated PrP showed strikingly different neuropathological features compared with those expressing diglycosylated PrP.[97]

PrP chain contains in its sequence three Cys residues which are conserved in all mammalian sequences.[1,2] Two of these Cys residues are located in the *C*-terminal globular domain and their engagement in an intramolecular disulfide bond is critical for the maintenance of the PrP^C α-fold.[1,2,24] Early studies showed that this bond is also intramolecular in the solubilized fraction of PrP^{Sc}.[97] However the non-quantitative character of this trend and the importance of the disulfide bond as a folding keeper has lead to the repetitive consideration of a disulfide isomerization as a suitable mechanism for conversion.[97–99] Removal of the disulfide bridge in a C2-like chain of PrP (PrPΔ23–88 C178A) is tolerated in mice without inducing a spontaneous phenotype.[82] Notwithstanding this, the expression of this form reduces the susceptibility for conversion, indicating either mislocation effects or indeed an activity role.[82]

11.5 Inherited Disease and Structurally-based Mutations: Susceptibility to Misfolding and Sensitization to Prion Infection

A basic principle of PrP-related diseases is that the misfolding of PrP^C induced by a mutation linked to an inherited disorder or involving an essential residue is sufficient to trigger the pathology, causing both a lesion and the *de novo* generation of the transmissible agent.[2,100] However, the presence of such a mutant PrP could also increase the sensitivity to pre-existing prions that otherwise would remain silent and accelerate their propagation.[6,7,13,101–103] Indeed, approximately 15% of human PrP-related diseases are associated with auto-somal-dominant pathogenic mutations in the *PRNP* gene coding for PrP; these are classified as CJD, GSS or FFI, mainly due to historical reasons (Figure 11.3). In addition, sequence differences in the chain of PrP from different species displaying distinct susceptibility to suffer these disorders might be acting as activating mutations.[104] Lastly, the availability of 3D structures for the PrP chains from distinct species can also provide an initial scenario for the construction of new chain variants with high propensity to conversion.

11.5.1 Genetic CJD and its Mouse Models: Non-transmissible Spontaneous Degeneration

Genetic forms of CJD (gCJD) occur with insertions in the octarepeat region, chain truncating at 145 and single substitutions in the region comprised

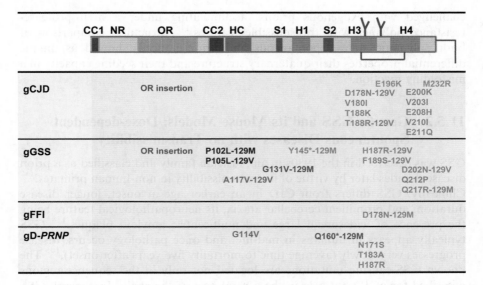

Figure 11.3 Human PrP inherited diseases and the *PRNP* mutations. Genetic forms of CJD (gCJD), GSS (gGSS), FFI (gFFI) and of some familiar neuropsychiatric disorders (gD-*PRNP*) are caused by mutations in *PNRP*. Mutations are located below and in the colour of the affected region.

between α2 and α3. In all the cases the inheritance is autosomal dominant and with a very high penetrance. As a hallmark of PrP-related diseases, the clinicopathological phenotype varies as a function of the mutation, the polymorphisms at codon 129, and most likely on a collection of yet unknown factors.[105]

Among the inherited PrP-diseases, gCJD resulting from E200K mutations in the *PRNP* gene is the commonest form.[106] This mutation is linked to a disease phenotype that resembles that of the typical sporadic CJD and for which the carriers of the mutation remain asymptomatic for several decades.[107] Onset of the disease correlates with the dose of expression of the mutant protein, with a mean age of 58 for heterozygous and 50 for homozygous.[108] Transgenic mice carrying the E200K mutation have been generated both on mouse–human chimeric and on human backgrounds.[109,110] Of these, the second group has been reported to be free of spontaneous disease but to undergo disease upon inoculation with human prions from E200K cases and sporadic forms of CJD.[110]

Transgenic mice expressing the mouse homologue of a nine OR insertional PrP mutant associated with gCJD have also been generated.[111,112] These animals suffer a spontaneous and progressive neurodegeneration with symptoms of ataxia, cerebellar granule cell loss, gliosis and PrP deposition.[113–115] In these mice the mutant protein is accumulated in brain as a highly neurotoxic, insoluble and weakly protease-resistant aggregate that lack infectious properties.[115] However, the mutant protein can convert into infectious forms when challenged with exogenous prions, accumulating under a high protease-resistance polymer with transmissible features.[115] Structural comparison of both polymeric states, spontaneous and prion-induced, showed as unique differential properties their quaternary structure and their seeding capacity in a misfolding reaction.[115,116]

11.5.2 Genetic GSS and its Mouse Models: Dose-dependent Spontaneous Diseases with no Transmissibility

GSS was described in the 1930s in an Austrian family and classified as a prion disease decades later by virtue of its transmissibility in non-human primates.[117] Clinically, GSS differs from CJD in an earlier age at onset, longer disease duration, and prominent cerebellar ataxia, its neuropathological feature being the presence of widespread large and multicentric amyloid plaques.[118] GSS typically appears in humans in mid-life, and once pathology occurs, disease progresses very slowly (average time to mortality five years after onset).[105] The known GSS-causing mutations are located not only in the sequence regions described for gCJD but also in the central part of the chain. In general, GSS is transmissible but with a very low efficiency, being in some cases never assessed.[117–120]

Transgenic mice expressing the equivalent of the human gGSS P102L mutation (MoPrP P101L) were first constructed using a MoPrP background and a randomly integrating approach.[121] Of these mice, the line expressing high levels of the mutant protein developed neurological dysfunction at 166 days of age with very low or even undetectable levels of PrP^{res}.[121] Lower levels of expression also resulted in spontaneous disease but at a more advanced age.[122,123] Brain extracts from affected mice did not transmit the neurodegeneration to wild-type mice, but accelerated the neurodegeneration in lower MoPrP P101L expressing mice.[11,123,124] The use of a knock-in approach with the same mutation showed that the P101L substitution in PrP does not cause disease but increases the susceptibility to exogenous prions.[101,102] These findings have been confirmed in models expressing the HuPrP P102L in a MoPrP$^{[0/0]}$ background.[110]

11.5.3 FFI and its Mouse Models: Transmissible Spontaneous Disease with Undetectable PrP^{res}

FFI refers to a disease identified in 1986 featured by insomnia and dysautonomia and which was later linked to the D178N mutation with positive transmission to mice.[119,125,126] In this case, the core clinical features are the disruption of the normal sleep–wake cycle, a sympathetic overactivity, endocrine abnormalities and the impairment of attention which correlates with the affection of the thalamus.[118] It must be stressed that FFI was initially thought to be controlled not only by the mutation in 178 but also by the polymorphism at codon 129 of the *PRNP* gene (129MM); however, a larger number of cases has shown that such genotype–phenotype association is not absolute.[127,128]

Randomly integration approaches and MoPrP transgenes carrying the D177N mutation, the 3F4 epitope and the M128V polymorphism have resulted in mice with sleep abnormalities.[129] For these mice, the occurrence of either spontaneous disease or the spontaneous generation of a transmissible agent has not yet been reported.[129,130] Knock-in strategies with a MoPrP carrying an intrinsic barrier to mouse prions and the D177N substitution has lead to a unique FFI animal model.[131] In this mouse, the mutant PrP chain reproduces the biochemical peculiarities described for the human chain (reduced expression levels, similar trafficking and post-translational modifications), and at 16 months, many homozygotic mice displayed both the behavioural and the CNS alterations of the human disease.[131,132] As in FFI, which characteristically produces very little PrP^{res} and is often not detected, in the brain of these mice PrP^{res} could not be easily detected under conventional conditions and, only on ageing, was a small amount of PrP^{res} found.[126,128,132] Moreover, the transmissibility of the disease is also recapitulated in this mouse model.[118]

11.5.4 Structural-designed Mutations for Spontaneous Disorder: Spontaneous Disease with Full Transmissibility and PrP^{res}

Comparison of the different 3D structures of mammalian PrP has identified the β2–α2 loop as differential structural element among species. This loop is highly disordered in human, cattle, mouse, dog and cat, whereas it adopts a rigid and defined structure in elk.[56,133] This structural diversity is dictated by the residues at positions 170 and 174, which are serine (S) and asparagine (N) in flexible states and N and T in those rigid.[24,56,133] Contrary to the transgenic mice expressing MoPrP wt at a similar dose, mice expressing the MoPrP S170N-N174T rigid mutant display a progressive transmissible spongiform neurodegeneration.[104] This spontaneous disease was characterized by ataxia, hindlimb paresis, weight loss, lethargy, kyphosis and a variable onset (145–637 days). Histological examination of the brains revealed spongiform changes, astrogliosis, microgliosis and the presence of multicentric PrP deposits. This central lesion pattern was accompanied by a peripheral myopathy similar to the MoPrP wt overexpression models.[69] Importantly the disease was transmitted with a 100% attack rate to mice overexpressing wt MoPrP, and from the latter to wt mice, but not to PrP-deficient mice.

11.5.5 A Molecular Switch Defining Cross-species Barriers for Prion Infections

Prions from different species vary considerably in their transmissibility to xenogeneic hosts. The variable transmission barriers depend on sequence differences between incoming PrP^{Sc} and host PrP^{C}, and there is evidence for correlations with strain-dependent conformational properties of PrP^{Sc} aggregates. The 3D structure of the β_2–α_2 loop region varies significantly between species and is influenced by the residue types in the two amino acid sequence positions 170 (most commonly S or N) and 174 (N or T). Prions from five different species were inoculated into transgenic mice expressing either flexible-loop or rigid-loop PrP^C variants. Similar β_2–α_2 loop conformations enabled transmission between otherwise resistant species, whereas dissimilar loops correlated with strong transmission barriers. When classified according to the 170S/N polymorphism, historical data indicated that transmission barriers are low between species with the same amino acid residue in position 170, and high between those with different residues. These findings point to a triggering role of the local β_2–α_2 loop amino acid sequence for interspecies prion transmissibility.[134]

Acknowledgements

M. Gasset is supported by grants from the Ministerio de Ciencia e Innovación (BFU2009-07971) and from Fundación Cien. A. Aguzzi is supported by the European Union (PRIORITY and LUPAS), the Swiss National Foundation, the Stammbach Foundation, the Novartis Research Foundation, and holds a European Research Council Advanced Investigator Grant.

References

1. S. B. Prusiner, *N. Engl. J. Med.*, 2001, **344**, 1516.
2. A. Aguzzi and A. M. Calella, *Physiol. Rev.*, 2009, **89**, 1105.
3. O. Chakrabarti, A. Ashok and R. S. Hegde, *Trends Biochem. Sci.*, 2009, **34**, 287.
4. O. Chakrabarti and R. S. Hegde, *Cell*, 2009, **137**, 1136.
5. S. B. Prusiner, *Adv. Virus Res.*, 1984, **29**, 1.
6. R. A. Somerville, *Trends Biochem. Sci.*, 2002, **27**, 606.
7. B. Chesebro, *Br. Med. Bull.*, 2003, **66**, 1.
8. C. Weissmann and E. Flechsig, *Br. Med. Bull.*, 2003, **66**, 43.
9. G. Legname, I. V. Baskakov, H. O. Nguyen, D. Riesner, F. E. Cohen, S. J. DeArmond and S. B. Prusiner, *Science*, 2004, **305**, 673.
10. J. Castilla, P. Saa, C. Hetz and C. Soto, *Cell*, 2005, **121**, 195.
11. J. Collinge and A. R. Clarke, *Science*, 2007, **318**, 930.
12. N. R. Deleault, B. T. Harris, J. R. Rees and S. Supattapone, *Proc. Nat. Acad. Sci. U. S. A.*, 2007, **104**, 9741.
13. L. Manuelidis, *J. Cell Biochem.*, 2007, **100**, 897.
14. B. Chesebro, M. Trifilo, R. Race, K. Meade-White, C. Teng, R. LaCasse, L. Raymond, C. Favara, G. Baron, S. Priola, B. Caughey, E. Masliah and M. Oldstone, *Science*, 2005, **308**, 1435.
15. B. Chesebro, B. Race, K. Meade-White, R. Lacasse, R. Race, M. Klingeborn, J. Striebel, D. Dorward, G. McGovern and M. Jeffrey, *PLoS Pathog.*, 2010, **6**, e1000800.

16. D. Westaway, P. A. Goodman, C. A. Mirenda, M. P. McKinley, G. A. Carlson and S. B. Prusiner, *Cell*, 1987, **51**, 651.
17. K. Hsiao, H. F. Baker, T. J. Crow, M. Poulter, F. Owen, J. D. Terwilliger, D. Westaway, J. Ott and S. B. Prusiner, *Nature*, 1989, **338**, 342.
18. Z. Y. Ning, D. M. Zhao, J. M. Yang, Y. L. Cui, L. P. Meng, C. D. Wu and H. X. Liu, *Anim. Biotechnol.*, 2005, **16**, 55.
19. B. Caughey and G. S. Baron, *Nature*, 2006, **443**, 803.
20. R. González-Iglesias, M. A. Pajares, C. Ocal, J. C. Espinosa, B. Oesch and M. Gasset, *J. Mol. Biol.*, 2002, **319**, 527.
21. V. G. Ostapchenko, N. Makarava, R. Savtchenko and I. V. Baskakov, *J. Mol. Biol.*, 2008, **383**, 1210.
22. M. Gasset, M. A. Baldwin, D. H. Lloyd, J. M. Gabriel, D. M. Holtzman, F. E. Cohen, R. Fletterick and S. B. Prusiner, *Proc. Natl Acad. Sci. U. S. A.*, 1992, **89**, 10940.
23. G. Forloni, N. Angeretti, R. Chiesa, E. Monzani, M. Salmona, O. Bugiani and F. Tagliavini, *Nature*, 1993, **362**, 543.
24. K. Wüthrich and R. Riek, *Adv. Protein Chem.*, 2001, **57**, 55.
25. N. Stahl, D. R. Borchelt, K. Hsiao and S. B. Prusiner, *Cell*, 1987, **51**, 229.
26. M. Rogers, A. Taraboulos, M. Scott, D. Groth and S. B. Prusiner, *Glycobiology*, 1990, **1**, 101.
27. M. S. Palmer and J. Collinge, *J. Hum. Mutat.*, 1993, **2**, 168.
28. K. M. Green, S. R. Browning, T. S. Seward, J. E. Jewell, D. L. Ross, M. A. Green, E. S. Williams, E. A. Hoover and G. C. Telling, *J. Gen. Virol.*, 2008, **89**, 598.
29. M. E. Juanes, G. Elvira, A. García-Grande, M. Calero and M. Gasset, *J. Biol. Chem.*, 2009, **284**, 2787.
30. R. Yadavalli, R. P. Guttmann, T. Seward, A. P. Centres, R. A. Williamson and G. C. Telling, *J. Biol. Chem.*, 2004, **279**, 21948.
31. B. Vincent, M. A. Cisse, C. Sunyach, M. V. Guillot-Sestier and F. Checler, *Curr. Alzheimer Res.*, 2008, **5**, 202.
32. M. Dron, M. Moudjou, J. Chapuis, M. K. Salamat, J. Bernard, S. Cronier, C. Langevin and H. Laude, *J. Biol. Chem.*, 2010, **285**, 10252.
33. J. B. Oliveira-Martins, S. Yusa, A. M. Calella, C. Bridel, F. Baumann, P. Dametto and A. Aguzzi, *PLoS One*, 2010, **5**, e9107.
34. E. T. Parkin, N. T. Watt, A. J. Turner and N. M. Hooper, *J. Biol. Chem.*, 2004, **279**, 11170.
35. D. R. Taylor, E. T. Parkin, S. L. Cocklin, J. R. Ault, A. E. Ashcroft, A. J. Turner and N. M. Hooper, *J. Biol. Chem.*, 2009, **284**, 22590.
36. R. Diaz-Espinoza and C. Soto, *Prion*, 2010, **4**, 53.
37. I. V. Baskakov, *FEBS Lett.*, 2009, **583**, 2618.
38. C. Weissmann, *Folia Neuropathol.*, 2009, **47**, 104.
39. I. V. Baskakov and L. Breydo, *Biochim. Biophys. Acta*, 2007, **1772**, 692.
40. I. H. Solomon, J. A. Schepker and D. A. Harris, *Curr. Issues Mol. Biol.*, 2009, **12**, 51.
41. F. E. Cohen, K. M. Pan, Z. Huang, M. Baldwin, R. J. Fletterick and S. B. Prusiner, *Science*, 1994, **264**, 530.

42. J. Safar, H. Wille, V. Itri, D. Groth, H. Serban, M. Torchia, F. E. Cohen and S. B. Prusiner, *Nat. Med.*, 1998, **4**, 1157.
43. R. K. Meyer, M. P. McKinley, K. A. Bowman, M. B. Braunfeld, R. A. Barry and S. B. Prusiner, *Proc. Natl. Acad. Sci. U. S. A.*, 1986, **83**, 2310.
44. M. P. McKinley, R. K. Meyer, L. Kenaga, F. Rahbar, R. Cotter, A. Serban and S. B. Prusiner, *J. Virol.*, 1991, **65**, 1340.
45. A. Taraboulos, D. Serban and S. B. Prusiner, *J. Cell. Biol.*, 1990, **110**, 2117.
46. F. Tagliavini, F. Prelli, J. Ghiso, O. Bugiani, D. Serban, S. B. Prusiner, M. R. Farlow, B. Ghetti and B. Frangione, *EMBO J.*, 1991, **10**, 513.
47. G. Giaccone, L. Verga, O. Bugiani, B. Frangione, D. Serban, S. B. Prusiner, M. R. Farlow, B. Ghetti and F. Tagliavini, *Proc. Natl. Acad. Sci. U. S. A.*, 1992, **89**, 9349.
48. S. Tzaban, G. Friedlander, O. Schonberger, L. Horonchik, Y. Yedidia, G. Shaked, R. Gabizon and A. Taraboulos, *Biochemistry*, 2002, **41**, 12868.
49. N. Makarava, G. G. Kovacs, O. Bocharova, R. Savtchenko, I. Alexeeva, H. Budka, R. G. Rohwer and I. V. Baskakov, *Acta Neuropathol.*, 2010, **119**, 177.
50. D. W. Colby, K. Giles, G. Legname, H. Wille, I. V. Baskakov, S. J. DeArmond and S. B. Prusiner, *Proc. Natl. Acad. Sci. U. S. A.*, 2009, **106**, 20417.
51. D. W. Colby, R. Wain, I. V. Baskakov, G. Legname, C. G. Palmer, H. O. Nguyen, A. Lemus, F. E. Cohen, S. J. DeArmond and S. B. Prusiner, *PLoS Pathog.*, 2010, **6**, e1000736.
52. F. Wang, X. Wang, C. G. Yuan and J. Ma, *Science*, 2010, **327**, 132.
53. D. Shmerling, I. Hegyi, M. Fischer, T. Blattler, S. Brandner, J. Gotz, T. Rulicke, E. Flechsig, A. Cozzio, C. von Mering, C. Hangartner, A. Aguzzi and C. Weissmann, *Cell*, 1998, **93**, 203.
54. C. Wolschner, A. Giese, H. A. Kretzschmar, R. Huber, L. Moroder and N. Budisa, *Proc. Natl. Acad. Sci. U. S. A.*, 2009, **106**, 7756.
55. Z. Huang, J. M. Gabriel, M. A. Baldwin, R. J. Fletterick, S. B. Prusiner and F. E. Cohen, *Proc. Natl. Acad. Sci. U. S. A.*, 1994, **91**, 7139.
56. H. M. Schätzl, M. Da Costa, L. Taylor, F. E. Cohen and S. B. Prusiner, *J. Mol. Biol.*, 1995, **245**, 362.
57. H. Büeler, M. Fischer, Y. Lang, H. Fluethmann, H.-P. Lipp, S. J. DeArmond, S. B. Prusiner, M. Aguet and C. Weissmann, *Nature*, 1992, **356**, 577.
58. P. Tremblay, Z. Meiner, M. Galou, C. Heinrich, C. Petromilli, T. Lisse, J. Cayetano, M. Torchia, W. Mobley, H. Bujard, S. J. DeArmond and S. B. Prusiner, *Proc. Natl. Acad. Sci. U. S. A.*, 1998, **95**, 12580.
59. G. R. Mallucci, S. Ratté, E. A. Asante, J. Linehan, I. Gowland, J. G. Jefferys and J. Collinge, *EMBO J.*, 2002, **21**, 202.
60. J. Collinge, M. A. Whittington, K. C. Sidle, C. J. Smith, M. S. Palmer, A. R. Clarke and J. G. Jefferys, *Nature*, 1994, **370**, 295.

61. I. Tobler, S. E. Gaus, T. Deboer, P. Achermann, M. Fischer, T. Rülicke, M. Moser, B. Oesch, P. A. McBride and J. C. Manson, *Nature*, 1996, **380**, 639.
62. S. B. Colling, M. Khana, J. Collinge and J. G. Jefferys, *Brain Res.*, 1997, **755**, 28.
63. J. Herms, T. Tings, S. Gall, A. Madlung, A. Giese, H. Siebert, P. Schürmann, O. Windl, N. Brose and H. Kretzschmar, *J. Neurosci.*, 1999, **19**, 8866.
64. P. M. Lledo, P. Tremblay, S. J. DeArmond, S. B. Prusiner and R. A. Nicoll, *Proc. Natl. Acad. Sci. U. S. A.*, 1996, **93**, 2403.
65. J. Bremer, F. Baumann, C. Tiberi, C. Wessig, H. Fischer, P. Schwarz, A. D. Steele, K. V. Toyka, K. A. Nave, J. Weis and A. Aguzzi, *Nat. Neurosci.*, 2010, **13**, 310.
66. C. Korth, B. Stierli, P. Streit, M. Moser, O. Schaller, R. Fischer, W. Schulz-Schaeffer, H. Kretzschmar, A. Raeber, U. Braun, F. Ehrensperger, S. Hornemann, R. Glockshuber, R. Riek, M. Billeter, K. Wüthrich and B. Oesch, *Nature*, 1997, **390**, 74.
67. M. Polymenidou, R. Moos, M. Scott, C. Sigurdson, Y. Z. Shi, B. Yajima, I. Hafner-Bratkovic, R. Jerala, S. Hornemann, K. Wuthrich, A. Bellon, M. Vey, G. Garen, M. N. James, N. Kav and A. Aguzzi, *PLoS One*, 2008, **3**, e3872.
68. T. Canello, R. Engelstein, O. Moshel, K. Xanthopoulos, M. E. Juanes, J. Langeveld, T. Sklaviadis, M. Gasset and R. Gabizon, *Biochemistry*, 2008, **47**, 8866.
69. D. Westaway, S. J. DeArmond, J. Cayetano-Canlas, D. Groth, D. Foster, S. L. Yang, M. Torchia, G. A. Carlson and S. B. Prusiner, *Cell*, 1994, **76**, 117.
70. S. Huang, J. Liang, M. Zheng, X. Li, M. Wang, P. Wang, D. Vanegas, D. Wu, B. Chakrabarty, A. P. Hays, K. Chen, S. G. Chen, S. Booth, M. Cohen, P. Gambetti and Q. Kong, *Proc. Natl. Acad. Sci. U. S. A.*, 2007, **104**, 6800.
71. V. Askanas, M. Bilak, W. K. Engel, R. B. Alvarez, F. Tome and A. Leclerc, *Neuroreport*, 1993, **5**, 25.
72. E. Sarkozi, V. Askanas and W. K. Engel, *Am. J. Pathol.*, 1994, **145**, 1280.
73. G. Zanusso, G. Vattemi, S. Ferrari, M. Tabaton, E. Pecini, T. Cavallaro, G. Tomelleri, M. Filosto, P. Tonin, E. Nardelli, N. Rizzuto and S. Monaco, *Brain Pathol.*, 2001, **11**, 182.
74. H. Büeler, A. Aguzzi, A. Sailer, R. A. Greiner, P. Autenried, M. Aguet and C. Weissmann, *Cell*, 1993, **73**, 1339.
75. A. Sailer, H. Büeler, M. Fischer, A. Aguzzi and C. Weissmann, *Cell*, 1994, **77**, 967.
76. S. Brandner, S. Isenmann, A. Raeber, M. Fischer, A. Sailer, Y. Kobayashi, S. Marino, C. Weissmann and A. Aguzzi, *Nature*, 1996, **379**, 339.
77. H. Büeler, A. Raeber, A. Sailer, M. Fischer, A. Aguzzi and C. Weissmann, *Mol. Med.*, 1994, **1**, 19.

78. J. Castilla, A. Gutiérrez-Adán, A. Brun, B. Pintado, B. Parra, M. A. Ramírez, F. J. Salguero, F. Díaz San Segundo, A. Rábano, M. J. Cano and J. M. Torres, *J. Neurosci.*, 2004, **24**, 2156.

79. C. J. Sigurdson, G. Manco, P. Schwarz, P. Liberski, E. A. Hoover, S. Hornemann, M. Polymenidou, M. W. Miller, M. Glatzel and A. Aguzzi, *J. Virol.*, 2006, **80**, 12303.

80. C. J. Sigurdson, M. Heikenwalder, G. Manco, M. Barthel, P. Schwarz, B. Stecher, N. J. Krautler, W. D. Hardt, B. Seifert, A. J. MacPherson, I. Corthesy and A. Aguzzi, *J. Infect. Dis.*, 2009, **199**, 243.

81. M. Fischer, T. Rülicke, A. Raeber, A. Sailer, M. Moser, B. Oesch, S. Brandner, A. Aguzzi and C. Weissmann, *EMBO J.*, 1996, **15**, 1255.

82. T. Muramoto, S. J. DeArmond, M. Scott, G. C. Telling, F. E. Cohen and S. B. Prusiner, *Nat. Med.*, 1997, **3**, 750.

83. E. Flechsig, D. Shmerling, I. Hegyi, A. J. Raeber, M. Fischer, A. Cozzio, C. von Mering, A. Aguzzi and C. Weissmann, *Neuron*, 2000, **27**, 399.

84. T. Muramoto, M. Scott, F. E. Cohen and S. B. Prusiner, *Proc. Natl. Acad. Sci. U. S. A.*, 1996, **93**, 15457.

85. R. S. Hegde, J. A. Mastrianni, M. R. Scott, K. A. Defea, P. Tremblay, M. Torchia, S. J. DeArmond, S. B. Prusiner and V. R. Lingappa, *Science*, 1998, **279**, 827.

86. I. Radovanovic, N. Braun, O. T. Giger, K. Mertz, G. Miele, M. Prinz, B. Navarro and A. Aguzzi, *J. Neurosci.*, 2005, **25**, 4879.

87. F. Baumann, M. Tolnay, C. Brabeck, J. Pahnke, U. Kloz, H. H. Niemann, M. Heikenwalder, T. Rülicke, A. Bürkle and A. Aguzzi, *EMBO J.*, 2007, **26**, 538.

88. A. Li, H. M. Christensen, L. R. Stewart, K. A. Roth, R. Chiesa and D. A. Harris, *EMBO J.*, 2006, **26**, 548.

89. E. Flechsig and C. Weissmann, *Curr. Mol. Med.*, 2004, **4**, 337.

90. J. Ma, R. Wollmann and S. Lindquist, *Science*, 2002, **298**, 1781.

91. H. Stimson, J. Hope, A. Chong and A. L. Burlingame, *Biochemistry*, 1999, **38**, 4885.

92. P. M. Rudd, T. Endo, C. Colominas, D. Groth, S. F. Wheeler, D. J. Harvey, M. R. Wormald, H. Serban, S. B. Prusiner, A. Kobata and R. A. Dwek, *Proc. Natl. Acad. Sci. U. S. A.*, 1999, **96**, 13044.

93. P. M. Rudd, A. H. Merry, M. R. Wormald and R. A. Dwek, *Curr. Opin. Struct. Biol.*, 2002, **12**, 578.

94. S. A. Priola and V. A. Lawson, *EMBO J.*, 2001, **20**, 6692.

95. E. Cancellotti, F. Wiseman, N. L. Tuzi, H. Baybutt, P. Monaghan, L. Aitchison, J. Simpson and J. C. Manson, *J. Biol. Chem.*, 2005, **280**, 42909.

96. E. Cancellotti, B. M. Bradford, N. L. Tuzi, R. D. Hickey, D. Brown, K. L. Brown, R. M. Barron, D. Kisielewski, P. Piccardo and J. C. Manson, *J. Virol.*, 2010, **84**, 3464.

97. E. Turk, D. B. Teplow, L. E. Hood and S. B. Prusiner, *Eur. J. Biochem.*, 1988, **176**, 21.

98. E. Welker, L. D. Raymond, H. A. Scheraga and B. Caughey, *J. Biol. Chem.*, 2002, **277**, 33477.
99. S. Lee and D. Eisenberg, *Nat. Struct. Biol.*, 2003, **10**, 725.
100. A. Aguzzi and M. Polymenidou, *Cell*, 2004, **116**, 313.
101. J. C. Manson, E. Jamieson, H. Baybutt, N. L. Tuzi, R. Barron, I. McConnell, R. Somerville, J. Ironside, R. Will, M. S. Sy, D. W. Melton, J. Hope and C. Bostock, *EMBO J.*, 1999, **18**, 6855.
102. R. M. Barron, V. Thomson, E. Jamieson, D. W. Melton, J. Ironside, R. Will and J. C. Manson, *EMBO J.*, 2001, **20**, 5070.
103. C. Weissmann and E. Flechsig, *Br. Med. Bull.*, 2003, **66**, 43.
104. C. J. Sigurdson, N. P. Nilsson, S. Hornemann, M. Heikenwalder, G. Manco, P. Schwarz, D. Ott, T. Rülicke, P. P. Liberski, C. Julius, J. Falsig, L. Stitz, K. Wüthrich and A. Aguzzi, *Proc. Natl. Acad. Sci. U. S. A.*, 2009, **106**, 304.
105. G. G. Kovács, M. W. Head, I. Hegyi, T. J. Bunn, H. Flicker, J. A. Hainfellner, L. McCardle, L. László, C. Jarius, J. W. Ironside and H. Budka, *Brain Pathol.*, 2002, **12**, 1.
106. K. Hsiao, Z. Meiner, E. Kahana, C. Cass, I. Kahana, D. Avrahami, G. Scarlato, O. Abramsky, S. B. Prusiner and R. Gabizon, *N. Engl. J. Med.*, 1991, **324**, 1091.
107. Z. Meiner, R. Gabizon and S. B. Prusiner, *Medicine*, 1997, **76**, 227.
108. E. S. Simon, E. Kahana, J. Chapman, T. A. Treves, R. Gabizon, H. Rosenmann, N. Zilber and A. D. Korczyn, *Ann. Neurol.*, 2000, **47**, 257.
109. H. Rosenmann, G. Talmor, M. Halimi, A. Yanai, R. Gabizon and Z. Meiner, *J. Neurochem.*, 2001, **76**, 1654.
110. E. A. Asante, I. Gowland, A. Grimshaw, J. M. Linehan, M. Smidak, R. Houghton, O. Osiguwa, A. Tomlinson, S. Joiner, S. Brandner, J. D. Wadsworth and J. Collinge, *J. Gen. Virol.*, 2009, **90**, 546.
111. L. W. Duchen, M. Poulter and A. E. Harding, *Brain*, 1999, **116**, 55.
112. S. Krasemann, I. Zerr, T. Weber, S. Poser, H. Kretzschmar, G. Hunsmann and W. Bodemer, *Brain Res. Mol. Brain Res.*, 1995, **34**, 173.
113. R. Chiesa, P. Piccardo, B. Ghetti and D. A. Harris, *Neuron*, 1998, **21**, 1339.
114. R. Chiesa, B. Drisaldi, E. Quaglio, A. Migheli, P. Piccardo, B. Ghetti and D. A. Harris, *Proc. Natl. Acad. Sci. U. S. A.*, 2000, **97**, 5574.
115. R. Chiesa, P. Piccardo, E. Quaglio, B. Drisaldi, S. L. Si-Hoe, M. Takao, B. Ghetti and D. A. Harris, *J. Virol.*, 2003, **77**, 7611.
116. E. Biasini, A. Z. Medrano, S. Thellung, R. Chiesa and D. A. Harris, *J. Neurochem.*, 2008, **104**, 1293.
117. S. J. Collins and C. L. Masters, *Sci. Prog.*, 1995, **78**, 217.
118. S. Collins, C. A. McLean and C. L. Masters, *J. Clin. Neurosci.*, 2001, **8**, 387.
119. J. Tateishi and T. Kitamoto, *Brain Pathol.*, 1995, **5**, 53.
120. P. Brown, C. J. Gibbs, P. Rodgers-Johnson, D. M. Asher, M. P. Sulima, A. Bacote, L. G. Goldfarb and D. C. Gajdusek, *Ann. Neurol.*, 1994, **35**, 513.
121. K. K. Hsiao, M. Scott, D. Foster, D. F. Groth, S. J. DeArmond and S. B. Prusiner, *Science*, 1990, **250**, 1587.

122. P. Tremblay, K. L. Ball, K. Kaneko, D. Groth, R. S. Hegde, F. E. Cohen, S. J. DeArmond, S. B. Prusiner and J. G. Safar, *J. Virol.*, 2004, **78**, 2088.
123. K. E. Nazor, F. Kuhn, T. Seward, M. Green, D. Zwald, M. Purro, J. Schmid, K. Biffiger, A. M. Power, B. Oesch, A. J. Raeber and G. C. Telling, *EMBO J.*, 2005, **24**, 2472.
124. K. K. Hsiao, D. Groth, M. Scott, S. L. Yang, H. Serban, D. Rapp, D. Foster, M. Torchia, S. J. Dearmond and S. B. Prusiner, *Proc. Natl. Acad. Sci. U. S. A.*, 1994, **91**, 9126.
125. E. Lugaresi, R. Medori, P. Montagna, A. Baruzzi, P. Cortelli, A. Lugaresi, P. Tinuper, M. Zucconi and P. Gambetti, *N. Engl. J. Med.*, 1986, **315**, 997.
126. R. Medori, H. J. Tritschler, A. LeBlanc, F. Villare, V. Manetto, H. Y. Chen, R. Xue, S. Leal, P. Montagna, P. Cortelli P, P. Tinuper, P. Avoni, M. Mochi, A. Baruzzi, J. J. Hauw, J. Ott, E. Lugaresi, L. Autilio-Gambetti and P. Gambetti, *N. Engl. J. Med.*, 1992, **326**, 444.
127. L. G. Goldfarb, R. B. Petersen, M. Tabaton, P. Brown, A. C. LeBlanc, P. Montagna, P. Cortelli, J. Julien, C. Vital and W. W. Pendelbury *et al.*, *Science*, 1992, **258**, 806.
128. J. J. Zarranz, A. Digon, B. Atares, A. B. Rodriguez-Martinez, A. Arce, N. Carrera, I. Fernández-Manchola, M. Fernández-Martínez, C. Fernández-Maiztegui, I. Forcadas, L. Galdos, J. C. Gómez-Esteban, A. Ibáñez, E. Lezcano, A. López de Munain, J. F. Martí-Massó, M. M. Mendibe, M. Urtasun, J. M. Uterga, N. Saracibar, F. Velasco and M. M. de Pancorbo, *J. Neurol. Neurosurg. Psychiatry*, 2005, **76**, 1491.
129. S. Dossena, L. Imeri, M. Mangieri, A. Garofoli, L. Ferrari, A. Senatore, E. Restelli, C. Balducci, F. Fiordaliso, M. Salio, S. Bianchi, L. Fioriti, M. Morbin, A. Pincherle, G. Marcon, F. Villani, M. Carli, F. Tagliavini, G. Forloni and R. Chiesa, *Neuron*, 2008, **60**, 598.
130. P. Gambetti, P. Parchi, R. B. Petersen, S. G. Chen and E. Lugaresi, *Brain Pathol.*, 1995, **5**, 43.
131. W. S. Jackson, A. W. Borkowski, H. Faas, A. D. Steele, O. D. King, N. Watson, A. Jasanoff and S. Lindquist, *Neuron*, 2009, **63**, 438.
132. P. Parchi, R. B. Petersen, S. G. Chen, L. Autilio-Gambetti, S. Capellari, L. Monari, P. Cortelli, P. Montagna, E. Lugaresi and P. L. Gambetti, *Brain Pathol.*, 1995, **8**, 539.
133. A. D. Gossert, S. Bonjour, D. A. Lysek, F. Fiorito and K. Wüthrich, *Proc. Natl. Acad. Sci. U. S. A.*, 2005, **102**, 646.
134. C. J. Sigurdson, P. R. Nilsson, S. Hornemann, G. Manco, N. Fernández-Borges, P. Schwarz, J. Castilla, K. Wüthrich and A. Aguzzi, *J. Clin. Invest.*, 2010, **120**, 2590.
135. I. V. Baskakov, *FEBS Lett.*, 2007, **274**, 3756.

CHAPTER 12

Mouse Models of Ischemia

DAVID C. HENSHALL[*a] AND ROGER P. SIMON[b]

[a] Department of Physiology and Medical Physics, Royal College of Surgeons in Ireland, Dublin, Ireland; [b] Morehouse School of Medicine, Atlanta, GA, USA

12.1 Introduction

The brain receives ~15% of cardiac output and has a very high dependence on the continual delivery of oxygen and glucose. Local cerebral blood flow (CBF) in mice ranges from 48–96 ml /100 g/min in white matter and from 86–198 ml/ 100 g/min in grey matter.[1] Interruption of the blood supply to the entire brain or a region (ischemia—in effect the brain's equivalent of a heart attack) results in reduced delivery of these substrates. Once below certain thresholds, reduced CBF to a brain region causes cessation of normal brain function and, with bi-hemispheric involvement, loss of consciousness in a few seconds. The downstream effects of ischemia depend on the duration of the episode and the brain region(s) affected. There can be full recovery of function or a varying degree of permanent brain damage with attendant neurologic deficits. Not all neurons or non-neuronal cells are equally tolerant of an interrupted blood supply. After global ischemia, for example, there is a selective death of a population of hippocampal pyramidal neurons while neighbouring cells are spared.

Our understanding of the cell and molecular pathogenesis of ischemic brain damage is comprehensive although not yet complete. Ischemia results in the collapse of cellular ion gradients and metabolic suppression. There is release of neurotransmitters, oxidative stress, oedema, necrosis and attendant inflammation. In regions where energy failure is less complete, coordinated signalling pathways are activated and cell death bears some of the hallmarks of apoptosis.

RSC Drug Discovery Series No. 6
Animal Models for Neurodegenerative Disease
Edited by Jesús Avila, Jose J. Lucas and Félix Hernández
© Royal Society of Chemistry 2011
Published by the Royal Society of Chemistry, www.rsc.org

Knowledge of these processes has informed the deployment of various therapeutic interventions which mitigate neuron loss and functional deficits following experimental ischemia; many have been advanced for clinical trials.

Animal models of stroke serve two crucial purposes. First, they provide a way to elucidate the pathophysiology of stroke; *in vivo* preparations model key elements common to the human condition while enabling variables to be controlled or modulated. Second, they provide the means to test potential therapeutic agents. *In vivo* stroke models in species including primates, cats and rats appeared in the 1960s and 1970s.[2-5] Core concepts of the pathophysiology of stroke—upper and lower limits of CBF which constituted ischemia, tissue viability thresholds and excitotoxicity—were established.[6-9] The introduction of stroke models in rats by Tamura *et al.*[10] and Koizumi *et al.*[11] made stroke research even more accessible. An exponential increase in experimental stroke research ensued, with models also used to evaluate therapeutic agents, beginning with drugs targeting glutamate neurotoxicity.[12] A plethora of agents targeting apoptosis, oxidative stress and inflammation have since been developed and evaluated.

The advent of mouse models of stroke allowed investigators to test specific gene effects in genetically altered mice. These also brought with them special technical challenges. Mouse models of ischemia have facilitated a breathtaking range of studies evaluating gene effects, extending our understanding of the pathophysiology of stroke in a manner not previously possible, and identifying novel routes to neuroprotection.

This chapter summarises the various mouse models of stroke. Particular emphasis is given to their origins and the strengths and weaknesses of each. We also include a special focus on mouse models of ischemic tolerance, which have been used as a route to identifying endogenous programmes of neuroprotection.

12.2   Stroke: Incidence, Causes and Definitions

Stroke is the second or third most common cause of death worldwide. It is also the leading cause of adult disability. It is a major human health and socioeconomic problem. The World Health Organization estimates that 15 million people every year suffer a stroke. One third of these people die, while another third are left with permanent disability. Indeed, while mortality is significant following stroke, it is the large proportion of patients with permanent disabilities which creates the greater socioeconomic burden. Stroke is uncommon in people under the age of 40 years, but its incidence thereafter increases exponentially with age. Rates are generally higher in men than women. Stroke incidence also shows considerable variation between different ethnic groups[13] and countries.[14]

12.2.1   Stroke: Terms and Conditions

12.2.1.1   *Ischemic Stroke*

Ischemic stroke involving occlusion of a major cerebral blood vessel represents the most common form of stroke; approximately 80% of all strokes are ischemic (Table 12.1). The cause is usually thromboembolic, arising following either the blockage of a vessel by an embolus (*e.g.* a piece of shed atherosclerotic plaque or a clot from the cardiac chambers) or due to local thrombus formation in the vessel itself. Focal ischemia is rarely complete, with some residual blood flow to the territory of the vessel either from the vessel itself or collateral circulation.

12.2.1.2 Hemorrhagic Stroke

Hemorrhagic stroke refers to bleeding into the brain. This can be haemorrhagic conversion (bleeding into an area of acute brain ischemia) or intracerebral haemorrhage (bleeding from a ruptured blood vessel in brain parenchyma) (Table 12.1). When a blood vessel ruptures near the surface of the brain (usually from an aneurysm), bleeding is usually referred to as subarachnoid hemorrhage. While not as common as strokes following blockages, these carry a high risk of mortality.

12.2.1.3 Global Cerebral Ischemia

Global cerebral ischemia results from cardiorespiratory arrest. In this case, the blood supply to all or most of the brain is briefly interrupted. This is also a less common form of stroke (Table 12.1).

12.2.1.4 Infarct

An infarct refers to a volume of tissue in which all cell types—neurons, glia, vessels and nerve fibres—have died. A clear demarcation of this zone is usually evident within 1–2 days after an ischemic stroke. When observed long after the initial stroke, the infarcted tissue comprises cavities (encephalomalacia) containing fluid and certain cell types.[15]

12.2.1.5 Selective Neuronal Injury

Selective neuronal injury is found following global ischemia and affects particularly the CA1 pyramidal neurons of the hippocampus and cortical lamina. The tissue structure itself remains preserved.[15] Two explanations were originally advanced to explain selective vulnerability. A vascular theory was proposed by Spielmeyer (1925), while Vogt (1937) proposed that physico-chemical

Table 12.1 Major forms of human stroke.

Type	Features
Ischemic stroke	Thromboembolic occlusion of a major cerebral vessel
Hemorrhagic stroke	Ischemia resulting from bleeding into the brain
Global ischemia	Complete (brief) cessation of cerebral blood flow

properties were the basis.[8] The cell and molecular basis of selective neuronal vulnerability to ischemia remains incompletely understood.

12.2.2 Transient *vs.* Permanent Ischemia

Vessel occlusion may be permanent or transient. In the former case, CBF is fully obstructed such that no return of flow to a given perfusion territory occurs. The perfusion field downstream of a permanently occluded vessel will eventually infarct. However, some degree of vessel recannilization is normal in human stroke. Accordingly, models of temporary vessel occlusion are considered better models of stroke, whereby reperfusion of the previously ischemic territory is permitted at some later stage. Interestingly, while reperfusion is the only means to avoid infarction, there is evidence that re-introduction of blood flow into ischemic brain can sometimes worsen damage[16,17]—so-called reperfusion damage. The mechanism is not fully understood but may involve hyperperfusion, reinjection of clots to the brain, enhanced infiltration of pro-inflammatory cells, free radicals, or other molecules.[18]

12.3 Pathophysiology of Ischemic Brain Injury: Penumbra, Cell Death and Neuroprotection

Ischemic cell death is the direct result of reduced CBF. The degree and duration of the reduction are both critical; infarction typically results when CBF falls to ≤ 10 ml/100 g/min for a period of a few minutes to hours. When the cause of stroke is occlusion of a major cerebral artery, this produces a dense ischemic core which is surrounded by an area of constrained blood supply in which certain metabolic processes are maintained (termed the penumbra). If blood flow is not restored to this region it too becomes infarcted. This area represents tissue, however, which can be salvaged either by recannilization or by introduction of pharmacological treatments which protect the tissue by interrupting various aspects of the pathophysiology.[19] Moreover, distinct molecular penumbras are generated by focal ischemia whereby differential expression of protective and cell death-related genes emerge in a perfusion-dependent manner.[20] While global ischemia is not associated with an evolving penumbra, watershed zones where brain survival hangs in the balance are recognized.[21]

12.3.1 Cause of Tissue Injury in Ischemia

What causes cell death after ischemia? This differs according to the region. Within the ischemic core, which after a middle cerebral artery (MCA) occlusion in rodents is the striatum, there is a combination of total energy failure leading to electrical and membrane failure, loss of intracellular ion homeostasis, tissue acidosis due to anaerobic glycolysis, release of glutamate and other neurotransmitters leading to excitotoxicity, oedema, and cell membrane rupture and necrosis.[22–24]

Within the ischemic penumbra, partial preservation of energy facilitates a more controlled cell death process. However, this also involves loss of intracellular ion homeostasis, in particular a rise in intracellular calcium, which can lead to necrosis.[22] Acidosis activates selective cation channels which also promote calcium entry.[25] Neurotransmitter release proceeds either *via* reversal of glia re-uptake transporters or depolarization-induced synaptic release.

Specific signalling pathways such as those associated with apoptosis are activated within ischemic tissue and in the penumbra in particular. A causal role for these genes is evinced by the tissue-sparing effects of various gene deletions or delivery of inhibitors of enzymes such caspases in stroke models.[26–31]

Other factors which contribute to ischemic brain injury include spreading depression—recurring waves of depolarization which further expend cellular energy and may contribute to the gradual recruitment of penumbral tissue into the infarct.[32,33] Adherence of immune cells to the microvasculature probably exacerbates interruption of blood flow particularly during vessel recannilization. Immune cells also invade ischemic tissue and, along with activated resident microglia, contribute further to ischemic injury by pro-inflammatory processes.[34,35] However, protective roles for certain T cell populations have been identified.[36]

12.3.2 Post-stroke Repair

In the aftermath of a stroke, there begins cortical remodelling and reorganization. A number of processes are thought to occur which may contribute to functional recovery or, sometimes, development of co-morbidities such as epilepsy. Indeed, beyond functional deficits relating to the original stroke, epilepsy is a common consequence, and stroke-induced epilepsy has also been modelled.[37] Processes established as important for functional recovery include axon and dendritic sprouting,[38,39] and neurogenesis.[40]

12.3.3 Neuroprotection and Translation

Outside the immediate area of infarction in the ischemic core, there is significant salvageable tissue.[19] A multitude of agents have been studied and reviews of these can be found elsewhere.[33,41] They include glutamate receptor antagonists, calcium channel antagonists, acid-sensing ion channel inhibitors, free radical scavengers, vascular growth factors, proteasome inhibitors and anti-apoptotic agents.[33,41–43] Various physiological approaches have also shown protective efficacy, including hypothermia.[44]

Over 150 clinical trials of neuroprotection for stroke have been initiated.[41] Disappointingly, we still have no frontline neuroprotective agent with which to prevent brain damage from stroke in humans. Tissue plasminogen activator (t-PA), which dissolves blood clots, remains the only approved treatment in clinical use.[45] Discussion of the reasons for the failure of animal models and preclinical research to translate to the clinic can be found elsewhere.[46–48]

12.4   Mouse Models of Ischemia

12.4.1   General

The purpose of any animal model of brain ischemia is to reduce CBF to all or part of the brain. The major models are detailed in Table 12.2 and excellent reviews can be found elsewhere.[49–51] The MCA is the most commonly occluded vessel in human stroke and, appropriately, is the vessel most commonly targeted in animal models of focal ischemia (Figure 12.1). In rodents, the MCA feeds the caudate–putamen (striatum), globus pallidus, major parts of the cerebral cortex and the adjacent white matter. Stroke is also modelled with short periods of global ischemia, and other less-common but nonetheless important aetiologies have been modelled including lacunar stroke (Table 12.2).

Assessment of ischemic brain damage can be undertaken in a variety of ways including volumetric assessments based on histological staining of brain tissue sections. Particularly useful is the technique using 2,3,5-triphenyltetrazolium chloride (TTC), a vital dye, to stain fresh coronal slices.[52] A method for determining infarct size which took account of brain swelling was introduced later.[53]

The utility of mouse models lies with their relative genetic homogeneity and the ability to introduce or inactivate various genes, insert mutations, and even

Table 12.2 Major mouse models of ischemia and their strengths and weaknesses.

Model	Strengths	Weaknesses
Surgical transection/direct occlusion	Highly reproducible	Technically demanding, damage to facial muscles and MCA
Intraluminal filament occlusion	Simplicity, reproducibility, widespread use	Large ischemic infarcts, strong influence of collaterals, hypothermia, damage beyond MCA territory
Embolic/thromboembolic	Relevance to aetiology of human stroke	Variability/control
Photothrombotic	Technically simple, control over where infarct is placed, smaller cortical infarcts	Vascular damage, certain aspects of pathophysiology
Endothelin-1	Technical simplicity	Requirement for surgery (*e.g.* craniotomy), small size of infarct
Lacunar stroke	Relevance to a common form of human stroke	Variability/control
Global ischemia (2 or 4 VO)	Reasonable model of global ischemia	Neurosurgical skill, variability, mortality
Global ischemia (cardiac arrest)	More clinically relevant model of global ischemia	Striatum damage > hippocampus

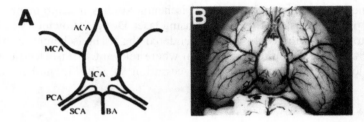

Figure 12.1   Cerebrovascular anatomy of the mouse. (**A**) Schematic of the major blood vessels at the base of the mouse brain. Depicted are the anterior cerebral artery (ACA), middle cerebral artery (MCA), internal carotid artery (ICA), posterior cerebral artery (PCA), superior cerebellar artery (SCA), and basilar artery (BA). (**B**) Photomicrograph showing the major blood vessels. Reprinted with permission from ref. 94. Copyright © 1998 Elsevier Science B.V.

Table 12.3   Common problems with mouse ischemia models.

Inter-animal and inter-strain infarct size variability
Infarction beyond striatum–neocortex field (*e.g.* hippocampus)
Variability of collateral circulation
Hypothermia
Subarachnoid hemorrhage
Retinal ischemia[a]

[a]Specific to the intraluminal filament occlusion model

introduce human versions of genes,[30] to evaluate pathogenic mechanisms underlying the cell death process.[50] Mice, being smaller than rats, are generally cheaper to purchase and house and they breed more quickly. There are also limitations with the use of mice, several of which are detailed in Table 12.3. Their small size necessitates a fine degree of surgical skill and adaptations of certain equipment. Mice, like rats, have lissencephalic brains (they lack the convoluted cortical surface of higher mammals) and this introduces unavoidable pathophysiological differences from human stroke.[50]

12.4.2   Tools of the Trade

In vivo modelling of stroke in mice necessitates measurement and maintenance of several physiologic parameters during experiments. Blood pressure should be maintained at ~100 mmHg and body and brain temperature at ~36–37 °C, which avoids undue influence of cerebral perfusion and hypo/hyperthermia, respectively. Control of respiration using small animal ventilators is required and arterial blood gases should be regularly monitored to ensure pCO_2 remains within normal physiologic range (35–45 mmHg). Blood oxygen should be maintained at ~100 mmHg. In focal ischemia studies measurement of local CBF or tissue perfusion, for example using laser Doppler flowmetry (LDF), is critical to confirm the occlusion has rendered the tissue ischemic (a fall of at least 50%), and to monitor reperfusion where relevant. An isoelectric electro-encephalogram is a means to gauge cessation of CBF in global ischemia models.

12.4.3 Surgical Occlusion of the Middle Cerebral Artery

Studies at the University of Glasgow by Tamura *et al.* introduced a popular rat model of permanent MCA occlusion.[10] Under anaesthesia and intubation, rats underwent a temporal bone craniectomy to expose the MCA. The MCA was then occluded and infarction developed within the cortex and basal ganglia.[10] A modified version of the Tamura model, involving MCA coagulation and bisection, and a simple neurologic exam to measure functional deficits, was later introduced by Bederson *et al.*[54] The Tamura model was also adapted to enable occlusion and reperfusion using a snare ligature around the MCA,[55] as well as distal occlusions combined with carotid artery occlusion.[56]

The Tamura model requires significant surgical skill (Table 12.2). Furthermore, feeding problems commonly arise because of damage to nerve and muscles supplying the jaw. This can result in poor recovery and difficulty with executing chronic studies. Reperfusion is possible with the model but physical damage to the MCA is a common problem.

Variations on the Tamura model approach have been introduced for use in mice by leading laboratories.[57–59] However, this approach has not been widely adopted. Instead, a model involving intraluminal filament/suture occlusion of the MCA, pioneered in rats, became the standard laboratory model of focal cerebral ischemia in mice.

12.4.4 Intraluminal Filament Occlusion

A simple and non-invasive method for occluding the MCA in rats was introduced by Koizumi *et al.*[11] and by Longa *et al.*[60] In this model, a nylon suture is introduced into the internal carotid artery and advanced just beyond the point where the MCA branches off from the circle of Willis (Figure 12.1), effectively occluding all blood flow to the MCA. The model overcomes several of shortcomings of the Tamura model. It is less technically demanding, and avoids the need for craniectomy which can cause changes to brain temperature and blood–brain–barrier integrity. However, CBF monitoring (*e.g.* using LDF) is critical to affirm the suture has effectively occluded the MCA. Infarction routinely develops in both striatum and cortex. Subsequent studies refined the technique by introducing suture coating.[61] The relationship between occlusion times and infarct volume is well characterized. For example, commonly employed occlusion times in rats are in the range 90–120 minutes. Occlusion times of less than 60 minutes typically affect only the striatum, while occlusion times of longer than three hours effectively represent a permanent occlusion.

The filament occlusion model was adapted for use in mice in the mid-1990s by multiple laboratories both as a permanent and reversible occlusion technique[62–67] (Table 12.2). A narrower, 5-0 to 8-0 suture is required and, as in rats, this is routinely coated to improve microvascular adherence. Intraluminal filament occlusion in mice causes an infarct encompassing the ipsilateral striatum and cortex (Figure 12.2). Occlusion times in mice were found to be necessarily shorter than in rats, and 60–90 minutes has been the most widely employed (Figure 12.2). Infarction of other brain regions including the hippocampus, however, is common in mice. The model has other shortcomings (Table 12.3) and can be associated with a high incidence of subarachnoid hemorrhage from suture penetration of the MCA.[68]

12.4.5 Thromboembolic Occlusion

Most human strokes are caused by either thrombosis or embolism. Both the surgical MCA route and the intraluminal filament model lack certain features of this pathophysiology, in particular occlusion by a thrombus or embolus which would in turn be amenable to dissolution with fibrinolysis. This deficiency may be important because, in clinical practice, future neuroprotectants will probably be 'piggy-backed' with t-PA-mediated fibrinolysis.

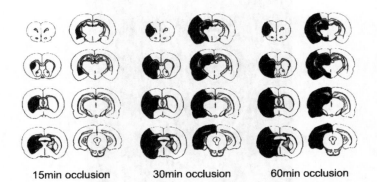

15min occlusion   30min occlusion   60min occlusion

Figure 12.2 Distribution of ischemic damage after intraluminal filament occlusion in mice. Line diagrams depict the extent of brain damage following 15, 30 and 60 min intraluminal filament occlusion at eight coronal levels of the mouse brain. A 15 minute occlusion restricted damage to the striatum. Both 30 and 60 minute occlusions result in damage to the cerebral cortex, but the hippocampus and thalamus which lie outside MCA territory are also damaged. Reprinted with permission from ref. 100. Copyright © 2003 Elsevier B.V.

A number of animal models were developed in the mid-1980s which used autologous or heterologous blood clots injected into vessels to produce focal cerebral ischemia.[69,70] Difficulty, however, in controlling where emboli lodged led groups to introduce improved models. Chopp's laboratory introduced models which addressed the previous shortcomings (Table 12.2). Zhang *et al.* used an intraluminal approach with a catheter to deposit fibrin-rich emboli in mice which blocked the MCA[71] (Figure 12.3). A similar model was also developed for use with rats.[72] The models proved capable of significantly reducing tissue perfusion within the MCA territory and producing infarcts which encompassed the striatum and cortex[71,72] (Figure 12.4). More recently, Chopp and colleagues employed the mouse embolic model for live brain imaging.[73]

12.4.6   Photothrombotic

The photothrombotic occlusion model of ischemia was introduced by Ginsberg's laboratory.[74] The model involves intravenous injection of Rose Bengal, a potent photosensitive dye which is then activated by irradiation using a halogen or xenon lamp exposed to the skull for a few minutes. The resultant highly reactive singlet oxygen causes microvascular damage and blood clots. This reduces vessel patency leading to localized ischemia. The model has certain

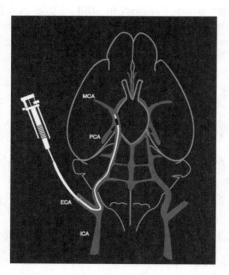

Figure 12.3   Embolic MCA occlusion stroke in mice using a directed approach. Figure shows a schematic drawing of the placement of the catheter through the external carotid artery (ECA) and the internal carotid artery (ICA) of the mouse, leaving the tip slightly proximal to the origin of the MCA. A clot is then injected which occludes flow to the MCA. Reprinted by permission from ref. 71. Copyright © 1997 Macmillan Publishers Ltd.

Figure 12.4   Ischemic brain injury after embolic stroke in the mouse. A series of photographs of TTC-stained coronal sections from a mouse 24 hours after embolization. Infarction, the regions in white, were located in the territory of the MCA. Reprinted by permission from ref. 71. Copyright © 1997 Macmillan Publishers Ltd.

strengths including reduced neurosurgical demands and it is reported to be highly reproducible. However, the model is challenging in terms of the care required to avoid brain temperature effects and entry of the dye into the parenchyma.[75]

By causing a thrombotic occlusion, the photothrombotic approach models an aspect of the pathophysiology of human stroke missing from either the surgical occlusion or intraluminal filament techniques. Photothrombotic occlusion produces small infarcts which are more proportionate to infarcts in common human strokes, as opposed to the very large infarcts produced by complete MCA occlusion.[51] However, the model fails to cause striatal infarcts and it is associated with a degree of vasogenic oedema which is more commonly associated with traumatic brain injury rather than ischemia. The model does not seem to produce an evolving penumbra.[51] The model has been used in mice (Table 12.2) and proven suitable to evaluate neuroprotective agents.[75]

12.4.7 Endothelin/vasoconstrictors

Fuxe *et al.* first demonstrated that intracerebral injection of the potent vasoconstrictor endothelin-1 (ET-1) produced lesions in the brain.[76] Later, McCulloch's group showed that when directly applied onto the exposed MCA, ET-1 could produce sufficient arterial constriction as to render the vascular territory of the rat MCA ischemic.[77] This meant that physical damage to the MCA could be completely avoided. Sharkey *et al.* adapted the model further, showing MCA occlusion could be produced by stereotaxic perivascular microinjection of ET-1 within brain parenchyma adjacent to the MCA.[78] This reduced the surgical demands and offered a means with which to induce MCA occlusion in conscious animals.

Due to the affinity of ET-1 for the ET-A receptor, attempts to control the duration of ET-1-induced MCA occlusion by application of ET-A receptor antagonists were unsuccessful. As an alternate approach to developing a reperfusion model using the ET-1 approach, ET-3 was shown to produce an effective MCA occlusion and infarct, and occlusion could be reversed by co-injection of an ET-A receptor antagonist.[79]

The ET-1 approach to focal ischemia has been investigated in different mice strains[80–82] (Table 12.2). Interestingly, single or even multiple intracerebral injections of ET-1 alone have been rather ineffective at producing ischemia in mice even at relatively high doses. Large ischemic lesions which encompass both striatum and cortex are only produced when ET-1 is combined with a nitric oxide synthase inhibitor to reduce vasodilation and ipsilateral common carotid artery occlusion.[82] This would suggest that the ET-1 model in mice is unsuitable until a more straightforward and reliable method is developed.

12.4.8 Global Ischemia Models

Brief episodes of global ischemia result in selective neuron loss which affects the CA1 pyramidal neurons, cortical layers 1 and 3, and the neurons of the striatum. Various models have been developed to model global ischemia with early versions involving decapitation or use of a neck tourniquet.[50] Cardiac arrest—sometimes combined with cardiopulmonary resuscitation—has also been employed in larger animals to model global ischemia and recently in mice.[83] However, this causes ischemia throughout the body, is associated with a significant burden of post-operative care, offers substantial problems with control of brain temperature, and in mice results in prominent injury in the striatum and relative sparing of the hippocampus.[50]

The first rodent model of global ischemia in which only cerebral arteries were occluded was introduced by Pulsinelli and Brierley.[84] The four vessel occlusion (4VO) model involved cauterization of both vertebral arteries followed by reversible ligation of both carotids for short periods of time. Ten-minute occlusions restricted damage almost exclusively to the CA1 sector of the hippocampus, while damage appeared within the striatum and thalamus when occlusion time was extended to 20 or 30 minutes.[84]

The basic 4VO model has undergone various modifications, including addition of hypotension and a 3VO version.[85–88] A 2VO occlusion involving bilateral common carotids was also proven successful in rats when combined with hypotension.[89] Global ischemia models require careful physiological monitoring (particularly brain temperature) and can be associated with high mortality and seizures.

A 2VO-only approach in gerbils without hypotension—possible because of their absent posterior communicating artery—became a popular model of global ischemia.[90,91] However, a simple and reliable mouse model of global ischemia was less forthcoming. Models first emerged in the mid-to-late 1990s. Mice have a posterior communicating artery (PcomA) so bilateral common carotid occlusion alone was typically insufficient to produce global ischemia, although prolonged occlusion times of 30 minutes were reported to produce histological injury.[92] More commonly, bilateral occlusions are performed accompanied by other manipulations such as controlled ventilation and systemic hypotension.[93,94] Major problems have been encountered including high mortality, cortical injury and occurrence of seizures.[94] The inter-animal variability in the PcomA seen between strains strongly influences outcome[93,95] (Figure 12.5). However, vascular anatomy alone may not be sufficient to explain differential histological outcome following global ischemia between different mouse strains.[95] Global ischemia models have been applied to genetically modified mice to evaluate effects of cell death–regulatory genes on the pathogenesis of ischemic brain injury.[96–98]

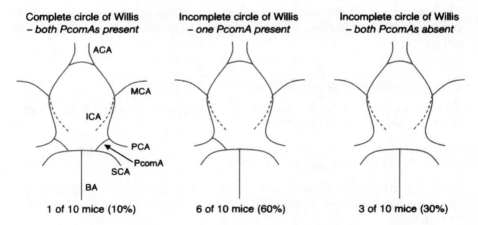

Complete circle of Willis
– both PcomAs present

ACA
MCA
ICA
PCA
PcomA
SCA
BA

1 of 10 mice (10%)

Incomplete circle of Willis
– one PcomA present

6 of 10 mice (60%)

Incomplete circle of Willis
– both PcomAs absent

3 of 10 mice (30%)

Figure 12.5 Natural variation in the completeness of the circle of Willis in mice. Illustrations of the variability in the circle of Willis derived from mice given carbon black infusions reveal that only one of 10 mice had a complete circle of Willis containing both posterior communicating arteries (PcomA). Other abbreviations as in Figure 12.1. Reprinted with permission from ref. 100. Copyright © 2004 Elsevier B.V.

12.5   Problems and Pitfalls Specific to Mouse Models of Ischemia

12.5.1   General Challenges for Modelling Stroke in Mice

Several general challenges are associated with stroke modelling in mice (Table 12.3).

12.5.1.1   *Collateral Circulation*

Wide variations in cerebral anatomy, including pial circulation, collateral length and completeness of the PcomA, are seen between mouse strains[92,93,95,99] (Figure 12.5). For example certain strains, in particular C57BL/6 mice, have poorly developed posterior communicating arteries which limit the extent of collateral blood flow. C57BL/6 mice typically develop larger ischemic lesions than other strains in both focal and global cerebral ischemia models.[95]

12.5.1.2   *Other Cerebrovascular Anomalies*

The circle of Willis has been found to be incomplete in up to 90% of C57BL/6 mice.[92,100]

12.5.1.3   *Ischemic Infarcts in Brain Regions outside the Territory of the MCA*

Hippocampal blood flow is reduced in the intraluminal filament model.[100,101] The hippocampus and other brain regions beyond the MCA vascular territory commonly infarct after focal ischemia in mice. This does not normally occur in rats.

12.5.1.4   *Subarachnoid Hemorrhage*

This is mainly a problem for the filament occlusion model where it may occur in up to one third of animals.[102]

12.5.1.5   *Hypothermia*

While hyperthermia is a common problem after filament occlusion in rats, the opposite problem (hypothermia) is common to mice subject to focal cerebral ischemia.[103]

12.5.1.6   *Retinal Ischemia*

Block *et al.* demonstrated that the filament occlusion model caused retinal ischemia in rats.[104] This was because the suture blocked the ophthalmic artery which branches from the circle of Willis adjacent to the MCA. Recently, retinal ischemia and histological correlates of damage were reported in mice using the filament occlusion model.[105] Since certain sensorimotor tests require visual cues, this has significant implications for the design and interpretation of post-stroke neurological testing.

12.5.2 Inter-strain Differences in Intrinsic Vulnerability to Excitotoxicity

The C57BL/6 mouse strain has been reported to be resistant to cell death after systemic kainic acid, a model of excitotoxicity.[106] This could introduce interpretational difficulties when comparing infarcts between strains or gene manipulations independently of any cerebrovascular anomalies. However, the importance of this reported phenotype as a confounder in ischemia is uncertain. C57BL/6 mice are highly vulnerable to ischemic damage, and this appears to relate only to cerebrovascular anatomy.[107] Second, the purported resistance to seizure-related excitotoxicity is model-dependent; when seizures are triggered by electrical stimulation or intra-amygdala rather than systemic kainic acid, cell death is readily induced in C57BL/6 mice.[108,109]

12.6 Other Problems and Pitfalls of Mouse Models

12.6.1 Stroke Size

Among several potential shortfalls in the way we model stroke is the volume of ischemic damage and its relevance to human stroke. Most (90%) human strokes affect ~5–15% of the ipsilateral hemisphere.[51] In the setting of very large (malignant) infarction, the affected area exceeds 39% of the ipsilateral hemisphere and there is attendant brain herniation and pan-hemispheric infarction.[51] In humans, such events are associated with very high mortality. In mouse filament occlusion models, infarct volumes are commonly in the range 21–45% of the ipsilateral hemisphere; most mouse models therefore reproduce an extreme form of human stroke, which itself comprises only ~10% of events in patients.[51] An interesting point of note is that in epilepsy, a field in which the authors are also active, a similar criticism has been levelled. Specifically, that most animal models of temporal lobe epilepsy do not in fact reproduce the typical human pattern of select hippocampal neuron loss but instead model widespread brain damage.[110] Closer attention to modelling infarcts relevant to the most common forms of human stroke may reduce the drop-off in translation of experimental findings when tested in the clinic.

12.6.2 Lacunar Strokes

Small lesions in subcortical grey or white matter (lacunar strokes) represent 25% of human strokes.[111] Their causes and effects are thought to differ from large-vessel occlusion-mediated stroke such as those resulting from MCA occlusion. Indeed, the aetiology of lacunar infarcts remains uncertain. Accordingly, modelling these strokes is an important but under-researched area. Several animal models of lacunar stroke have been developed or reported which produce small, deep lesions (Table 12.2). Technical approaches include injection of various emboli and intracerebral microapplication of ET-1.[112] Few of these models have yet been adapted to mice so future standardization and cross-comparisons of strain effects is required.

12.6.3 Narrow Temporal Relationship Between Occlusion Duration and Infarction

Generally, mice are more vulnerable to ischemic damage after MCA occlusion than are rats. For example, a 15-minute MCA occlusion in mice produces a similar size infarct to an occlusion twice as long in rats.[51] Furthermore, mice display a dramatic escalation in ischemic damage with extended occlusion times. For example, extending the duration of occlusion in the intraluminal filament model by just 15 minutes can increase infarct size by up to fivefold.[100] Dramatic escalation of damage also occurs with small increases in occlusion time in global ischemia models.[113] Accordingly, occlusion times are critically important variables in comparing outcomes and therapeutic interventions.

12.6.4 Animal Age, Co-morbidities and Other Variables

Stroke predominantly affects older people, but there is an overwhelming reliance on young animals in experimental modelling. This arises for obvious reasons such as cost and minimizing potential sources of variability. Nevertheless, this means that potentially important aspects of stroke pathophysiology are not represented. Indeed, studies show strong effects of aging on the pathogenesis of stroke in rodent models.[114,115] This may influence the later efficacy of therapeutic treatments developed in models that do not accurately capture key elements of human pathophysiology. Gender is also an under-reported variable which has been shown to affect certain outcomes in murine stroke models.[116] Co-morbidities such as hypertension, present in human patients, have been modelled by studies in hypertensive rats. However, despite the availability of hypertensive mice,[117,118] equivalent studies in mice have yet to be documented. Together, the influence of age, gender and normo- *versus* hypertensive animals should continue to be incorporated into mouse stroke models to ensure we generate data with relevance to the human conditions.

12.7 Ischemic Tolerance in Mice

Not all ischemic episodes are harmful to the brain. Clinical data show that strokes tend to be smaller in patients who had previously experienced transient ischemic attacks (mini-strokes) in the same vascular distribution as the stroke.[119] The mechanistic basis for this probably rests with the phenomena of ischemic tolerance—exposure of a tissue or organ to a low intensity of an otherwise harmful stressor temporarily generates a state resistant to damage caused by a subsequent and otherwise harmful insult. Tolerance represents an endogenous programme of tissue protection which likely evolved from mechanisms to cope with restricted substrate supply. Indeed, some of the deduced cell and molecular mechanisms underlying tolerance bear close resemblance to processes activated in hibernating animals which endure long periods of low oxygen/glucose without incurring permanent harm.[120–122]

Experiments in gerbils were the first to explore ischemic tolerance in the brain, demonstrating that ischemic damage after global ischemia could be reduced by prior brief ischemic challenges.[123,124] Roger Simon and colleagues later characterized the spatio-temporal profile of ischemic tolerance in a rat intraluminal filament model of stroke. These studies showed a window of tolerance where reduced brain injury followed a prolonged MCA occlusion lasted at least four days after the brief MCA occlusion.[125] Several molecular effectors of tolerance were subsequently elucidated, including the anti-apoptotic gene *bcl-2*.[126]

Stenzel-Poore *et al.* developed a mouse model of ischemic tolerance, showing that a 15-minute MCA occlusion using the filament model delivered three days before a 60-minute MCA reduced cortical injury by ~80%.[122] In the same study, tissue microarray analyses elucidated the transcriptome of ischemic tolerance. Remarkably, gene expression patterns differed almost completely in tolerant brain from brain subject to prolonged ischemia.[122] Most clearly, ischemic tolerance was associated with suppression of genes associated with energy-expensive functions including metabolism and transport.[122] The findings were taken as evidence of genomic reprogramming. That is, the preconditioning stress altered the gene repertoire normally expressed after an ischemic injury. This is a conserved feature of tolerance and is evident when non-ischemic preconditioning agents are used and in epileptic tolerance.[127,128]

More recently, Simon and colleagues have endeavoured to identify the processes set in motion by preconditioning which might explain how suppressed gene transcription is effected at the point of the second challenge. A mouse brain screen for changes to microRNAs (small non-coding RNA sequences which function mainly by blocking mRNA translation) revealed several changed by ischemic preconditioning targeted methyl CpG binding protein 2 (MeCP2), a transcriptional repressor. MeCP2 levels were found to increase after ischemic preconditioning and mice lacking *Mecp2* were unable to develop ischemic tolerance after preconditioning.[129] A proteomics screen by An Zhou's team also identified transcriptional repressors as a major group of proteins up-regulated by preconditioning.[130] Among several interesting leads, they identified polycomb group proteins (PcG). PcG proteins were shown to associate with the promoter regions of several predicted targets after ischemic preconditioning including potassium channel genes which had originally been identified among those mRNA transcripts down-regulated in ischemic tolerance.[122]

What are some of the next challenges in the field of ischemic tolerance? Further exploratory work can now begin on these new pathways representing the systems biology of regulators of the molecular pathogenesis of ischemic brain injury. As we gain the necessary tools to deploy and control their myriad functions, these epigenetic modifiers of gene transcription may provide the means to recapitulate the powerful neuroprotective effects of ischemic preconditioning. In turn, this may bring us closer to treatments which can be given after a stroke to protect the brain or, conceivably, delivered prophylactically to those at risk of developing a stroke.

12.8   Conclusions

Stroke is a major health problem which places a significant socioeconomic burden on society. Mouse models of ischemia have been a powerful resource in our efforts to unravel the pathophysiology of stroke and screen potential therapeutic agents for neuroprotection. However, their relatively small size and aspects of their genetics, in particular variable cerebrovascular anatomy, pose significant challenges to experimentalists evaluating drug and gene effects. Nevertheless, it is likely that mouse models of ischemia will continue to underpin our efforts to improve our understanding and develop therapeutic approaches to treat and cure stroke.

References

1. T. M. Jay, G. Lucignani, A. M. Crane, J. Jehle and L. Sokoloff, *J. Cereb. Blood Flow Metab.*, 1988, **8**, 121.
2. A. W. Brown and J. B. Brierley, *Experientia*, 1966, **22**, 546.
3. B. S. Meldrum, B. J. Excell, J. B. Brierley, A. W. Brown and M. A. McSheehy, *Electroencephalogr. Clin. Neurophysiol.*, 1968, **24**, 594.
4. J. B. Brierley, A. W. Brown and B. S. Meldrum, *Brain Res.*, 1971, **25**, 483.
5. K. A. Hossmann and K. Sato, *Electroencephalogr. Clin. Neurophysiol.*, 1971, **30**, 535.
6. J. Astrup, B. K. Siesjo and L. Symon, *Stroke*, 1981, **12**, 723.
7. K. A. Hossmann, *J. Cereb. Blood Flow Metab.*, 1982, **2**, 275.
8. D. I. Graham, *Br. J. Anaesth.*, 1985, **57**, 3.
9. B. S. Meldrum, M. C. Evans, J. H. Swan and R. P. Simon, *Med. Biol.*, 1987, **65**, 153.
10. A. Tamura, D. I. Graham, J. McCulloch and G. M. Teasdale, *J. Cereb. Blood Flow Metab.*, 1981, **1**, 53.
11. J. Koizumi, Y. Yoshida, T. Nakazawa and G. Oonedda, *Jpn. J. Stroke*, 1986, **8**, 1.
12. R. P. Simon, J. H. Swan, T. Griffiths and B. S. Meldrum, *Science*, 1984, **226**, 850.
13. J. P. Stansbury, H. Jia, L. S. Williams, W. B. Vogel and P. W. Duncan, *Stroke*, 2005, **36**, 374.

14. T. Truelsen, B. Piechowski-Jozwiak, R. Bonita, C. Mathers, J. Bogousslavsky and G. Boysen, *Eur. J. Neurol.*, 2006, **13**, 581.
15. M. Nedergaard, *Acta Neurol. Scand.*, 1988, **77**, 81.
16. J. Aronowski, R. Strong and J. C. Grotta, *J. Cereb. Blood Flow Metab.*, 1997, **17**, 1048.
17. S. Kuroda and B. K. Siesjo, *Clin. Neurosci.*, 1997, **4**, 199.
18. B. K. Siesjo and P. Siesjo, *Eur. J. Anaesthesiol.*, 1996, **13**, 247.
19. K. A. Hossmann, *Ann. Neurol.*, 1994, **36**, 557.
20. F. R. Sharp, A. Lu, Y. Tang and D. E. Millhorn, *J. Cereb. Blood Flow Metab.*, 2000, **20**, 1011.
21. A. Bizzi, A. Righini, R. Turner, D. Le Bihan, K. H. Bockhorst and J. R. Alger, *Magn. Reson. Imaging*, 1996, **14**, 581.
22. D. W. Choi, *Neuron*, 1988, **1**, 623.
23. B. Meldrum, *Prog. Clin. Biol. Res.*, 1990, **361**, 275.
24. U. Dirnagl, C. Iadecola and M. A. Moskowitz, *Trends Neurosci.*, 1999, **22**, 391.
25. Z. G. Xiong, X. P. Chu and R. P. Simon, *J. Membr. Biol.*, 2006, **209**, 59.
26. J. Chen, T. Nagayama, K. Jin, R. A. Stetler, R. L. Zhu, S. H. Graham and R. P. Simon, *J. Neurosci.*, 1998, **18**, 4914.
27. D. M. Rosenbaum, G. Gupta, J. D'Amore, M. Singh, K. Weidenheim, H. Zhang and J. A. Kessler, *J. Neurosci. Res.*, 2000, **61**, 686.
28. N. Plesnila, S. Zinkel, D. A. Le, S. Amin-Hanjani, Y. Wu, J. Qiu, A. Chiarugi, S. S. Thomas, D. S. Kohane, S. J. Korsmeyer and M. A. Moskowitz, *Proc. Natl. Acad. Sci. U. S. A.*, 2001, **98**, 15318.
29. G. Cao, W. Pei, H. Ge, Q. Liang, Y. Luo, F. R. Sharp, A. Lu, R. Ran, S. H. Graham and J. Chen, *J. Neurosci.*, 2002, **22**, 5423.
30. L. E. Kerr, A. L. McGregor, L. E. Amet, T. Asada, C. Spratt, T. E. Allsopp, A. J. Harmar, S. Shen, G. Carlson, N. Logan, J. S. Kelly and J. Sharkey, *Cell Death Differ.*, 2004, **11**, 1102.
31. C. Culmsee, C. Zhu, S. Landshamer, B. Becattini, E. Wagner, M. Pellecchia, K. Blomgren and N. Plesnila, *J. Neurosci.*, 2005, **25**, 10262.
32. J. A. Hartings, M. L. Rolli, X. C. Lu and F. C. Tortella, *J. Neurosci.*, 2003, **23**, 11602.
33. A. Durukan and T. Tatlisumak, *Pharmacol. Biochem. Behav.*, 2007, **87**, 179.
34. D. Amantea, G. Nappi, G. Bernardi, G. Bagetta and M. T. Corasaniti, *FEBS J.*, 2009, **276**, 13.
35. R. Jin, G. Yang and G. Li, *J. Leukoc. Biol.*, 2010, **87**, 779.
36. A. Liesz, E. Suri-Payer, C. Veltkamp, H. Doerr, C. Sommer, S. Rivest, T. Giese and R. Veltkamp, *Nat. Med.*, 2009, **15**, 192.
37. H. Karhunen, J. Jolkkonen, J. Sivenius and A. Pitkanen, *Neurochem. Res.*, 2005, **30**, 1529.
38. R. J. Nudo, B. M. Wise, F. SiFuentes and G. W. Milliken, *Science*, 1996, **272**, 1791.
39. J. Liauw, S. Hoang, M. Choi, C. Eroglu, G. H. Sun, M. Percy, B. Wildman-Tobriner, T. Bliss, R. G. Guzman, B. A. Barres and G. K. Steinberg, *J. Cereb. Blood Flow Metab.*, 2008, **28**, 1722.

40. A. Arvidsson, T. Collin, D. Kirik, Z. Kokaia and O. Lindvall, *Nat. Med.*, 2002, **8**, 963.
41. M. D. Ginsberg, *Neuropharmacology*, 2008, **55**, 363.
42. S. H. Graham and J. Chen, *J. Cereb. Blood Flow Metab.*, 2001, **21**, 99.
43. E. Martinez-Vila and P. I. Sieira, *Cerebrovasc. Dis.*, 2001, **11**, 60.
44. M. D. Ginsberg, L. L. Sternau, M. Y. Globus, W. D. Dietrich and R. Busto, *Cerebrovasc. Brain Metab. Rev.*, 1992, **4**, 189.
45. The National Institute of Neurological Disorders and Stroke rt-PA Stroke Study Group, *N. Engl. J. Med.*, 1995, **333**, 1581.
46. N. G. Wahlgren and N. Ahmed, *Cerebrovasc. Dis.*, 2004, **17**(Suppl 1), 153.
47. U. Dirnagl, *J. Cereb. Blood Flow Metab.*, 2006, **26**, 1465.
48. E. S. Sena, H. B. van der Worp, P. M. Bath, D. W. Howells and M. R. Macleod, *PLoS Biol.*, 2010, **8**, e1000344.
49. M. D. Ginsberg and R. Busto, *Stroke*, 1989, **20**, 1627.
50. R. J. Traystman, *ILAR J.*, 2003, **44**, 85.
51. S. T. Carmichael, *NeuroRx*, 2005, **2**, 396.
52. J. B. Bederson, L. H. Pitts, S. M. Germano, M. C. Nishimura, R. L. Davis and H. M. Bartkowski, *Stroke*, 1986, **17**, 1304.
53. R. A. Swanson, M. T. Morton, G. Tsao-Wu, R. A. Savalos, C. Davidson and F. R. Sharp, *J. Cereb. Blood Flow Metab.*, 1990, **10**, 290.
54. J. B. Bederson, L. H. Pitts, M. Tsuji, M. C. Nishimura, R. L. Davis and H. Bartkowski, *Stroke*, 1986, **17**, 472.
55. T. Shigeno, G. M. Teasdale, J. McCulloch and D. I. Graham, *J. Neurosurg.*, 1985, **63**, 272.
56. S. T. Chen, C. Y. Hsu, E. L. Hogan, H. Maricq and J. D. Balentine, *Stroke*, 1986, **17**, 738.
57. C. Backhauss, C. Karkoutly, M. Welsch and J. Krieglstein, *J. Pharmacol. Toxicol. Methods.*, 1992, **27**, 27.
58. H. Nawashiro, D. Martin and J. M. Hallenbeck, *J. Cereb. Blood Flow Metab.*, 1997, **17**, 229.
59. J. Lubjuhn, A. Gastens, G. von Wilpert, P. Bargiotas, O. Herrmann, S. Murikinati, T. Rabie, H. H. Marti, I. Amende, T. G. Hampton and M. Schwaninger, *J. Neurosci. Methods.*, 2009, **184**, 95.
60. E. Z. Longa, P. R. Weinstein, S. Carlson and R. Cummins, *Stroke*, 1989, **20**, 84.
61. L. Belayev, O. F. Alonso, R. Busto, W. Zhao and M. D. Ginsberg, *Stroke*, 1996, **27**, 1616.
62. P. H. Chan, H. Kamii, G. Yang, J. Gafni, C. J. Epstein, E. Carlson and L. Reola, *Neuroreport*, 1993, **5**, 293.
63. Z. Huang, P. L. Huang, N. Panahian, T. Dalkara, M. C. Fishman and M. A. Moskowitz, *Science*, 1994, **265**, 1883.
64. G. Yang, P. H. Chan, J. Chen, E. Carlson, S. F. Chen, P. Weinstein, C. J. Epstein and H. Kamii, *Stroke*, 1994, **25**, 165.
65. E. S. Connolly, Jr., C. J. Winfree, D. M. Stern, R. A. Solomon and D. J. Pinsky, *Neurosurgery*, 1996, **38**, 523.

66. W. M. Clark, N. S. Lessov, M. P. Dixon and F. Eckenstein, *Neurol. Res.*, 1997, **19**, 641.
67. R. Hata, G. Mies, C. Wiessner, K. Fritze, D. Hesselbarth, G. Brinker and K. A. Hossmann, *J. Cereb. Blood Flow Metab.*, 1998, **18**, 367.
68. D. Tsuchiya, S. Hong, T. Kayama, S. S. Panter and P. R. Weinstein, *Brain Res.*, 2003, **970**, 131.
69. J. A. Zivin, M. Fisher, U. DeGirolami, C. C. Hemenway and J. A. Stashak, *Science*, 1985, **230**, 1289.
70. S. M. Papadopoulos, W. F. Chandler, M. S. Salamat, E. J. Topol and J. C. Sackellares, *J. Neurosurg.*, 1987, **67**, 394.
71. Z. Zhang, M. Chopp, R. L. Zhang and A. Goussev, *J. Cereb. Blood Flow Metab.*, 1997, **17**, 1081.
72. Z. Zhang, R. L. Zhang, Q. Jiang, S. B. Raman, L. Cantwell and M. Chopp, *J. Cereb. Blood Flow Metab.*, 1997, **17**, 123.
73. Z. G. Zhang, L. Zhang, G. Ding, Q. Jiang, R. L. Zhang, X. Zhang, W. B. Gan and M. Chopp, *Stroke*, 2005, **36**, 2701.
74. B. D. Watson, W. D. Dietrich, R. Busto, M. S. Wachtel and M. D. Ginsberg, *Ann. Neurol.*, 1985, **17**, 497.
75. G. W. Kim, T. Sugawara and P. H. Chan, *J. Cereb. Blood Flow Metab.*, 2000, **20**, 1690.
76. K. Fuxe, A. Cintra, B. Andbjer, E. Anggard, M. Goldstein and L. F. Agnati, *Acta Physiol. Scand.*, 1989, **137**, 155.
77. M. J. Robinson, I. M. Macrae, M. Todd, J. L. Reid and J. McCulloch, *Neurosci. Lett.*, 1990, **118**, 269.
78. J. Sharkey, I. M. Ritchie and P. A. Kelly, *J. Cereb. Blood Flow Metab.*, 1993, **13**, 865.
79. D. C. Henshall, S. P. Butcher and J. Sharkey, *Brain Res.*, 1999, **843**, 105.
80. S. Zhang, J. Boyd, K. Delaney and T. H. Murphy, *J. Neurosci.*, 2005, **25**, 5333.
81. Y. Wang, K. Jin and D. A. Greenberg, *Brain Res.*, 2007, **1167**, 118.
82. N. Horie, A. L. Maag, S. A. Hamilton, H. Shichinohe, T. M. Bliss and G. K. Steinberg, *J. Neurosci. Methods*, 2008, **173**, 286.
83. J. Kofler, K. Hattori, M. Sawada, A. C. DeVries, L. J. Martin, P. D. Hurn and R. J. Traystman, *J. Neurosci. Methods*, 2004, **136**, 33.
84. W. A. Pulsinelli and J. B. Brierley, *Stroke*, 1979, **10**, 267.
85. E. Kagstrom, M. L. Smith and B. K. Siesjo, *J. Cereb. Blood Flow Metab.*, 1983, **3**, 170.
86. M. Kameyama, J. Suzuki, R. Shirane and A. Ogawa, *Stroke*, 1985, **16**, 489.
87. H. D. Mennel, D. Sauer, C. Rossberg, G. W. Bielenberg and J. Krieglstein, *Exp. Pathol.*, 1988, **35**, 219.
88. R. Schmidt-Kastner, W. Paschen, B. G. Ophoff and K. A. Hossmann, *Stroke*, 1989, **20**, 938.
89. M. L. Smith, G. Bendek, N. Dahlgren, I. Rosen, T. Wieloch and B. K. Siesjo, *Acta Neurol. Scand.*, 1984, **69**, 385.

90. T. Kirino, *Brain Res.*, 1982, **239**, 57.
91. B. M. Djuricic, W. Paschen, H. J. Bosma and K. A. Hossmann, *J. Neurol. Sci.*, 1983, **58**, 25.
92. M. Fujii, H. Hara, W. Meng, J. P. Vonsattel, Z. Huang and M. A. Moskowitz, *Stroke*, 1997, **28**, 1805.
93. K. Kitagawa, M. Matsumoto, G. Yang, T. Mabuchi, Y. Yagita, M. Hori and T. Yanagihara, *J. Cereb. Blood Flow Metab.*, 1998, **18**, 570.
94. K. Murakami, T. Kondo, M. Kawase and P. H. Chan, *Brain Res.*, 1998, **780**, 304.
95. J. C. Wellons, 3rd, H. Sheng, D. T. Laskowitz, G. Burkhard Mackensen, R. D. Pearlstein and D. S. Warner, *Brain Res.*, 2000, **868**, 14.
96. K. Kitagawa, M. Matsumoto, Y. Tsujimoto, T. Ohtsuki, K. Kuwabara, K. Matsushita, G. Yang, H. Tanabe, J. C. Martinou, M. Hori and T. Yanagihara, *Stroke*, 1998, **29**, 2616.
97. M. Kawase, K. Murakami, M. Fujimura, Y. Morita-Fujimura, Y. Gasche, T. Kondo, R. W. Scott and P. H. Chan, *Stroke*, 1999, **30**, 1962.
98. D. Tsuchiya, S. Hong, Y. Matsumori, T. Kayama, R. A. Swanson, W. H. Dillman, J. Liu, S. S. Panter and P. R. Weinstein, *Neurosurgery*, 2003, **53**, 1179.
99. H. Zhang, P. Prabhakar, R. Sealock and J. E. Faber, *J. Cereb. Blood Flow Metab.*, 2010, **30**, 923.
100. B. W. McColl, H. V. Carswell, J. McCulloch and K. Horsburgh, *Brain Res.*, 2004, **997**, 15.
101. Y. G. Ozdemir, H. Bolay, E. Erdem and T. Dalkara, *Brain Res.*, 1999, **822**, 260.
102. R. Schmid-Elsaesser, S. Zausinger, E. Hungerhuber, A. Baethmann and H. J. Reulen, *Stroke*, 1998, **29**, 2162.
103. P. A. Barber, L. Hoyte, F. Colbourne and A. M. Buchan, *Stroke*, 2004, **35**, 1720.
104. F. Block, C. Grommes, C. Kosinski, W. Schmidt and M. Schwarz, *Neurosci. Lett.*, 1997, **232**, 45.
105. E. C. Steele, Jr., Q. Guo and S. Namura, *Stroke*, 2008, **39**, 2099.
106. P. E. Schauwecker and O. Steward, *Proc. Natl. Acad. Sci. U. S. A.*, 1997, **94**, 4103.
107. F. C. Barone, D. J. Knudsen, A. H. Nelson, G. Z. Feuerstein and R. N. Willette, *J. Cereb. Blood Flow Metab.*, 1993, **13**, 683.
108. G. Mouri, E. Jimenez-Mateos, T. Engel, M. Dunleavy, S. Hatazaki, A. Paucard, S. Matsushima, W. Taki and D. C. Henshall, *Brain Res.*, 2008, **1213**, 140.
109. F. Kienzler, B. A. Norwood and R. S. Sloviter, *J. Comp. Neurol.*, 2009, **515**, 181.
110. R. S. Sloviter, *Epilepsia*, 2009, **50**(Suppl 12), 11.
111. C. L. Sudlow and C. P. Warlow, *Stroke*, 1997, **28**, 491.
112. E. L. Bailey, J. McCulloch, C. Sudlow and J. M. Wardlaw, *Stroke*, 2009, **40**, e451.

113. R. P. Simon, H. Cho, R. Gwinn and D. H. Lowenstein, *J. Neurosci.*, 1991, **11**, 881.
114. Y. Liu, D. M. Jacobowitz, F. Barone, R. McCarron, M. Spatz, G. Feuerstein, J. M. Hallenbeck and A. L. Siren, *J. Cereb. Blood Flow Metab.*, 1994, **14**, 348.
115. M. Nagayama, T. Aber, T. Nagayama, M. E. Ross and C. Iadecola, *J. Cereb. Blood Flow Metab.*, 1999, **19**, 661.
116. H. Kitano, J. M. Young, J. Cheng, L. Wang, P. D. Hurn and S. J. Murphy, *J. Cereb. Blood Flow Metab.*, 2007, **27**, 1377.
117. M. F. Elias, R. N. Sorrentino, C. A. Pentz, 3rd and J. R. Florini, *Exp. Aging Res.*, 1975, **1**, 251.
118. P. Hamet, D. Malo and J. Tremblay, *Hypertension*, 1990, **15**, 904.
119. M. Weih, K. Kallenberg, A. Bergk, U. Dirnagl, L. Harms, K. D. Wernecke and K. M. Einhaupl, *Stroke*, 1999, **30**, 1851.
120. K. U. Frerichs, C. Kennedy, L. Sokoloff and J. M. Hallenbeck, *J. Cereb. Blood Flow Metab.*, 1994, **14**, 193.
121. P. W. Hochachka, L. T. Buck, C. J. Doll and S. C. Land, *Proc. Natl. Acad. Sci. U. S. A.*, 1996, **93**, 9493.
122. M. P. Stenzel-Poore, S. L. Stevens, Z. Xiong, N. S. Lessov, C. A. Harrington, M. Mori, R. Meller, H. L. Rosenzweig, E. Tobar, T. E. Shaw, X. Chu and R. P. Simon, *Lancet*, 2003, **362**, 1028.
123. K. Kitagawa, M. Matsumoto, M. Tagaya, R. Hata, H. Ueda, M. Niinobe, N. Handa, R. Fukunaga, K. Kimura, K. Mikoshiba and T. Kamada, *Brain Res.*, 1990, **528**, 21.
124. K. Kitagawa, M. Matsumoto, K. Kuwabara, M. Tagaya, T. Ohtsuki, R. Hata, H. Ueda, N. Handa, K. Kimura and T. Kamada, *Brain Res.*, 1991, **561**, 203.
125. J. Chen, S. H. Graham, R. L. Zhu and R. P. Simon, *J. Cereb. Blood Flow Metab.*, 1996, **16**, 566.
126. S. Shimizu, T. Nagayama, K. L. Jin, L. Zhu, J. E. Loeffert, S. C. Watkins, S. H. Graham and R. P. Simon, *J. Cereb. Blood Flow Metab.*, 2001, **21**, 233.
127. M. P. Stenzel-Poore, S. L. Stevens, J. S. King and R. P. Simon, *Stroke*, 2007, **38**, 680.
128. E. M. Jimenez-Mateos, S. Hatazaki, M. B. Johnson, C. Bellver-Estelles, G. Mouri, C. Bonner, J. H. Prehn, R. Meller, R. P. Simon and D. C. Henshall, *Neurobiol. Dis.*, 2008, **32**, 442.
129. T. A. Lusardi, C. D. Farr, C. L. Faulkner, G. Pignataro, T. Yang, J. Lan, R. P. Simon and J. A. Saugstad, *J. Cereb. Blood Flow Metab.*, 2010, **30**, 744.
130. M. Stapels, C. Piper, T. Yang, M. Li, C. Stowell, Z. G. Xiong, J. Saugstad, R. P. Simon, S. Geromanos, J. Langridge, J. Q. Lan and A. Zhou, *Sci. Signal.*, 2010, **3**, ra15.

CHAPTER 13

A Non-transgenic Rat Model of Sporadic Alzheimer's Disease

KHALID IQBAL*, XIAOCHUAN WANG, JULIE BLANCHARD AND INGE GRUNDKE-IQBAL

New York State Institute for Basic Research in Developmental Disabilities, Staten Island, New York, USA

13.1 Introduction

Alzheimer's disease (AD) is the major cause of dementia in middle- and old-age individuals in most of the world. The current worldwide estimates range from 20–30 million suffering from this disease and this number is expected to reach around 100 million by 2050 (see www.alz.org). AD is histopathologically characterized by numerous intraneuronal neurofibrillary tangles and extracellular amyloid β (Aβ) plaques and clinically by progressive dementia. Only a sparse number of neurofibrillary tangles and/or Aβ plaques has been reported in aged primates and polar bears;[1–3] aged rabbits, guinea pigs, rats or mice do not show plaques or tangles. A modest to moderate number of diffuse Aβ plaques and neurofibrillary tangles and associated neurodegeneration have been observed in ~20% of aged lemurian primate, *Microcebus murinus* (also referred to as the mouse lemur); subiculum and entorhinal cortex are only involved occasionally in those animals more than eight years old.[4,5] Because of the lack of any practical animal model, research in the AD field remained very limited till the early 1990s. The discoveries of Aβ[6] and abnormally hyperphosphorylated tau[7] as the major protein subunits of plaque amyloid and neurofibrillary tangles, respectively, exploded research on the development of experimental animal models.

RSC Drug Discovery Series No. 6
Animal Models for Neurodegenerative Disease
Edited by Jesús Avila, Jose J. Lucas and Félix Hernández
© Royal Society of Chemistry 2011
Published by the Royal Society of Chemistry, www.rsc.org

The first generation of experimental animal models were the transgenic mice overexpressing or knocking out of βAPP or tau and animals directly injected/perfused with Aβ.[8–10] While the latter approach demonstrated the neurotoxicity of Aβ, in the former case animals exhibited minimal pathology and were not employed for further studies.

The discovery of disease-causing mutations, first in βAPP[11,12] and then in presenilin-1,[13,14] presenilin-2,[15] and finally in tau in frontotemporal dementia with parkinsonism linked to chromosome 17 (FTDP-17), though not in AD, led to the generation of many transgenic mice expressing one or a combination of these mutations in one or more human genes in the same animal (see www.alzforum.org). These and other transgenic mice expressing other human genes such as APOE and a-synuclein or knock-out models of these endogenous genes have been extensively employed to understand disease mechanisms, generate disease biomarkers, and for preclinical studies in the development of therapeutic drug candidates. However, most of these experimentally altered animals model the familial form of the disease which, in the case of AD, represents less than 1% of all cases.

Surprisingly, to date very little effort has been made to develop animal models of sporadic AD, which accounts for over 99% of all cases of this disease. Two closest such models are:

- the inducible transgenic mouse model which expresses constitutively active glycogen synthase kinase 3β (GSK-3β);[16]
- the transgenic rat model which expresses the truncated human tau, $tau_{151-391}$.[17]

The inducible transgenic GSK-3β mouse model shows abnormal hyper-phosphorylation of tau, neurodegeneration and associated cognitive impairment but does not show any Aβ pathology. The human $tau_{151-391}$ transgenic rats show many neurofibrillary tangles and neurodegeneration, mostly in the brainstem and associated motor impairment. These animals, hemizygotes of which survive around seven months, show an increase in tau hyperphosphorylated at Thr181 in their cerebrospinal fluid (CSF), levels of which increase with the severity of the disease.[18] However, Aβ pathology has not been reported in this transgenic rat model.

AD is multifactorial and probably involves several different etiopathogenic mechanisms.[19] Elucidation of these different disease mechanisms is crucial to the development of potent disease-modifying drugs. Thus, for rational drug development there is an enormous need for appropriate animal models. Generation of transgenic animals that can model various mechanisms of sporadic AD is very costly in time, effort and resources, and is beyond the reach of most research labs. An approach which is relatively inexpensive, rapid and can be successfully employed in most research labs is the use of viral vectors for the expression of desired gene(s) in a temporal and spatial selective manner.[20] In this chapter we describe a disease-relevant rat model of sporadic AD we have generated using an adeno-associated virus (AAV) expression vector.[21]

13.2   A Mechanism of AD Involving Cleavage of SETα

Neurofibrillary degeneration of abnormally hyperphosphorylated tau is one of the two hallmarks of AD.[7,22] A cause of abnormal hyperphosphorylation of tau is most probably the selective decrease in the activity of phosphoseryl/phosphothreonyl protein phosphatase (PP) 2A in the brains of sporadic AD cases.[23,24] PP2A, which accounts for around 70% of the total PP activity in human brain,[25] is a major phospho-tau phosphatase.[26–32] The activity of PP2A, a trimeric holoenzyme, is regulated:

  (i) positively by binding to its regulatory and variable B subunit[33] and by methylation of PP2Ac,[34] its catalytic C subunit;
 (ii) negatively by phosphotyrosinylation of PP2Ac at tyrosine 307;[35]
(iii) by two endogenous inhibitors, I_1^{PP2A} and I_2^{PP2A}. I_2^{PP2A},[36] which is also known as SETα, has a nuclear localization signal and is primarily localized in the cell nucleus.

In AD brain both the mRNA and protein expressions of I_2^{PP2A}/SETα are elevated and this protein is selectively cleaved at N175 into I_{2NTF} (*N*-terminal fragment) and I_{2CTF} (*C*-terminal fragment) and is translocated from the neuronal nucleus to the cytoplasm in the affected areas of the brain.[37] The translocated I_2^{PP2A}/SETα localizes with abnormal hyperphosphorylation of tau in the neurofibrillary tangles in the AD brain. Moreover, transfection of PC12 as well as tau stably transfected PC12 cells with I_2^{PP2A} results in abnormal hyperphosphorylation of tau and associated degeneration of cells.[38,39]

The cleavage products of I_2^{PP2A}/SETα, I_{2NTF} and I_{2CTF}, each 20 kDa or less, because of their small size can easily diffuse between the neuronal nucleus and the cytoplasm and we had found that both fragments could inhibit PP2A (Chen *et al.*, unpublished observation).

13.3   Generation of an AAV1-I_{2CTF} Rat Model of AD

To replicate in rat brain the findings of I_2^{PP2A} we found in AD brain, we packaged I_{2CTF} in recombinant AAV serotype 1, purified, concentrated and titred the virus (Figure 13.1). We also engineered enhanced green fluorescent protein (GFP) in the vector so that both AAV1-I_{2CTF} animals and vector-alone control animals expressed GFP (AAV1-GFP). In this way the expression of the inserted gene(s) can be easily followed. We infected with 8×10^9 AAV1 genomic equivalents in 4 µl (ICV, bilaterally with 2 µl on each side) pups of Wistar rats within 24 hours of birth, *i.e.* p 0.5. The infection of newborn animals with AAV1-induced expression is known to be widespread and long-lasting.[40]

We studied the AAV1-induced expression at three weeks, nine weeks, five months, eight months and 14 months and found expression at all the ages examined in the ventricular area, hippocampus and the cerebral cortex (Figure 13.2).

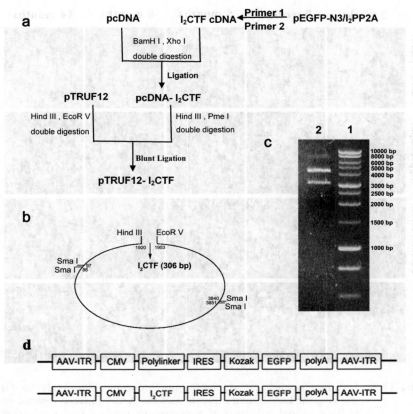

Figure 13.1 (a) Construct of pTRUF12-I$_{2CTF}$ plasmid. I$_{2CTF}$ cDNA was generated by using pEGFP-NC/12PP2A (wt) as a template by PCR (primer 1: 5'-GAT<u>GGATCC</u>AAAGCCAGCAGGAAGA (BamH I site underlined), primer 2: 5' GAT<u>CTCGAG</u>TTAGTCATCTTCTC (Xho I site underlined). pcDNA3.1 vector (Invitrogen, Carlsbad, CA) linked with I$_{2CTF}$ cDNA via BamH I site and Xho I double digestion to construct into pcDNA-I$_{2CTF}$. (b) Then, I$_{2CTF}$ DNA was cut out of pcDNA via Hind III and Pme I double digestion. After Hind III and EcoR V double digestion, pTRUF12 acted as a vector to link with the insert gene I$_{2CTF}$ with a blunt ligation. (c) As a key restriction site to be packed with an adeno-associated virus (AAV), Sma I was identified in the constructed plasmid *via* agarose gel electrophoresis. 1. DNA marker; 2. Fragments from pTRUF12-I$_{2CTF}$ cleaved by Sma I. (d) Linear maps of the AAV vector plasmids based on pTRUF12.

While the expression of I$_{2CTF}$ had no effect on the level of the expression of PP2Ac, the phosphatase activity towards hyperphosphorylated tau in the brains of AAV1-I$_{2CTF}$ animals, however, was inhibited (Figure 13.3). The inhibition of the PP2A activity in AAV1-I$_{2CTF}$ rats compared with the AAV1-GFP control animals was inhibited in the ventricular area, in the cerebral cortex and in the hippocampus in eight-month animals. Furthermore, like in

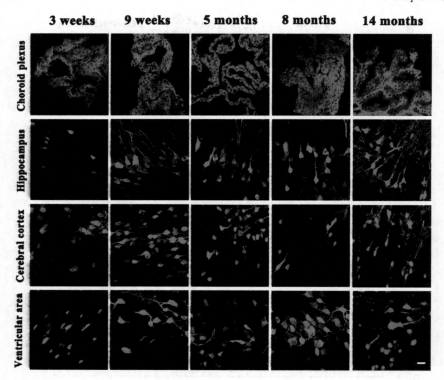

3 weeks   9 weeks   5 months   8 months   14 months

Choroid plexus

Hippocampus

Cerebral cortex

Ventricular area

Figure 13.2   Immunohistochemical staining using monoclonal antibody to GFP showing the long-lasting expression of the AAV1-induced inserted gene in the neuronal cell layers of hippocampus, cerebral cortex, ventricular area, and choroid plexus at different ages. Scale bar: 25 mm.

AD brain, the AAV1-I$_{2CTF}$ rats showed a marked abnormal hyperphosphorylation of tau in the hippocampus, cerebral cortex and ventricular areas at Ser199, Thr205, Thr212, Ser214, Thr231/Ser235, Ser262, Ser396 and Ser422 (Figure 13.4). The level of total tau was decreased in the AAV1-I$_2^{PP2A}$ animals. The decrease in total tau probably represented neurodegeneration. The AAV1-I$_{2CTF}$ animals also showed intraneuronal accumulation of both Aβ_{x-40} and Aβ_{x-42} in all the above brain areas studied. In addition to tau and Aβ pathologies, the AAV1-I$_{2CTF}$ rats showed a marked decrease in neuronal plasticity as determined by MAP2, synapsin 1 and synaptophysin levels. Neurodegeneration, as measured by Fluoro Jade B staining and decreased levels of neuron-specific βIII tubulin normalized with total tubulin, was found in AAV1-I$_{2CTF}$ animals.

Finally, the AAV1-I$_{2CTF}$ rats were impaired in spatial learning both at five months and eight months, and showed impairment to maintain or reactivate the long term memory trace of spatial information at nine months in Morris water maze (Figure 13.5). These impairments are characteristic of hippocampal

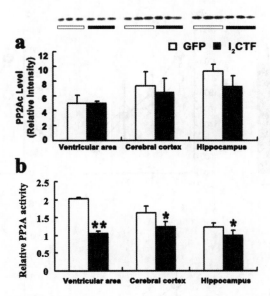

Figure 13.3 Level and activity of PP2A in AAV1-I_{2CTF} and AAV1-GFP control rats. (a) Level of PP2A catalytic subunit, PP2Ac, and (b) PP2A activity towards phospho Ser199 tau normalized with the PP2Ac level. *p < 0.05; **p < 0.01.

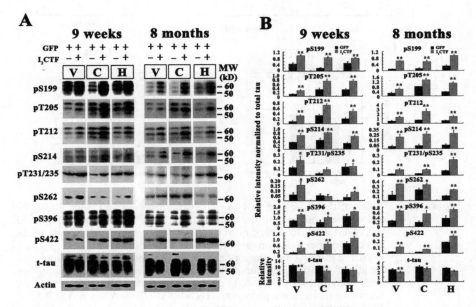

Figure 13.4 Western blots and quantitative analysis of the blots showing a decrease in the level of total and abnormal hyperphosphorylation of tau in AAV1-I_{2CTF} rats in comparison with the AAV1-GFP control animals. *p < 0.05; **p < 0.01.

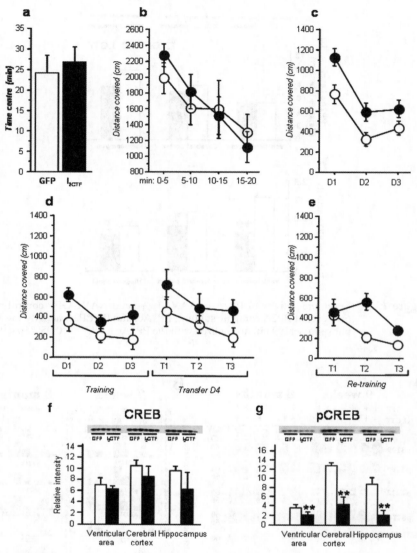

Figure 13.5   Effect of the expression of I$_{2\mathrm{CTF}}$ on spatial learning and memory and its
association with the brain level of pSer133-CREB. While compared with the
AAV1-GFP rats (control group), the AAV1-I$_{2\mathrm{CTF}}$ rats showed no change in
anxiety (**a**) or exploration (**b**) in the open field, these animals were impaired
in spatial learning and memory at age five months; $p < 0.005$ (**c**) and eight
months; $p < 0.01$ (**d**, left curves) in Morris water maze during training but
not in the transfer test (**d**, right curves). This delayed learning of the task
suggests impairment to process and encode spatial information. The AAV1-
I$_{2\mathrm{CTF}}$ animals also showed long-term deficit of retention when retested one
month after the previous water maze test; $p < 0.05$. (**e**) This deficit reflects
retrograde amnesia which is a specific symptom of hippocampal dysfunction. These impairments in AAV1-I$_{2\mathrm{CTF}}$ rats were associated with a selective
decrease in the brain level of pSer133-CREB (**f**, **g**). **$p < 0.01$.

dysfunction and mimic the early cognitive deficits observed in AD. The cognitive impairment in AAV1-I_{2CTF} animals was associated with a marked decrease in the level of pSer133-CREB, especially in the hippocampus and the cerebral cortex. The AAV1-I_{2CTF} rat model essentially recapitulated key features of AD. The AAV1-I_{2CTF} rat is not only one of the most disease relevant animal models of sporadic AD but can also be replicated without considerable expense and effort generally required for generating a transgenic animal line.

13.4 Conclusions

The AAV1-induced expression of an inserted gene(s) offers a direct and relatively inexpensive model where any number of animals with the designed gene expression in a specific brain area can be generated. This technology enables a direct cause and effect relationship *in vivo* as well as for testing diagnostic biomarkers and therapeutic compounds. Employing this strategy and based on our previous findings in AD brain which showed an upregulation and cleavage of I_2^{PP2}/SETα, we generated AAV1-I_{2CTF} rats. The AAV1-$_{I2CTF}$ rats showed all key features of early AD pathology, *i.e.* accumulation of intraneuronal Aβ and abnormally hyperphosphorylated tau, neurodegeneration, upregulation of GSK-3β activity, a decrease in phospho CREB and impairment in spatial learning and long-term traces of spatial information. Thus, AAV1-I_{2CTF} rats represent a model of sporadic AD, and the pathology and cognitive decline observed in these animals demonstrate a novel etiopathogenic mechanism of the disease which is initiated with the cleavage and translocation of I_2^{PP2A} from the neuronal nucleus to the cytoplasm.

Acknowledgements

Secretarial assistance, including the preparation of this manuscript, was provided by Janet Murphy. Studies described from our lab in this manuscript were supported in part by the New York State Office of People with Developmental Disabilities, and by NIH/NIA grant AG019158.

References

1. L. C. Cork, R. E. Powers, D. J. Selkoe, P. Davies, J. J. Geyer and D. L. Price, *J. Neuropathol. Exp. Neurol.*, 1988, **47**, 629–641.
2. R. G. Struble, D. L. Price, Jr., L. C. Cork and D. L. Price, *Brain Res.*, 1985, **361**, 267–275.
3. H. M. Wisniewski, B. Ghetti and R. D. Terry, *J. Neuropathol. Exp. Neurol.*, 1973, **32**, 566–584.
4. N. Bons, F. Rieger, D. Prudhomme, A. Fisher and K. H. Krause, *Genes Brain Behav.*, 2006, **5**, 120–130.
5. P. Giannakopoulos, S. Silhol, V. Jallageas, J. Mallet, N. Bons, C. Bouras and P. Delaere, *Acta Neuropathol.*, 1997, **94**, 131–139.

6. G. G. Glenner and C. W. Wong, *Biochem. Biophys. Res. Commun.*, 1984, **122**, 1131–1135.
7. I. Grundke-Iqbal, K. Iqbal, Y. C. Tung, M. Quinlan, H. M. Wisniewski and L. I. Binder, *Proc. Natl. Acad. Sci. U. S. A.*, 1986, **83**, 4913–4917.
8. B. A. Yankner, L. R. Dawes, S. Fisher, L. Villa-Komaroff, M. L. Oster-Granite and R. L. Neve, *Science*, 1989, **245**, 417–420.
9. D. Quon, Y. Wang, R. Catalano, J. M. Scardina, K. Murakami and B. Cordell, *Nature*, 1991, **352**, 239–241.
10. A. Harada, K. Oguchi, S. Okabe, J. Kuno, S. Terada, T. Ohshima, R. Sato-Yoshitake, Y. Takei, T. Noda and N. Hirokawa, *Nature*, 1994, **369**, 488–491.
11. M. C. Chartier-Harlin, F. Crawford, H. Houlden, A. Warren, D. Hughes, L. Fidani, A. Goate, M. Rossor, P. Roques, J. Hardy and M. Mullan, *Nature*, 1991, **353**, 844–846.
12. E. Levy-Lahad, W. Wasco, P. Poorkaj, D. M. Romano, J. Oshima, W. H. Pettingell, C. E. Yu, P. D. Jondro, S. D. Schmidt and K. Wang *et al.*, *Science*, 1995, **269**, 973–977.
13. R. Sherrington, E. I. Rogaev, Y. Liang, E. A. Rogaeva, G. Levesque, M. Ikeda, H. Chi, C. Lin, G. Li, K. Holman, T. Tsuda, L. Mar, J.-F. Foncin, A. C. Bruni, M. P. Montesi, S. Sorbi, I. Rainero, L. Pinessi, L. Nee, I. Chumakov, D. Pollen, A. Brookes, P. Sanseau, R. J. Polinsky, W. Wasco, H. A. R. Da Silva, J. L. Haines, M. A. Pericak-Vance, R. E. Tanzi, A. D. Roses, P. E. Fraser, J. M. Rommens and P. H. St George-Hyslop, *Nature*, 1995, **375**, 754–760.
14. S. Sorbi, B. Nacmias, P. Forleo, S. Piacentini, R. Sherrington, E. Rogaev, P. St George Hyslop and L. Amaducci, *Lancet*, 1995, **346**, 439–440.
15. R. Sherrington, S. Froelich, S. Sorbi, D. Campion, H. Chi, E. A. Rogaeva, G. Levesque, E. I. Rogaev, C. Lin, Y. Liang, M. Ikeda, L. Mar, A. Brice, Y. Agid, M. E. Percy, F. Clerget-Darpoux, S. Piacentini, G. Marcon, B. Nacmias, L. Amaducci, T. Frebourg, L. Lannfelt, J. M. Rommens and P. H. St George-Hyslop, *Hum. Mol. Genet.*, 1996, **5**, 985–988.
16. J. J. Lucas, F. Hernandez, P. Gomez-Ramos, M. A. Moran, R. Hen and J. Avila, *EMBO J.*, 2001, **20**, 27–39.
17. N. Zilka, P. Filipcik, P. Koson, L. Fialova, R. Skrabana, M. Zilkova, G. Rolkova, E. Kontsekova and M. Novak, *FEBS Lett.*, 2006, **580**, 3582–3588.
18. N. Zilka, M. Korenova, B. Kovacech, K. Iqbal and M. Novak, *Acta Neuropathol.*, 2010, **119**, 679–687.
19. K. Iqbal, M. Flory, S. Khatoon, H. Soininen, T. Pirttila, M. Lehtovirta, I. Alafuzoff, K. Blennow, N. Andreasen, E. Vanmechelen and I. Grundke-Iqbal, *Ann. Neurol.*, 2005, **58**, 748–757.
20. I. M. Verma and M. D. Weitzman, *Annu. Rev. Biochem.*, 2005, **74**, 711–738.
21. X. Wang, J. Blanchard, E. Kohlbrenner, N. Clement, R. M. Linden, A. Radu, I. Grundke-Iqbal and K. Iqbal, *FASEB J.*, 2010, **24**, 4420–4432.

22. K. Iqbal, I. Grundke-Iqbal, T. Zaidi, P. A. Merz, G. Y. Wen, S. S. Shaikh, H. M. Wisniewski, I. Alafuzoff and B. Winblad, *Lancet*, 1986, **2**, 421–426.

23. C. X. Gong, S. Shaikh, J. Z. Wang, T. Zaidi, I. Grundke-Iqbal and K. Iqbal, *J. Neurochem.*, 1995, **65**, 732–738.

24. C. X. Gong, T. J. Singh, I. Grundke-Iqbal and K. Iqbal, *J. Neurochem.*, 1993, **61**, 921–927.

25. F. Liu, I. Grundke-Iqbal, K. Iqbal and C. X. Gong, *Eur. J. Neurosci.*, 2005, **22**, 1942–1950.

26. C. X. Gong, T. J. Singh, I. Grundke-Iqbal and K. Iqbal, *J. Neurochem.*, 1994, **62**, 803–806.

27. C. X. Gong, J. Wegiel, T. Lidsky, L. Zuck, J. Avila, H. M. Wisniewski, I. Grundke-Iqbal and K. Iqbal, *Brain Res.*, 2000, **853**, 299–309.

28. M. Bennecib, C. Gong, J. Wegiel, M. H. Lee, I. Grundke-Iqbal and K. Iqbal, *Alzheimers Rep.*, 2000, **3**, 295–304.

29. M. Bennecib, C. X. Gong, I. Grundke-Iqbal and K. Iqbal, *FEBS Lett.*, 2001, **490**, 15–22.

30. J. Z. Wang, I. Grundke-Iqbal and K. Iqbal, *Eur. J. Neurosci.*, 2007, **25**, 59–68.

31. L. Y. Cheng, J. Z. Wang, C. X. Gong, J. J. Pei, T. Zaidi, I. Grundke-Iqbal and K. Iqbal, *Neurochem. Res.*, 2000, **25**, 107–120.

32. L. Y. Cheng, J. Z. Wang, C. X. Gong, J. J. Pei, T. Zaidi, I. Grundke-Iqbal and K. Iqbal, *Neurochem. Res.*, 2001, **26**, 425–438.

33. M. Mumby, *Cell*, 2007, **130**, 21–24.

34. M. Floer and J. Stock, *Biochem. Biophys. Res. Commun.*, 1994, **198**, 372–379.

35. Y. Shi, *Cell*, 2009, **139**, 468–484.

36. M. Li and Z. Damuni, *Methods Mol. Biol.*, 1998, **93**, 59–66.

37. H. Tanimukai, I. Grundke-Iqbal and K. Iqbal, *Am. J. Pathol.*, 2005, **166**, 1761–1771.

38. I. Tsujio, T. Zaidi, J. Xu, L. Kotula, I. Grundke-Iqbal and K. Iqbal, *FEBS Lett.*, 2005, **579**, 363–372.

39. M. O. Chohan, S. Khatoon, I. G. Iqbal and K. Iqbal, *FEBS Lett.*, 2006, **580**, 3973–3979.

40. P. A. Lawlor, R. J. Bland, P. Das, R. W. Price, V. Holloway, L. Smithson, B. L. Dicker, M. J. During, D. Young and T. E. Golde, *Mol. Neurodegener.*, 2007, **2**, 11.

Subject Index

References to figures are given in *italic* type. References to tables are given in **bold** type.